STATE MONOPOLIES UNDER EC LAW

Wiley Series in EUROPEAN LAW

Barnard/EC Employment Law, Revised Edition
0-471-93917-X 587 Pages

Burrows/European Social Law
0-471-96537-5 451 Pages

Emiliou/The European Union and World Trade Law
0-471-95552-3 390 Pages

Caiger/1996 Onwards: Lowering the Barriers Further
0-471-95768-2 300 Pages

Cross/Electric Utility Regulation in the European Union
0-471-95793-3 357 Pages

Ehlermann/Proceedings of the European Competition Forum
0-471-96668-1 248 Pages

Farr/Harmonisation of Technical Standards, Second Edition
0-471-95925-X 502 Pages

Goh/European Air Transport Law and Competition
0-471-96159-0 295 Pages

Goyder/EC Distribution Law, Second Edition
0-471-96122-1 336 Pages

Hervey/Sex Equality Law in the European Union
0-471-96436-0 427 Pages

Holder/The Impact of EC Environmental Law in the United Kingdom
0-471-97535-4 398 Pages

Yaqub/European Travel Law
0-471-94354-1 761 Pages

Forthcoming

Khan/EC Anti-Dumping Law
0-471-97483-8 528 Pages

233805

GB1 # 24276

£12.95

To Miko & Angela

With very best wishes

from Anne.

Maybe you'll find a little space for this on your crowded bookshelves!

STATE MONOPOLIES UNDER EC LAW

Françoise Blum

Solicitor, Avocat à la Cour

Clifford Chance, Paris

Anne Logue

Solicitor

Prior Associates, Brussels and Dublin

JOHN WILEY & SONS

Chichester • New York • Weinheim • Brisbane • Singapore • Toronto

Published in 1998 by John Wiley & Sons Ltd,
Baffins Lane, Chichester,
West Sussex PO19 1UD, England

National 01243 779777
International (+44) 1243 779777
e-mail (for orders and customer service enquiries): cs-books@wiley.co.uk
Visit our Home Page on http://www.wiley.co.uk
or http://www.wiley.com

Other Wiley Editorial Offices

John Wiley & Sons, Inc., 605 Third Avenue,
New York, NY 10158-0012, USA

Wiley-VCH Verlag GmbH, Pappelallee 3,
D-69469 Weinheim, Germany

Jacaranda Wiley Ltd, 33 Park Road, Milton,
Queensland 4064, Australia

John Wiley & Sons (Asia) Pte Ltd, 2 Clementi Loop #02-01,
Jin Xing Distripark, Singapore 129809

John Wiley & Sons (Canada) Ltd, 22 Worcester Road,
Rexdale, Ontario M9W 1L1, Canada

Library of Congress Cataloging-in-Publication Data

Blum, Françoise.
State monopolies under EC law / Françoise Blum, Anne Logue.
p. cm.
Includes bibliographical references and index.
ISBN 0-471-94353-3 (cloth)
1. Government monopolies—Law and legislation—European Union countries. I. Logue, Anne. II. Title.
KJE6497.B57 1998
341.7'53—dc21 97–34884
CIP

British Library Cataloguing in Publication Data

A catalogue record for this book is available from the British Library

ISBN 0-471-94353-3

Typeset in 10½/12pt Baskerville by Dorwyn Ltd, Rowlands Castle, Hants.
Printed and bound in Great Britain by Bookcraft (Bath) Ltd, Midsomer Norton, Somerset.
This book is printed on acid-free paper responsibly manufactured from sustainable forestry, in which at least two trees are planted for each one used for paper production.

CONTENTS

Foreword *xi*

Acknowledgements *xv*

Select bibliography *xvii*

Stop press *xxi*

Tables *xxv*

National Cases *xxv*
Cases before the European Court of Justice *xxv*
Treaties *xxx*
Commission Communications *xxxii*
Commission Decisions *xxxii*
Commission Directives *xxxiv*
Commission Documents *xxxv*
Commission Notices *xxxvi*
Commission Regulations *xxxvi*
Council Directives *xxxvi*
Council Regulations *xxxvii*
European Parliament *xxxviii*
National Legislation *xxxviii*

Introduction *1*

The new interpretation of Article 90 *1* Structure of the book *4*

PART I ARTICLE 90 *5*

Chapter 1 Article 90(1): Member States' obligations *7*

General meaning *7* Meaning of specific phrases: 'Member States' *8*; 'public undertaking' *8*; 'special and exclusive rights' *9*; 'measures' *12*

Chapter 2 Article 90(2): permitted derogation from Treaty rules *17*

General meaning *17* Original scope of application *18* Recent scope of application *18* Application of Article 90(2) to the mere existence of a monopoly *19* Link between Article 90(2) and Article 90(1) *19* *Corbeau* and the necessity test *20* Specific phrases: 'services of general economic interest' *21*; 'Does not obstruct in law or in fact the particular tasks assigned to them' *25*; 'The development of trade must not be affected to such an extent as would be contrary to the interest of the Community' *35* Burden of proof *36* Direct effect of Article 90(2) *38* Possible defences *39*

Chapter 3 Article 90(3): Commission power to issue Directives or Decisions *41*

General meaning *41* Commission's supervisory power *41* Commission's enforcement power *42* Differences between Decisions and Directives under Article 90(3) *43* Article 90(3) Directives *48* Nature of Commission Directives *48* Applicable procedure *48* Commission's power to take normative measures by way of Directives only *50* Commission's power to legislate in respect of Article 90(1) measures *50* Examples of Commission Directives pursuant to Article 90(3) *51* Principal features of Article 90(3) Directives leading to the abolition of monopolies *51* Article 90(3) Decisions *53* Right of defence in Article 90(3) Decisions *54* Limited rights of complainants in Article 90(3) cases *55*

Chapter 4 State-related undertakings and the competition rules: Articles 85 and 86 and Article 90 *59*

Article 86 *59* Summary *59* Article 86 criteria *60* Traditional Article 86 abuses *65* Article 86 abuses characteristic of 'state-related' undertakings *73* Article 85 *93*

Chapter 5 State-related undertakings and the rules on the free movement of goods: Articles 30 to 37 and Article 90 *99*

Articles 30 to 34 *99* Article 36 *99* Article 37 *99* General rules on free movement of goods: Articles 30 to 36 *100* The traditional case-law of the Court *100* The recent case-law on monopolies over goods *105* Article 30 type restrictions *109* Conditions for exemption: justification and proportionality *113* Link between Article 30 infringements and Article 86 infringements *118* Rules governing State monopolies of a commercial character: Article 37 *119* Monopolies covered by Article 37 *120* Purpose and scope of Article 37 *121* Comparison between the scope of Article 30 to 36 and Article 37 *124* Link between Article 37 and Article 90 *126*

Chapter 6 State-related undertakings and the rules on the freedom to provide services: Articles 59 and 60 and Article 90 *129*

Article 59 *129* Article 60 *129* Article 56(1) *129* General application of Article 59 *130* Services *130* Services rendered for remuneration *130* Cross-border element *131* Direct effect and direct applicability *132* Justification of restrictions on the freedom to provide services *133* Article 56 justification *133* Public interest justification *133* The application of Article 59 to monopolies *136* Type of monopoly services affected by Article 59 *136* Restrictions caught by Article 59 *138* What are the possible justifications of restrictions on the freedom to provide services? *145*

Chapter 7 State-related undertakings and other Treaty Articles *153*

Article 222 *153* State aid rules *155* Article 52 *159* Article 90(2) and the Merger Regulation *161* The application of the Merger Regulation to State-related enterprises *162*

Chapter 8 Key Article 90(1) concepts *165*

Access to an essential facility *165* The concept of access *165* The concept of essential facility *165* Origin of the concept of access *167* The implications of the concept of access in competition law terms *167* Lessons from the case-law and legislation *174* Universal service *175* Universal service and service of general economic interest *181* Public service *181* Proportionality *182* Transparency *184* Non-discrimination *185* Unbundling *185* Liberalization/deregulation *186* Inter-operability *187*

PART II REGULATORY MEASURES COMPLEMENTING ARTICLE 90 IN THE UTILITY SECTOR *189*

Chapter 9 Competition and exclusive rights in the European Union energy market *191*

Introduction *191* Setting the scene: creation of an internal energy market *192* Regulation of the energy sector in the European Union: phase one *195* Regulation of the energy sector in the European Union: phase two *196* Electricity and gas proposals *196* Main provisions of the electricity Directive *203* Application of competition enforcement under Article 90 and other Treaty Articles to the energy sector *210* Conclusion *217*

Chapter 10 Competition and exclusive rights in the telecommunications sector *219*

Introduction *219* Background to Commission measures *220* Case-law and legislation *220* Importance of the sector *220* The preparatory work:

the various green papers *221* Article 90(3) Liberalization Directives *227* Introduction *227* Terminal equipment *228* Services *229* Satellite *231* Alternative infrastructure *233* Mobile *237* Full competition *240* Harmonization Directives *245* Introduction *245* ONP Framework Directive *246* ONP voice telephony *250* European Parliament and Council Directive on a common framework for general authorization and individual licences in the telecommunication services field (the Licensing Directive) *252* Directive 97/33/EC of the European Parliament and of the Council of 30 June 1997 on interconnection in telecommunications with regard to ensuring universal service and interoperability through application of ONP *254* Establishment of a coherent telecoms regime at Community level *257* Introduction: ONP and Article 90 *257* Interconnection and interoperability *258* Licensing *258* Universal service *259* Application of the EC competition rules *261* Article 90(3) Decisions *262* Mergers *262* Articles 85 and 86 Decisions *263* Competition rules and the telecommunications sector: Guidelines and a draft Notice *266* Conclusion *270*

Selective list of significant EU documents in the telecommunications sector
(appendix to chapter 10) *272*

Chapter 11 Competition and exclusive rights in the broadcasting sector *277*

Introduction *277* Regulation of broadcasting in the European Union *278* The Television Without Frontiers Directive *278* Application of competition rules under the Merger Regulation to broadcasting *289* Merger cases in the media field *289* Development of EU policy in the broadcasting sector *297* Green papers, Communications, Resolutions and Directives *297* Application of competition enforcement under Article 90 to the broadcasting sector and ancillary services *303* Key cases *303* Conclusion *315*

Chapter 12 Competition and exclusive rights in the postal sector *317*

Introduction *317* Green paper on the development of the single market for postal services *318* Case-law Express delivery service cases *322* The *Corbeau* case *324* Legislative measures on postal services *327* Proposed Directive on common rules for the development of the internal market of Community postal services and the improvement of quality of service *327* Commission draft Notice on the application of the competition rules to the postal sector and in particular on the assessment of certain State measures relating to postal services *333* Conclusion *337*

Chapter 13 Competition and exclusive rights in the EU transport sector *339*

Introduction *339* Air transport *340* The regulatory framework *340* Application of Article 90 to air transport *348* Ancillary services *353* Sea transport *360* Abuse of monopoly rights *360* Rail transport *365* The regulatory framework *365* Application of the competition rules to rail transport *370* Conclusion *382*

Index *383*

FOREWORD

There are many factors which explain why the discussion on Public Service and the market should have developed in Europe in the mid-1990s. Our economic environment has undergone rapid change during the last fifteen years. This period of change has heralded a crisis in adapting the traditional structures of our European societies. There is deep uneasiness as the nation states find it increasingly difficult to deal with some of the now global phenomena, involved in tackling inequalities and exclusion.

The contrast with the economy in the first half of the century is striking. International trade in goods was limited then, and trade in services was virtually non-existent. In many sectors, national economies were relatively sheltered. The scale of competition was determined by government decisions. The State decided on the establishment of customs duties or quotas in order to protect national production against imports. The State also decided to establish monopolies for certain activities, often in order to achieve better control of the establishment of infrastructure that was essential for the development of the country (telephone, transport, energy). Competition authorities, if existing, found that the scope of competition was limited by government decisions.

In the last fifty years, the internationalisation of trade and the interdependence of economies have increased rapidly, and indeed at a faster rate since the early 1980s, spreading like a contagion throughout the world, imposing on our firms and consequently on our public services ever-greater requirements of competitiveness. The arrival of the single market has certainly made this process more visible, but it has also acted as a shock-absorber, dampening the effect of an adjustment which would inevitably have been more harsh if the Member States had had to adjust on their own and in an uncoordinated way to an economy which had globalised without them.

Confronted with these changes, often transcending the national level, the European Union's task is to take joint action in situations where individual action by Member States is not sufficient, and to provide us with common rules in situations where States can no longer, on their own, set limits to market forces. These changes also mean that our public services must be adjusted so as to make them more competitive while at the same time ensuring that they still provide equality of access, continuity and universality, and that they continue to play their role in the pursuit of greater social cohesion and more satisfactory living conditions in Europe.

Community policy seeks to achieve this difficult balance while respecting the different concepts of Public Service that exist within the Union. It is based on the provisions of the Treaty, Article 90 of which lays down the principle of an evolving balance between Public Service and competition. This principle may be summarised as follows: restrictions of competition must be accepted if they are necessary and proportionate to the Public Service objectives which Member States are pursuing. Personally speaking, and contrary to a widespread but incorrect belief, I am of the opinion that the balance which the Commission has struck in determining this proportionality, subject to review by the Court of Justice, is favourable to services of general economic interest. European liberalisation policy is in no way comparable to all-out deregulation. On the contrary, liberalisation in Europe is based on the search for a balance whose guiding principle is to place the economy at the service of people and not vice-versa and on the conviction that there is an alternative to the move towards greater precariousness, inequality and exclusion.

It is possible and desirable in certain areas to introduce mechanisms to offset reliance on market forces alone. The Commission is aware of this. It is developing an active policy to this end, through its liberalisation policy, by endeavouring in particular to define and guarantee a universal service in all the public services subject to monopoly arrangements, but also through its action under structural policies.

Nevertheless, the Commission's policy and the judgements delivered by the Court of Justice and the Court of First Instance are at the centre of an ongoing and irrational controversy. Sentiments, traditions and sometimes even myths often prevail over legal analysis and rational argument. Of course, I accept that the irrational must have a place in political debate, but since this is also a legal question, I have always had the impression that, once the debate has been rationalised, many of the current difficulties could be resolved.

Bearing this in mind, this book by Anne Logue and Françoise Blum has impressed me both in terms of the quality of the research work carried out by the authors, and in terms of the rigour of their analysis of the gradual development of Commission decision-making practice and Court of Justice case-law on liberalisation. This rigorous approach gets away from the inaccuracies and polemics which too often characterise the debate on this subject and puts the question of the relationships between public services and Community law in its proper legal context. This book therefore makes a timely contribution to putting the discussion back on a firm and up-to-date technical basis.

It also presents an interesting interpretation of the development of the case-law on Article 90 and of the effect of Article 90(3) Directives even if I personally do not fully share the view of the authors on these two points. It seems to me that the case-law of the Court of Justice has fluctuated less than the authors seem to think. I also believe that Article 90(3) Directives do not establish new obligations but enforce existing rules, namely the rules laid down in the Treaty. They do not therefore introduce new legislation, but apply and define the obligations enshrined in the Treaty, subject to review by the Court of Justice.

There are, in an area which is still relatively new, such as the application of Article 90 of the Treaty of Rome, many points on which differences of interpretation are liable to arise between lawyers. The strict and always clear distinction between facts and analysis which the authors have endeavoured to maintain will, at all events, allow everyone to form his own judgement, particularly in the areas where the lessons to be drawn from case-law still leave a wide margin of interpretation in policy matters.

In an area in which politics and law are as intricately mixed as here, let me conclude by expressing the wish that this rigorous legal analysis will generate an open and dispassionate political discussion on the relationships between Services of General Economic Interest and competition, and on the, as it seems to me, close links between improving competitiveness and reinforcing solidarity. I am not one of those who think that the machinery for economic and social solidarity imposes a handicap on Europe. On the contrary, I think that if we are able to modernise the mechanisms, they will give the European economy a tremendous comparative advantage in the competition which has now developed at world level. We, and in particular the European Commission, must for this purpose point the way for the development of services of general economic interest in Europe.

Commissioner Karel Van Miert
December 1997

Anne Logue's dedication:

for Hugh
with love as always

ACKNOWLEDGEMENTS

The authors wish to express their extreme gratitude to Ulick Bourke (partner at Clifford Chance) for his invaluable help throughout the preparation of this book. The authors are particularly grateful for his enlightened comments on the case law of the Court and the Commission as well as for his continuous encouragement.

The authors also wish to thank the following people for their helpful contribution to specific sectors: Nick Elverston (broadcasting); Natasha Hobday (telecoms); Nicola Northway (energy); Ulick Bourke (postal services); Alex Nourry (transport) and to Michael Leonard (partner at Macmillan, White & Wheeler).

Finally, they wish to express their thanks to Sandra Raguier for her great help and immense patience in setting and correcting their manuscript.

SELECT BIBLIOGRAPHY

Arsac, Magali. L'appréhension du service public ferroviaire par le droit communautaire. Paris: Librairie générale de droit et de jurisprudence, 1997. xii, 138 pp. (Travaux et recherches Pantheon-Assas (Paris II). Droit-économie-sciences sociales).
Also: Mémoire-Université Panthéon-Assas (Paris-II), 1996.

Bredsgaard, Benedikte. Article 90 of the EC treaty as a legal basis for liberalisation. In: *Justitia*. no. 1 (1997), d. II, pp. 1–40.

Bright, Christopher. Article 90, economic policy and the duties of member states. In: *European competition law review*. 1993; vol. 14, no. 6, pp. 263–272.

Burke, Jarleth. Telecommunication, Article 90 EC and special and exclusive rights. In: *Irish competition law reports*. 1996, pp. 2.27–2.35.

Charbit, Nicolas. L'article 90 du Traité CE; I. Développements des années 1994–1995. In: *Revue de droit des affaires internationales*. No. 7 (1995), pp. 853–863.

Charbit, Nicolas. L'article 90 du Traité CE; II. Revirement sur la doctrine de l'abus automatique vers une reconnaissance des services publics. In: *Revue de droit des affaires internationales*. No. 8 (1995), pp. 988–997.

Chavrier, Henri. La jurisprudence de la Cour de justice sur l'application de l'article 90 du Traité de Rome. In: *Cahiers juridiques d'électricité et du gaz*. No. 509 (1995), pp. 125–139.

Chung, Chan-Mo. Article 90 of the Treaty of Rome in telecommunications, post, and air transport: a brief comparison. In: *Revue européenne de droit public*. No. 1 (1997), pp. 41–60.

Debéne, Marc et Raymundie, Olivier. Sur le service universel: renouveau du service public ou nouvelle mystification? In: *Critique de la raison communautaire*. 1996, pp. 109–126.

Deringer, Arved. Les articles 90 et 37 dans leurs relations avec un regime de concurrence non falsifie; les incidences des regles de concurrence et de l'article 222 sur les possibilites de nouvelles nationalisations ou socialisations de secteurs economiques. In: *L'entreprise publique et la concurrence*. 1969; 1969, pp. 388–408.

Dewost, Jean-Louis. Service public et droit communautaire: principes de base. In: *Europe, concurrence et service public.* 1995, pp. 153–162.

Dommering, E.J. Article 90 of the EEC Treaty and the telecommunications, broadcasting and postal services sectors. In: *State entrepreneurship, national monopolies and European Community law*; European monographs. 1993; vol. 6, pp. 53–72.

Dubois, Jean. L'approche de la Commission Européenne face à la libéralisation et aux obligations de service universel. In: *Managing universal service obligations in public utilities in the European Union.* 1996, pp. 17–25.

Edward, David Alexander Ogilvy and Hoskins, Mark. Article 90: deregulation and EC laws: reflections arising from the XVI FIDE Conference. In: *Common Market law review.* No. 1 (1995), pp. 157–186.

Ferrari Bravo, Luigi. Les articles 90 et 37 dans leurs relations avec un regime de concurrence non falsifie; les incidences des regles de concurrence et de l'article 222 sur les possibilites de nouvelles nationalisations ou socialisations de secteurs economiques. In: *L'entreprise publique et la concurrence.* 1969, pp. 410–438.

Frison-Roche, Marie-Anne. Libres propos sur le 'service public marchand' dans la perspective de la régulation des réseaux d'infrastructure essentielle. In: *Revue de jurisprudence de droit des affaires.* No. 10 (1995), pp. 847–852.

Gardner, Anthony L. The velvet revolution: Article 90 and the triumph of the free market in Europe's regulated sectors. In: *European competition law review.* No. 2 (1995), pp. 78–86.

Graff, Michèle. L'article 90-2 et son interprétation. In: *Revue de la concurrence et de la consommation.* No. 88 (1995), pp. 5–11.

Grard *et al.* Vers un service public européen / sous la direction de Loïc Grard, Jacques Vandamme et Francois van der Mensbrugghe; avant-propos de Christian Stoffaes 637 pp. In Aspe Europe, 1996 (Collection TEPSA).

Gyselen, Luc. National monopolies – an evolving state action doctrine under article 90 EEC. In: *State entrepreneurship, national monopolies and European Community law*; European monographs. 1993; vol. 6, pp. 1–13.

Hancher, Leigh and Slot, Piet Jan. Article 90. In: *European competition law review.* 1990; vol. 11, no 1, pp. 35–39.

Hocepied, Christian. Les directives article 90 paragraphe 3: une espèce juridique en voie de disparition? In: *Revue des affaires européennes.* No. 2 (1994), pp. 49–63.

Huglo, Jean-Guy. L'article 90 du Traité de Rome. In: *Tendances actuelles et évolution de la jurisprudence de la Cour de justice et du Tribunal de première instance des Communautés européennes.* Vol. 2, 1995, pp. 29–38.

IRENE. Service public et droit communautaire: les grands textes / IRENE, Institut de recherche et d'études sur les normativités européennes; sous la direction de Robert Kovar, Denys Simon. Paris: La documentation francaise, 1996. 395 pp. (Travaux de la CEDECE. Série Textes et documents).

Kovar, Robert. Droit communautaire et service public: espirit d'orthodoxie ou pensée laïcisée. In: *Revue trimestrielle de droit européen.* No. 2 (1996), pp. 215–242.

Kovar, Robert. Droit communautaire et service public: espirit d'orthodoxie ou pensée laïcisée. In: *Revue trimestrielle de droit européen.* No. 3 (1996), pp. 493–533.

Lesquins, Jean-Louis. Réconcilier service public et intégration européene: un effort symétrique indispensable et une stratégie diversifiée. In: *La mission d'intérêt économique général dans l'Union européenne.* 1996, pp. 65–72.

Masson. Europe, concurrence et service public: avril 1995 / CEEP. Paris: Masson: Colin, 1995. ix, 196 pp.

Monet, Hubert. A propos de l'article 90 du Traité CEE relatif aux monopoles publics. In: *Revue de jurisprudence de Liège, Mons et Bruxelles.* 1994; 101ème année/1994, no. 11, pp. 373–379.

Murphy, Frances. The jurisdiction of national courts under article 90(2) EEC. In: *European business law review.* 1993; vol. 4, no. 1, pp. 3–5.

Pappalardo, Aurelio. Regime de l'article 90 du Traité CEE: les aspects juridiques. In: *L'entreprise publique et la concurrence.* 1969; pp. 76–103.

Pappalardo, Aurelio. State measures and public undertakings; article 90 of the EEC treaty revisited. In: *European competition law review.* 1991; vol. 12, no. 1, pp. 29–39.

Rantala, Meri and Kronenberger, Vincent. Les droits des tiers dans la procédure d'adoption des décisions de l'article 90, par. 3, du traité instituant la Communauté européenne. In: *Journal de tribunaux. Droit européen.* No. 42 (1997), pp. 169–173.

Rodrigues, Stéphane. Prospective du service public en Europe: le marché intérieur, entre concurrence et utilité publique. In: *Revue des affaires européennes.* No. 2 (1994), pp. 65–79.

Rodrigues, Stéphane. Comment intégrer les principes du service public dans le droit positif communautaire: quelques propositions. In: *Revue française de droit administratif.* no. 2 (1995), pp. 335–342.

Rundegren, Hans. Developments in case law concerning Article 90 of the EC Treaty. In: *Managing universal service obligations in public utilities in the European Union.* 1996, pp. 63–71.

L'entreprise publique et la concurrence: les articles 90 et 37 du Traité CEE et leurs relations avec la concurrence: *Semaine de Bruges* 1968 – Public enterprises and competition. Bruges : De Tempel, 1969. 516 pp. (Cahiers de Bruges. NS; 22).

Slot, Piet Jan. Les obligations de service public (OSP) aux Pays-Bas en relation à la législation de la Communauté européenne (CE).

Slot, Piet Jan. Competition and public service obligation: the example of the energy sector. In: *Current and future perspectives on EC competition law.* 1997, pp. 109–120.

Sutherland, Peter D. Progress to European Union: a challenge for the public service. In: *Administration.* 1993; vol. 41, no. 2, pp. 105–119.

Tesauro, Guiseppe. Intervento pubblico nell'economia e art. 90, n. 2, del Trattato CE. In: *Il diritto dell'Unione europea.* No. 3 (1996) pp. 719–740.

Van Miert, Karel. Les missions d'intérêt économique général et l'article 90 paragr. 2 du Traité CE dans la politique de la Commission. In: *Il diritto dell'economia.* No. 2 (1997), pp. 277–288.

Voisset, Michèle. Le service public autrement: de quelques effets du droit communautaire sur le droit français des services publics industriels et commerciaux. In: *Revue française de droit administratif.* No. 2 (1995), pp. 304–319.

STOP PRESS

Since this book has gone to press, the Court has delivered judgments in two important cases.

First, what is referred to in this book as *the gas and electricity cases* involving the Commission against France (C-159/94), the Netherlands (C-157/94), Spain (C-160/94) and Italy (C-158/94).

Second, the *Franzén case* between Allmänna Aklagaren and Harry Franzén relating to the Swedish monopoly over the retail sale of alcohol.

The judgments were issued on the same day, 23 October 1997. They provide a significant contribution to the case law of the Court on State monopolies.

The gas and electricity cases

The Commission had started proceedings against four Member States in relation to their monopoly over the import and export of electricity and, in one case, over gas as well. The Commission considered that these monopolies were contrary to Article 37 and had to be abolished. It stated that the derogation contained in Article 90(2) did not apply to them; and, in any event, that the Member States had not proven that the monopoly was necessary to enable the enterprises concerned to remain economically viable.

The Court's judgments are particularly interesting for the following reasons:

(i) *Article 37*: the Court agreed with the Commission that the import and export monopolies constituted a discrimination contrary to Article 37.

The Court held that it was therefore not necessary to examine whether these monopolies were contrary to Articles 30-36.

An important implication is that Article 36 does not apply to justify a monopoly contrary to Article 37.

The Court, however, contrary to the Commission, held that the derogation contained in Article 90(2) might apply and that it was necessary to examine whether the criteria contained in Article 90(2) were met.

(ii) *Article 90(2) and the Corbeau test*: the Court confirmed that the validity of the monopolies in question depended on whether they were necessary to enable the gas and electricity enterprises to carry out their task of general economic interest.

The Court rejected the Commission's narrow interpretation of its case-law in the *Corbeau* case: according to the Commission the monopolies would only be valid if they were absolutely necessary to enable the enterprises concerned to remain *commercially viable*. Conversely, if these enterprises could continue to operate and remain financially sound under conditions of competition, this would show that their monopoly was not necessary and was therefore contrary to the Treaty. The Court disagreed and held that there could be other circumstances when the monopolies could be valid. One should also examine whether the monopolies are necessary to enable the enterprises concerned to carry out all the *specific obligations* imposed upon them by the State as part of their task of general economic interest.

One may consider that the Court slightly widens or, at least, clarifies the test set out in its judgment in the *Corbeau* case of 1993, particularly in the light of its definition of specific obligations.

(iii) *Specific obligations*: the Court explained that the obligations must be *specific and directly* linked to the task of general economic interest. These obligations may vary between Member States. For example, the Court held that the obligation to provide electricity at the lowest possible price was a specific obligation in the Netherlands, but not in France. It followed that the examination of whether the monopoly was necessary to satisfy this obligation would only be carried out in respect of the Netherlands. In France, other specific obligations would be taken into account, namely that supply of electricity to all citizens, on a continuous and equal basis.

The Court explained that the monopolies could be compatible with the Treaty if they were absolutely necessary to meet such specific obligations.

(iv) *Burden of proof*: the Court clarified the question of who bears the burden of proof. It held that insofar as Article 90(2) constitutes a derogation from Treaty rules, it is *for the Member States seeking to rely on it* to prove that they meet all the conditions contained in the Article.

However, once a Member State has explained in detail why it considers that a monopoly is necessary to enable the enterprise concerned to carry out all its specific obligations, it does not have to go further and prove that there could be no other less restrictive way to carry out these obligations. It is then for the Commission when it attacks the monopoly to prove that the monopoly is not necessary and, in particular, that there may be other less restrictive ways to carry out the specific task entrusted to the monopoly enterprise. The Commission should prove its case using all relevant economic, social and financial data. The Court stated that the Commission had failed to produce this evidence in the present case. It was therefore impossible for the Court to rule on whether the monopolies in question went no further than was necessary for them to carry out their tasks in economically acceptable conditions.

It is submitted that this case follows the previous line of Court cases. It further clarifies the application of Article 90(2) and the issue of the burden to proof.

Franzén

In this case Mr Franzén was prosecuted for selling alcohol in Sweden in breach of the State company's monopoly over the retail sale of alcohol. The national court stayed the proceedings to ask the Court whether the Swedish monopoly was compatible with Articles 30 and 37 of the Treaty. The Court made a distinction between the national rules which governed the existence and operation of the monopoly and which had to be examined under Article 37 alone and those which could be separated from the existence and operation of the monopoly itself and which should be examined under Articles 30 and 36.

The most interesting part of the judgment for the understanding of the compatibility of state monopolies with EC laws relates to the Court's analysis under Article 37. The Court held that Article 37 presupposes the existence of commercial monopolies: it stressed that these monopolies do not have to be abolished but only adjusted to ensure that they do not lead to discrimination towards foreign goods. Once it has been demonstrated that the monopoly has been established to further a public interest, it is only necessary to check that it does not lead to discrimination against foreign goods.

The Court held, at paragraph 39, that Article 37 aims at eliminating restrictions on the free movement of goods, *except restrictions* on trade which are *inherent in the existence of the monopoly itself*.

Two lessons may be drawn from this:

(i) the validity of the existence and operation of a commercial monopoly may be examined under Article 37 alone. If the Court finds that the existence and operation of the monopoly are valid under Article 37, it will not examine them also under Article 30. The Court found that the Swedish monopoly over the retail sale of alcohol was compatible with Article 37 because it did not discriminate against foreign goods more than against domestic goods. It therefore did not examine whether it might be contrary to Article 30.

(ii) the Court did not carry out the balancing test that it normally carries out in Article 90 cases: i.e. it did not ask whether the monopoly was *absolutely necessary* to carry out the task of public interest entrusted to the enterprise by the State.

In this case and contrary to *the gas and electricity cases* and other cases involving State monopolies, the Court does not regard the monopoly as an exception to the Treaty which must be justified in accordance with the criteria of Article 90(2). It regards the monopoly in question as being one of those which the draftsmen of the Treaty must have wished to allow, irrespective of its inherent restrictions on trade.

The absence of examination under Articles 30-36 and Article 90(2) is surprising. The Court appears to accept the validity – *per se* – of certain commercial monopolies falling under Article 37. It seems that this case goes against the general trend of cases developed since the *Corbeau* judgment.

Should this case be limited to monopolies falling under Article 37 and, then, only to 'certain' of those monopolies? Or, is this a new approach from the Court which might extend to Article 30 and 59 monopolies? The question remains open.

It is submitted that the case does not invalidate the Court's approach in other State monopoly cases (for example in *the gas and electricity cases* of the same date). It might, however, open new possibilities to defend certain State monopolies.

TABLES

NATIONAL CASES

Belgium

La Poste v SPRL Delta Express and SA Hays Business Services (A.C./5.884/94 and A.C./7.829/94) Tribunal de Commerce de Bruxelles (22 May 1995) 89

France

Arret Cour de Cassation (Chambre Commerciale), 12 December 1995 (Meteorologie Nationale) 63
Conseil de la Concurrence Avis No.96-A-10 (25 June 1966) 32
SARL Pro-Mess and Direction Departementale de la Poste de Seine-Maritime, T.A. de Rouen (22 February 1995) 89

CASES BEFORE THE EUROPEAN COURT OF JUSTICE

Ahmed Saeed Flugreisen, Silver line Reisburo GmbH v Zentrale zur Bekampfung unlauteren Wettbewerbs eV C-66/86 [1989] ECR 803 10, 11, 12, 23, 24, 26, 28, 38, 67, 69, 70, 184, 267, 349
AKZO v European Commission C-62/86 [1991] ECR 3359 . . . . .92, 269
Alpine Investments BV v Minister van Financien C-384/93 [1995] ECR I-1141 . . . . .131, 132, 135
Bachmann C-204/90 [1992] ECR I-249 135
Banchero C-387/93 [1995] ECR I-4663 . . . . .79, 82, 103
Banco de Credito Industrial SA (now Banco Exterior de Espana SA) v Ayuntamiento de Valencia C-387/92 [1994] ECR I-877 . .24, 155, 156, 157
Bavaria Fluggesellschaft and Germanair v Eurocontrol [1977] ECR 1517. .61

BBC v European Commission T-70/89 [1991] ECR II-535 . . . . .86, 313
Belgishe Radio en Televisie and Societe belge des auteurs, compositeurs et editeurs v SV SABAM and NV FONIOR (BRT v SABAM) C-127/73 [1974] ECR 313 . . . . .22, 23
Bodson (Corinne) v Pompes Funebres des regions liberees SA C-30/87 [1988] ECR 2479 . . . . . . .8, 12, 23, 27, 64, 65, 67, 68, 69, 70, 105, 106, 121
Bond Van Adverteerders C-352/85 [1988] ECR 2085 136
British Leyland C-226/84 [1986] ECR 3263. 70
Bundesverband der Bilanzbuchhalter eV v European Commission T-84/94 [1995] ECR II-101 . . . . .56, 57
CBEM/CLT and IPB C-311/84 [1985] ECR 3261. . . . . .63, 323
Campus Oil v Minister for Industry and Energy C-72/83 [1984] ECR 2727 . . . . .13, 105, 109, 111, 116, 118, 214–15
Centre belge d'etudes de marche – Telemarketing (CBEM) SA v Compagnie Luxembourgeoise de telediffusion SA and information publicite Benelux (Telemarketing) SA C-311/84 [1985] ECR 3261 60, 77, 78. 79, 80, 85, 86, 167, 168, 173, 312–13, 357
Coditel C-62/79 [1980] ECR 881. 134
Commercial Solvents v European Commission C-6/73 and C-7/73 [1974] ECR 223. . . . . .77, 78, 79, 86, 167, 169
Corbeau (Paul) C-320/91 [1993] ECR I-2533 . . . . .3, 4, 19, 20, 23, 28, 29, 30, 31, 32, 33, 34, 35, 37, 60, 71, 73, 76, 82, 87, 88, 89, 90, 91, 96, 117, 137, 150, 151, 175, 180, 181, 183, 212, 215, 318, 322, 323, 324–7, 330, 337
Corsica Ferries Italia SRL v Corpo dei Piloti del Porto di Genova C-18/93 [1994] ECR I-1783. . . . . .12, 28, 69, 137, 139, 140, 148, 364
Costa v Enel C-6/64 [1964] ECR 614 . . . . .106, 120
Debauve C-52/79 [1980] ECR 833. 131
Decoster C-69/91 [1993] ECR I-5335 113
Ellinike Radiophonia Tileorassi Anonimi Etairia (ERT AE) v Dimotiki Etairia Pliroforissis (DEP) C-260/89 [1991] ECR I-2925 3, 13, 20, 27, 28, 37, 38, 73, 74, 75, 86, 106, 131, 136, 137, 139, 140, 149, 172, 185, 304, 305–7, 318
European Commission v Belgium (cable television broadcasts) C-11/95 [1997] 2 CMLR 289. 286
European Commission v Belgium (Dassonville) C-2/78 [1979] ECR 1761 . . . . .100, 101, 102, 105
European Commission v Denmark (Danish co-insurance) C-252/83 [1986] ECR 3713 . . . . .131, 133, 136
European Commission v France (French co-insurance) C-220/83 [1986] ECR 3663 . . . . .131, 133, 136
European Commission v France (French tourist guide) C-154/89 [1991] ECR I-659 . . . . .132, 134, 136
European Commission v Germany (German co-insurance) C-205/84 [1986] ECR 3755 . . . . .131, 133, 136

European Commission v Greece (Greek tourist guide) C-198/89 [1991] ECR I-727 . . . . 132, 134, 136
European Commission v Ireland (Irish co-insurance) C-206/84 [1986] ECR 3817 . . . . 131, 133, 136
European Commission v Italy C-78/82 [1983] ECR 1955 . . . . 122, 127
European Commission v Italy (Italian tourist guide) C-180/89 [1991] ECR I-709 . . . . 132, 134, 136
European Commission v Italy (Transparency) C-118/85 [1987] ECR 2599 61
European Commission v The Netherlands (Mediawet) C-353/89 [1991] ECR I-4069 131, 134, 136, 137, 139, 140, 141, 142, 143, 144, 145, 149, 182, 185, 307–8
European Commission v The Netherlands, Italy, France, Spain (Electricity and Gas) C-157/94 to C-160/94 [1997] ECR 5699 34, 35, 36, 56, 105, 125, 126, 194, 213, 216
European Commission v United Kingdom C-222/94 [1996] ECR I-4025 . . . . 282, 284–6
European Commission v Union Internationale de Chemins de Fer (UIC) C-264/95 P (11 March 1997) 381
Federation Francaise des Societes d'Assurance v European Commission T-106/95 [1997] ECR II-229 . . . . 24, 157, 158, 182
Federation Francaise des Societes d'Assurance v Ministry of Agriculture and Fisheries C-244/94 [1995] ECR I-4013 61
SAT Fluggesellschaft v Eurocontrol C-364/92 [1994] ECR I-0043 61, 130, 352–3
France v European Commission (Terminal Equipment) C-202/88 [1991] ECR I-1223 2, 3, 4, 10, 14, 18, 19, 20, 27, 37, 43, 44, 45, 46, 48, 50, 68, 74, 75, 105, 106, 107, 108, 109, 111, 114, 115, 117, 121, 123, 124, 137, 144, 187, 210, 228, 230, 235, 266, 308
France, Italy, United Kingdom, Ireland v European Commission (Transparency Directive) C-188/80 to C-190/80 [1982] ECR 2545 7, 8, 9, 14, 44, 47, 158, 162, 184
France v European Commission (C-325/91) [1993] ECR I-3283 50
Gemeente Almelo v Energiebedriff Ijsselmij NV C-393/92 [1994] ECR 1477 . . . . 3, 19, 23, 30, 32, 33, 34, 35, 37, 71, 87, 93, 94, 96, 106, 109, 111, 115, 118, 120, 121, 147, 150, 151, 175, 183, 201, 210, 211–13
General Motors v European Commission C-26/75 [1975] ECR 1367 70, 312
Gesellschaft zur Vertwertung von Leistungsschutz rechten GmbH (GVL) v European Commission C-7/82 [1983] ECR 483 23
Greek oil monopoly C-347/88 [1990] ECR 4747 . . . . 122, 124, 125, 127
Grimaldi v Fond des Maladies Professionelles [1989] ECR 4407 333
Hansen [1979] ECR 935 122
Hoffman la Roche v European Commission [1979] ECR 461 63
Hunermund v Landesapothekerkammer Baden-Wuttenberg C-292/92 [1993] ECR 6787 103

Italy v European Commission (British Telecommunications Plc) C-41/83 [1985] ECR 873 ....22, 23, 26, 115, 220, 222, 223
Independent Television Publications (ITP) Ltd v European Commission T-76/89 [1991] ECR II-575 ....86, 313
Keck and Mithouard C-267/91 and C-268/91 [1993] ECR 6097 100, 102, 103, 104, 105
Klaus Höfner Fritz Elser v Macrotron GmbH (Höfner) C-41/90 [1991] I-1979 10, 14, 23, 34, 60, 62, 63, 64, 73, 75, 76, 81, 82, 83, 108, 117, 130, 131, 137, 146, 210, 318, 335, 337
Ladbroke Racing Limited v European Commission T-32/93 [1994] ECR II-1015 ....24, 56, 57
Lagauche C-46/90 and C-93/91 [1993] ECR I-5267 ....113, 114, 120
LTU v Eurocontrol [1976] ECR 1541 61
Merci Convenzionali Porto di Genova SpA v Siderurgica Gabrielli (Port of Genoa) C-179/90 [1991] ECR I-5889 14, 34, 60, 64, 70, 71, 73, 76, 118, 137, 318, 335, 337, 361, 363–4, 373
Metropole Television SA and Reti Televisive Italiane v European Commission T-528/93, T-542/93, T-543/93, T-546/93 [1996] ECR II-649 ....93, 94, 95, 96, 310–12
Michelin v European Commission [1983] ECR 3461 ....63, 64
Ministere public v Asjes et al (Nouvelles Frontieres) C-209/84 to C-213/84 [1986] ECR II-1425 340
Miritz C-91/75 [1976] ECR 217.................................122, 123
Muller-Hein C-10/71 [1971] ECR 723.................22, 23, 38, 181, 267
Netherlands-Koninklijk PTT Nederland NV and PTT Post BV v European Commission (Dutch messenger services) C-48/90 and C-66/90 [1992] ECR I-565...................................72, 80, 137, 318, 324
Night Services T-374/94 ....86, 170
NV GB-Inno-BM v Vereniging van de Kleinhandelaars in Tabak (ATAB) C-13/77 [1977] ECR 2115.........11, 13, 73, 75, 105, 111, 117, 118, 312
Poucet et Pistre C-159/91 and C-160/91 [1993] ECR 637 ....61, 62, 130
Publico Ministerio v Manghera C-59/75 [1976] ECR 91.....3, 122, 123, 124
R. v Bouchereau C-30/77 [1977] ECR 1999..........................133
R. v Secretary of State for Transport Ex p. Factortame C-221/89 [1991] ECR I-3905...282
Radio Telefis Eireann (RTE) v European Commission T-76/89R, T-77/89R, T-91/89R [1989] ECR 1141 313–15
Radio Telefis Eireann (RTE) v European Commission T-69/89 [1991] ECR II-485...86, 313–15
Radio Telefis Eireann (RTE), Independent Television Publications Ltd (ITP) supported by Intellectual Property Owners Inc (IPO) v European Commission supported by Magill TV Guide Ltd C-241/91 P and C-252/91 P [1995] ECR I-743........................173, 313–15, 330
Regie des Telegraphes et des Telephones (RTT) v GB-Inno-BM SA C-18/88 [1991] ECR I-5941 14, 23, 27, 28, 60, 76, 78, 79, 80, 86, 112, 113, 114, 115, 118, 119, 140, 171, 183, 185, 186, 223–4, 318, 357

Rendo NV, Centraal Overijsselse Nutsbedrijven NV, NV Regional Energiebedrijf Salland v European Commission (Rendo I) T-16/91 [1992] ECR II-2417 . . . . .38, 39, 55, 56, 97, 106, 211, 213–14
(Rendo II) C-19/93 P [1995] ECR I-3319 . . . . .56, 211
REWE-Zentral v Bundesmonopolverwaltung fur Brauntwein (Cassis de Dijon) C-120/78 [1979] ECR 649 . . . . .100, 101, 102, 110, 112
RTL, Vereniging Veronica Omroep Organisatie and Commissariat voor de Media C-148/91 [1993] ECR I-0487 299
Rush Portugesa C-113/89 [1990] ECR I-1417 134
Sacchi C-155/73 [1974] ECR 409 1, 2, 3, 4, 12, 13, 18, 19, 23, 24, 65, 66, 73, 74, 106, 120, 126, 142, 143, 147, 185, 303–5
Sager v Dennemeyer C-76/90 [1991] ECR I-4221 136
Societe Civile Agricole du Centre d'Insemination de la Crespelle v Cooperative d'Elevage est d'Insemination Artificielle du Departement de la Mayenne (Crespelle) C-323/93 [1994] ECR 5077 27, 33, 34, 36, 69, 70, 71, 73, 110, 111, 114, 117
Societe d'Importation Edouard Leclerc-Siplec v TFI Publicite SA and M6 Publicite SA C-412/93 [1995] ECR I-179 283
Spain, Belgium, Italy v European Commission (Telecommunications Services) C-271/90, C-281/90, C-289/90 [1992] ECR 1–5833 9, 10, 12, 14, 20, 44, 50, 109, 131, 136, 137, 139, 140, 143, 144, 210, 266
Srl Ufficio Henry van Ameyde v Srl Ufficio Centrale Italiano di Assistenza Assicurativa Automobilisti in Circolazione Internazionale (UCI) C-90/76 [1977] ECR 1091 . . . . .13, 137, 138, 139
Steinike and Weinlig v Germany C-78/76 [1977] ECR 595 156
Stichting Collectieve Antenne Voorziening Gouda v Commissariat voor de Media C-288/89 [1991] ECR I-4007 307–8
Stoke on Trent v B&Q C-169/91 [1992] ECR 6635 103
Syndicat Francais de l'Express International v La Poste C-39/94 [1996] ECR I-3547 . . . . .155, 159
Syndicat Francais de l'Express International, DHL International Service CRIE, May Courier v European Commission T-77/95 [1997] ECR II-1 . . . . .92, 159
Syndicat National des Frabricants Raffineurs d'Huile de Graissage v Groupement d'Interet Economique 'inter-Huiles' (Waste Oils) C-172/82 [1983] ECR 555 . . . . .10, 13, 23, 109, 110, 111, 115
Taillandier C-92/91 [1993] ECR I-5383 . . . . .113, 114
Tetra Pak II T-83/91 [1994] ECR II-755 92
Thierry Tranchant v Telephone Store SARL C-91/94 [1995] ECR I-3911 113
Tilly v European Commission C-41/83 [1985] ECR 873 312
Torfaen v B&Q C-145/88 [1989] ECR 3851 103
Van Binsbergen C-33/74 [1974] ECR 1299 . . . . .132, 134, 135
Van Schinjdel (Jeroen) and Stichting Pensioenfonds voor Fysiotherapeuten C-430/93 and C-431/93 [1995] ECR I-4705 57

Van Wesemael C-110/78 and C-111/78 [1979] ECR 35 . . .132, 134, 135, 149
Volvo v Veng C-238/87 [1988] ECR 6211 . . . . .81, 173
Vereniging van Vlaamse Reisbureau C-311/85 [1987] ECR 3801. 69
Webb C-279/80 [1981] ECR 3305. . . . . .132, 134
Zuchner v Bayerische Vereinsbank AG C-172/80 [1981] ECR 2021 . . .22, 24

TREATIES

European Convention on Human Rights 1950 Article 10(1) 288
European Convention on Transfrontier Television 280–1, 285
Single Europe Act 1987. 332
Treaty of Rome 1957
Article 2 109
Article 2(3) 162
Article 3b 300
Article 3c 109
Article 3g 109
Article 5. . . . . .26, 69, 349, 364
Article 5(2) 266
Article 6 8
Article 7. . . . . .304, 325, 363
Article 7a 299
Article 7d 21
Article 9 105
Article 19(3) 65, 216
Article 30 13, 14, 35, 52, 56, 75, 99, 100, 101, 103, 104, 105, 106, 107, 108, 109, 110, 111, 112, 113, 118, 119, 121, 124, 125, 126, 137, 154, 192, 194, 210, 215, 306
Article 31. .99, 100, 105, 111, 113, 119, 121, 125, 210
Article 32. .99, 100, 105, 111, 113, 119, 121, 125, 210
Article 33. .99, 100, 105, 111, 113, 119, 121, 125, 210
Article 34 . . . . .99, 100, 105, 110, 111, 113, 119, 121, 125, 210
Article 35 . . . . .100, 111, 119, 121, 125, 210
Article 36 99, 100, 101, 108, 110, 111, 114, 116, 117, 118, 119, 125, 126, 127, 182, 210, 215
Article 37 56, 99, 100, 107, 119, 120, 121, 122, 123, 124, 125, 126, 127, 192, 194, 210, 212
Article 37(1) 119–20, 121, 123
Article 37(6) 122
Article 52 . . . . .53, 153, 159, 160, 161, 299, 300
Article 53 159
Article 55 . . . . .52, 62, 63
Article 56. . . . . .52, 74, 130, 133, 139, 149, 182, 306
Article 56(1) 129
Article 57(2) 197, 200

Article 59 13, 52, 74, 120, 129, 130, 131, 132, 133, 135, 136, 137, 138, 139, 140, 141, 142, 143, 144, 146, 150, 151, 161, 185, 192, 210, 234, 285, 306, 307, 308
Article 60 .....62, 129, 130, 131, 132, 136
Article 63 132
Article 66 .....62, 197
Article 67 200
Article 77 182
Article 84(2) 353
Article 85 13, 42, 47, 51, 59, 60, 62, 63, 65, 67, 68, 86, 93, 94, 193, 211, 212, 227, 240, 261, 263–5, 266, 267, 268, 306, 315, 339, 340, 345, 349, 352, 360, 370, 376, 380
Article 85(1) 12, 26, 69, 94, 96, 265, 309, 310, 349, 371, 373, 376, 377, 381
Article 85(3) 94, 95, 96, 97, 170, 214, 263, 309, 310, 345, 355, 371
Article 85(3)a 311
Article 86 1, 3, 13, 18, 27, 36, 42, 47, 51, 52, 53, 54, 59, 60, 62, 63, 65, 66, 67, 68, 69, 70, 71, 72, 73, 75, 76, 78, 79, 80, 82, 92, 93, 115, 117, 118, 119, 137, 140, 143, 144, 146, 150, 151, 154, 156, 161, 168, 186, 211, 212, 216, 223, 224, 227, 234, 235, 240, 261, 263–5, 266, 267, 268, 305, 306, 307, 312, 313, 314, 315, 322, 323, 324, 337, 339, 340, 345, 349, 352, 353, 357, 360, 361, 362, 363, 364, 370, 376
Article 86b 73, 76
Article 86c 28, 69, 352
Article 87 .....43, 45, 47
Article 88 341
Article 89 .....42, 341
Article 90 1–4, 14, 44, 47, 54, 56, 57, 60, 63, 68, 78, 79, 80, 86, 105, 106, 108, 109, 115, 117, 136, 137, 138, 153, 159, 160, 165–217, 210, 219, 228, 234, 254, 257, 258, 261, 266, 267, 277, 303–16, 315, 322, 325, 326, 337, 348–60, 352, 353, 361, 362, 370, 371, 382
Article 90(1) 2, 3, 7–15, 19–21, 25, 37, 38, 41, 42, 43, 44, 47, 48, 53, 54, 67, 68, 69, 70, 71, 72, 73, 74, 75, 76, 97, 106, 107, 109, 119, 126, 154, 155, 156, 159, 168, 215, 224, 229, 234, 235, 306, 307, 323, 324, 325, 326, 327, 348, 349, 362, 363, 364
Article 90(2) 4, 17–40, 41, 42, 52, 54, 56, 62, 66, 67, 68, 69, 71, 72, 74, 82, 94, 95, 96, 97, 107, 109, 111, 115, 117, 119, 126, 127, 128, 136, 147, 150, 155, 156, 157, 158, 161, 162, 163, 182, 183, 184, 185, 204, 211, 212, 213, 214, 215, 216, 224, 235, 236, 241, 267, 304, 306, 309, 311, 313, 314, 323, 325, 336, 363, 371
Article 90(3) 2, 38, 41–57, 50, 75, 144, 153, 156, 157, 158, 160, 166, 176, 210, 227–45, 261, 262, 263, 300, 327, 348, 349, 353, 365
Article 92 .....153, 155, 156, 158, 159
Article 92(1) 155
Article 92(2) 155, 157
Article 92(2)a 155

Article 92(3) 155, 156, 157
Article 93 . . . . .153, 155, 156, 158, 159
Article 93(3) 157
Article 100a. . .45, 47, 197, 200, 219, 241, 245, 257, 258, 268, 271, 327, 332
Article 128 288
Article 129b 220
Article 155 41
Article 169 . . . . .43, 45, 46, 56, 57, 288, 299, 307
Article 170 288
Article 173 . . . . .54, 57
Article 174 54
Article 175 57
Article 177 . . . . .63, 304
Article 189 . . . . .44, 48
Article 222 . . . . .2, 3, 106, 107, 108, 153, 154, 162
Article 235 300

COMMISSION COMMUNICATIONS

Application of the competition rules to access agreements in the telecommunications sector (draft), COM (96) 649 final. 268–70
Europe's way to the information society, COM (94) 347 final 220, 272
Progress report on the thinking and work done in the field of telecommunications and initial proposals for an action programme (December 1984) 221
Review of the situation in the telecommunications services sector (1992), SEC (92) 1048 . . . . .230, 272
Services of general economic interest, COM (96) 443 final 22, 60, 61, 175, 181
Towards a dynamic economy: green paper on the development of the Common Market for telecommunications services and equipment, COM (87) 290 final. 221–4, 271, 272
Universal service for telecommunications, COM (96) 419 final 244, 259

COMMISSION DECISIONS

ACRISS, OJ 1993 C149/9 357–8
Air France/Sabena (5 October 1992) 162
Amadeus/Sabre (2 August 1991), IP/91/784 .169, 358
Anseau Navewa OJ 1982 L167/39 . . . . .18, 23, 25
Atlas, OJ 1996 L239/23 . . . . .263, 264
B&I v Sealink [1992] 5 CMLR 255 . . . . .85, 86, 168
Belgacom/ITT Promedia, IP/97/292. . . . . .263, 265
British Midland v Aer Lingus, 92/213/EEC, OJ 1992 L96/34 85, 86, 170, 360

British Telecommunications Plc, 82/861/EEC, OJ 1982 L360/36. . . .66, 223
British Telecommunications Plc/MCI, IP/97/406 262–3
CEA Industrie/France Telecom/SGS Thomson (22 February 1993) 162, 164
Conditions imposed on the second operator of GSM radiotelephony services in Italy (4 October 1995) OJ 1995 L280/49. . .13, 53, 73, 239, 263
Conditions imposed on the second operator of GSM radiotelephony services in Spain, OJ 1996 L76/19. 263
Distribution of railway tickets by travel agents (25 November 1992), 92/568/EEC, OJ 1992 L366/47 380–2
EBU/Eurovision System (11 June 1993) OJ 1993 L179/23 . . .93, 95, 96, 310
Electricite de France (EDF) v ENEL (Grangevielle) (press release), IP/92/6680 (14 August 1992) 215–16
European Night Services (ENS) (21 September 1994) OJ 1994 L259/20 86, 91, 170, 375–8
Eurotunnel (13 December 1994), 94/894/EEC, OJ 1994 L354/66 373
Eurotunnel agreements on rail transport of goods: ACI 378
Eurotunnel usage agreements, OJ 1988 C292/2 .371–5
Global European Network, IP/97/242 . . . . .263, 265
HOV/SVZ/MCN, OJ 1994 L104/34 382
IBM, Bull. E.C. 10–1984 p.96 173
Ijsselcentrale (16 January 1991) OJ 1991 L28/32 . . . . .93, 95, 97, 98, 193
Insurance in Greece of public property and loans granted by Greek state-owned banks (24 April 1985) OJ 1985 L152/25 . . . . .53, 54, 55, 159
Iridium, IV/35.518. . . . . .263, 265
Jahrhundertvertrag (22 December 1992) OJ 1993 L50/14. . . .93, 95, 96, 214
Kali und Salz/MKD/Treuhand (14 December 1993). 162
Landing fees at Brussels airport (28 June 1995) OJ 1995 L216/8 12, 53, 69, 70, 348, 351–2
London European/Sabena (4 November 1988), 88/589/EEC, OJ 1988 L317/47 . . . . .86, 169, 354, 357
Lufthansa/Air Europe, report on competition policy (1990) point 106 . .360
Magill TV Guide/ITP, BBC and RTE (21 December 1988) OJ 1989 L78/43 313
Malpensa airport (press release), IP/93/684, (30 July 1993) 81, 169
MSG Media Service (9 November 1994), OJ 1994 L364/1 289–91
National Carbonizing Company, 76/185/EEC, OJ 1976 L35/6 86
Nordic Satellite Distribution (19 July 1995), OJ 1996 L53/20. 293–5
Omnitel Pronto Italia, OJ 1995 L280/49 262
Pechiney/Usinor (24 June 1991). 162, 164
Phoenix v Global One, OJ 1996 L239/57. . . . . .263, 264
Provision in The Netherlands of express delivery services (20 December 1989), 90/16/EEC, OJ 1990 L10/47. . . . . .43, 53, 54, 55, 72, 322
Provision in Spain of international express courier services (1 August 1990), 90/456/19, OJ 1990 L233/19. . . . . .42, 53, 72, 89, 322

Reductions in air and sea transport fares available only to Spanish nationals resident in the Canary Islands and Balearic Islands (22 June 1987) OJ 1987 L194/28; [1988] 1 CMLR 331 . . . . .53, 348
Refusal to grant access to the facilities of the port of Rødby in Denmark (21 December 1993) OJ 1994 L55/52 53, 72, 86, 166, 168, 181, 269, 360, 361–3
RTL/Veronica/Endemol (Holland Media Group – HMG) (20 September 1995), 96/346/EC. 292–3
Sabre/Air France and Iberia, report on competition policy (1993) point 238 358–9
Screensport/EBU members (19 February 1991) OJ 1991 L63/32 309–10
Sea Containers v Stena Sealink (interim measures) (21 December 1993) OJ 1994 L15/8 . . . . .84, 85, 86, 165, 166, 167, 168, 174, 269, 360–1, 373
Sterling Airways/SAS Denmark, Tenth report on competition policy (1981) point 135 349
Tariff structures in the combined transport of goods (24 February 1993), OJ 1993 L73/38 378–80
TAT-Paris (Orly)-London (27 April 1994), 94/290/EC, OJ 1994 L127/22 295–7
TAT-Paris (Orly)-Marseille and Paris (Orly)-Toulouse (27 April 1994), 94/291/EC, OJ 1994 L127/32. . . . . .54, 350–1
Telefonica/Canal Plus/Cablevision (8 October 1996), IP/96/897 295–7
Uniform Eurocheques (10 December 1984) OJ 1985 L35/43 . . . . .93, 94, 95
Vebacom 262
Viva Air (28 May 1993), 93/347/EEC, OJ 1993 L140/51 . . . . .344, 350

COMMISSION DIRECTIVES

Abolition of the restrictions on the use of cable television networks for the provision of telecommunications services (amending 90/388/EEC), 95/51/EC, OJ 1995 L256/49 . . . . .51, 233–6, 245
Competition in the markets for telecommunications services, 90/388/EEC, (28 June 1990) OJ 1990 L192/10 (amended 96/19/EC, OJ 1996 L73/13 . . . . .39, 51, 52, 53, 143, 151, 154, 171, 227, 229–30, 236, 247, 261, 267
Competition in the markets for telecommunications terminal equipment, 88/301/EEC, (16 May 1988) OJ 1988 L131/73 2, 45, 47, 51, 52, 53, 68, 106, 108, 114, 125, 154, 227–9, 261
Full competition in telecommunications markets (amending 90/388/EEC), 96/19/EC, OJ 1996 L74/13 . . . . .51, 52, 240–5, 246
Mobile and personal communications (amending 90/388/EEC), 96/2/EC, OJ 1996 L20/59. . . . . .51, 237–40, 246
Satellite communications, 94/46/EC, OJ 1994 L268/15 (amending 88/301/EEC and 90/388/EEC) 11, 12, 51, 52, 231–3
Transparency of financial relations between Member States and public undertakings, 80/723/EEC, (25 June 1980) OJ 1980 L195/35 8–9, 45, 51, 158

Transparency of financial relations between Member States and public undertakings, 85/413/EEC, OJ 1985 L229/20 51
Transparency of financial relations between Member States and public undertakings 93/84/EEC OJ 1993 L254/16 51

COMMISSION DOCUMENTS

XXIVth Report on Competition Policy (1994) 7
XXVth Report on Competition Policy (1995) 123
Bangemann Group Report on Europe and the Information Society, COM (94) 347 final 220, 272
Common approach in the field of satellite communications in the European Community, COM (90) 490 final 24–5, 272
Competition in the GSM mobile telecommunications market (press release), IP/96/791 (8 August 1996) 239
Digital video broadcasting communication (17 November 1993), COM (93) 557 final 278, 300–2
Electricity Market Organization (working paper), SEC (95) 464, (22 March 1995) 201
Green paper on a common approach to mobile and personal communications in the European Union, COM (94) 145 final . . .225, 272
Green paper on copyright and related rights in the information society (19 July 1995) COM (95) 382 final 316
Green paper on liberalization of telecommunications infrastructure and cable television networks, COM (94) 440 and COM (94) 682 final 225–7, 272, 236, 244, 245–6, 259
Green paper on the development of the single market for postal services, COM (91) 476 final 82, 87, 171, 317, 318–20
Green papers on broadcasting 278
Growth, competitiveness and employment (Delors White Paper), COM (93) 700 final 220, 272
Guidelines for the development of Community postal services, COM (93) 247 final 317, 320–1
Guidelines on the application of EEC competition rules in the telecommunications sector, OJ 1991 C233/2 .266–7
Internal Energy Market (Working Document), COM (88) 238 final, (2 May 1988) 192
Pluralism and media concentration in the Internal Market (23 December 1992) COM (92) 480 final 278, 298–300
Small and very small electricity systems in the internal market (working paper), SEC (95) 685, (10 May 1995) 201
Statement concerning Council Resolution on universal service in the telecommunications sector, 94/C/48/06, OJ 1994 C48/8 236
Strategy options to strengthen the European programme industry (6 April 1994) COM (94) 96 final 278, 297–8

COMMISSION NOTICES

Application of the competition rules to postal services (December 1995) (draft) OJ 1995 C322/3 . . . . .50, 53, 172, 333–7
Draft Directive on mobile and personal telecommunications, OJ 1995 C197/5. 237
Electricidad de Portugal (Pego), OJ 1993 C265/3 .216
Electricity in England and Wales, OJ 1990 C191/9. 216
Electricity in Northern Ireland, OJ 1992 C92/5 216
Electricity in Scotland, OJ 1990 C191/9 216
Independent power generators, OJ 1992 C92/4 216
International Energy Agency, OJ 1993 C300/8 216
Notion of concentration, 94/C385/02, OJ 1994 C385/5 . . . . . .161, 163, 164
Notion of undertakings concerned, 94/C385/03, OJ 1994 C385/12. 161
Reorganization of the electricity industry in Scotland, OJ 1990 C245/9 . . .65
Twinning program engineering, OJ 1992 C148/8 .216
Unisource v Telefonica, OJ 1997 C44/15. . . . . .263, 264
Uniworld, OJ 1997 C44/4. . . . . .263, 264

COMMISSION REGULATIONS

Agreements relating to computer reservation systems, (EEC) 3976/87, OJ 1987 L374/9 (as amended by (EEC) 2672/88, OJ 1988 L239/13; (EEC) 83/91, OJ 1991 L10/9; (EEC) 1618/93, OJ 1993 L155/23; (EEC) 3652/93, OJ 1993 L333/37) 354, 356
(EEC) 1617/93, OJ 1993 L155/18 . . . . .85, 168

COUNCIL DIRECTIVES

Advanced television standards, 95/47/EC, OJ 1995 L281 302–3
Allocation of railway infrastructure capacity and the charging of infrastructure fees, 95/19/EC, OJ 1995 L143/75. . . . . .170, 365
Annual accounts of limited liability companies, 78/660/EEC, OJ 1978 L222/11. 206
Application of open network provision to leased lines, 92/44/EEC, OJ 1992 L165/27 . . . . .276, 247–9
Application of open network provision to voice telephony, 95/62/EC, OJ 1995 L321/6 . . . . .249–50, 259, 260
Application of open network provision to voice telephony and on universal service replacing 95/62/EC (draft) 251–2, 260
Common framework for general authorizations and individual licences in the telecommunications services field, 97/13/EC, OJ 1997 L117 . . .252–4
Common rules for the development of Community postal services (16 December 1994) (draft). 160, 176–7

Common rules for the development of Community postal services (29 April 1997) (Common Position) OJ 1997 C-188/9 . . . . 177–81, 186, 327–33
Common rules for the internal market in electricity (draft), OJ 1992 C65/4 196–7
Common rules for the internal market in electricity (amended proposal), OJ 1994 C123/1 . . . . 199, 202–9
Common rules for the internal market in natural gas (draft), OJ 1992 C65/14 196–7
Common rules for the internal market in natural gas (amended proposal) OJ 1994 C123/26, COM (93) 643 final 199
Community procedure to improve the transparency of gas and electricity prices charged to industrial end users, 90/377/EEC, OJ 1990 L185/16 . . . . 191, 194, 196
Development of the Community railways (access), 91/440/EEC, OJ 1991 L237/25 . . . . 49, 92, 170, 186, 365–70, 372, 373, 376, 378
Establishment of the internal market for telecommunications services through the implementation of open network provision, 90/387/EEC, OJ 1990, L192/1 . . . . 225, 247, 248, 276, 257
Fares for scheduled air services between Member States, 87/601/EEC, OJ 1987 L374/12 . . . . 67, 341
General authorizations and individual licences in the field of telecommunications services (draft), COM (95) 545 final as amended by Common Position 9944/96 241–2
Ground handling, 96/67/EEC, OJ 1996 L272/36 . . . . 168, 353–4
Interconnection in telecommunications, 97/33/EC, OJ 1997 L199/32 243, 254–6, 259
Licensing of railway undertakings, 95/18/EC, OJ 1995 L143/70 . . . 170, 365
Public procurement relating to the excluded sectors, 93/38/EEC, OJ 1993 L199/84 Art.1(2) 9
Television without frontiers, 89/552/EEC, OJ 1989 L298/23 277–89
Transit of electricity through transmission grids, 90/547/EEC, OJ 1990 L313/30 . . . . 191, 194, 196, 216
Transit of natural gas through grids, 91/296/EEC, OJ 1991 L147/37 191, 194, 195–6

COUNCIL REGULATIONS

Access for air carriers to scheduled intra Community air service routes, (EEC) 2343/90, OJ 1990 L217/8 342
Access for Community air carriers to intra Community air routes, (EEC) 2408/92, OJ 1992 L240/8. . . . . 54, 182, 343, 344, 350, 351
Application of competition rules to air transport sector, (EEC) 3975/87, OJ 1987 L374/1 (amended by 2410/92, OJ 1992 L240/18) 339, 340, 341, 343, 345
Block exemption (Agreements and concerted practices in the air transport sector), (EEC) 3976/87, OJ 1987 L374/9 (amended by (EEC) 2344/90,

OJ 1990 L217/15 and (EEC) 2411/92, OJ 1992 L240/19) 341, 342, 343, 345, 346, 354
Block exemption (computerized reservation systems) (EEC) 83/91, OJ 1991 L10/9. . . . . .169, 184, 339
Block exemption (rail, road and inland waterways) (EEC) 1017/68, OJ 1968 L175/68 . . . . .339, 371, 372, 373, 375, 376, 378, 380, 381
Block exemption (sea transport) (EEC) 4056/86, OJ 1986 L378/4 339
Code of conduct for computerized reservation systems, (EEC) 2299/89, OJ 1989 L220/1 (as amended by 3089/93, OJ 1993 L278/1) 355, 356
Common rules for the allocation of slots at Community airports, (EEC) 95/93 (18 January 1993) OJ 1993 L14/1 . . . . .85, 168, 346, 347
Control of concentrations between undertakings (Merger) (EEC) 4064/89, OJ 1990 L257/14 . . . . .71, 153, 161, 162, 163, 261, 268, 289–97, 315
Coordination of airline capacity, (EEC) 84/91, OJ 1991 L10/14. 359
(EEC) 13/204, OJ 1962 13/204, 1959–62. 340
(EEC) 17/62, OJ 1962 13/204, 1959–62 42, 56, 94, 95, 263, 339, 340, 373, 381
(EEC) 141/62, OJ 1962 124/2752, 1959–62 339
Fares and rates for air services, (EEC) 2409/92, OJ 1992 L240/15 343
Fares for scheduled air services, (EEC) 2342/90, OJ 1990 L217/1 342
Freedom to provide services to maritime transport (EEC) 4055/86, OJ 1986 L378/1 . . . . .139, 364
Groundhandling, (EEC) 82/91, OJ 1991 L10/7. 339
Licensing of air carriers, (EEC) 2407/92, OJ 1992 L240/1. 343
Public service in transport by rail, road and inland waterway, (EEC) 1191/69, OJ 1969 L156/1 (as amended by 1893/91, OJ 1991 L169/1) 367

EUROPEAN PARLIAMENT

Desama Report EN/PR/225/225592 A3–0000/93 (14 April 1993) 198

NATIONAL LEGISLATION

Broadcasting Act 1990 s.43. . . . . .284, 285
Decree of the Council of the French Community on Audio-visual Communications 1987 (Belgium). 287
Sherman Act (United States) 86

INTRODUCTION

State monopolies have come under the limelight of the Community authorities and in particular the European Commission for about ten years.

The purpose of this book is to explain the fundamental changes which have taken place in the application of Treaty rules to state monopolies, i.e. to undertakings which have a close relationship with the State and have been granted certain privileges by it. These undertakings generally include the utility companies but may also extend to companies in other sectors.

Changes have occurred following Court judgments on Article 90 and complementary EC regulatory measures in the utility sector. It is important for practitioners and business to understand the change, so as to make best use of Article 90 and the new EC regulations.

The legal environment keeps changing to take into account technical, economic and social developments. This obviously makes it difficult to give definite legal answers to problems which companies and their advisers may encounter at any one time. However, an understanding of the key principles established in recent leading cases of the Court is essential to anticipate new trends and resolve any question in this area.

The new interpretation of Article 90

The traditional application of the Article can be found in the *Sacchi*[1] case of 1974. This concerned a monopoly granted to an enterprise by the Italian State over television broadcasting. The Court held that the fact that an undertaking to which the State grants exclusive rights has a monopoly is not as such incompatible with the Treaty rules on competition. The existence of a monopoly granted by the State is not contrary to Article 86. In its relations with third parties, however, the monopoly must behave in a fair manner consistent with the competition rules. In particular, it cannot impose unfair

[1] *Sacchi*: C-155/73, 30 April 1974, [1974] ECR 409.

charges or conditions on them or discriminate in favour of national suppliers against foreign suppliers.

Sacchi was in line with the views of the founders of the Treaty. They, consistent with the political mood, provided for the right of a State to reserve to itself certain sectors of the economy and thus exclude them from competition by other operators. Their line of thinking was expressed in Article 222 of the Treaty which stated that the Treaty should be neutral as far as public or private ownership was concerned. It was further expressed in Article 90(1) which refers to public undertakings and undertakings to which Member States grant special or exclusive rights and appears to take such exclusivity or monopoly for granted.

The fundamental change occurred with the *Terminal Equipment*[2] case. This concerned the monopoly then held by all national telecommunications organizations over terminal equipment such as telephone sets. The Commission had ruled in a Directive[3] based on Article 90(3) that this monopoly infringed Treaty rules on free movement of goods and was contrary to Article 90. Proceedings followed in which France and other Member States attacked the Commission Directive. Advocate General Tesauro advised the Court as follows (in paragraph 35 of the Opinion):

> I would reiterate that the fact that a monopoly – and often, in practice, other similar arrangements – is likely to impair the proper functioning of the common market and freedom of competition was duly considered by the draftsmen of the Treaty, who nonetheless chose to tolerate it, so that the Commission Directive . . . by simply abolishing the monopoly or the system of exclusive rights without taking the trouble of providing an appropriate statement of reasons, radically alters the basic framework set out in Article 90(1).

The Advocate General recommended that the Court should annul the articles of the Directive which required the abolition of the existing monopoly over terminal equipment. The Court did not follow the Advocate General and upheld the key parts of the Commission decision abolishing the monopoly. The Court held that:

> Even if Article 90 presupposes the existence of undertakings which have certain special or exclusive rights, it does not follow that all the special and exclusive rights are necessarily compatible with the Treaty. That depends on different rules, to which Article 90(1) refers.

The Court held that the monopoly was contrary to the rules on free movement of goods and was therefore illegal.

---

[2] *French Republic* v. *Commission* (the *Terminal Equipment* case) C-202/88, 19 March 1991, [1991] ECR 1223.

[3] Commission Directive of 16 May 1988 on competition in telecommunications terminal equipment OJ 1988 L131/73.

This case constitutes a turning point, which, if one uses the words of Advocate General Tesauro, radically alters the basic framework set out in Article 90(1). Monopolies can now potentially be illegal *per se.* Previously it was only in specific cases relating to the import of goods, as in *Manghera*[4], that the existence of monopolies could be contrary to the Treaty. It is now possible to challenge the validity of monopolies as such. In what circumstances? *Terminal Equipment* does not provide an answer. On the basis of *Terminal Equipment* all national monopolies, at least over products, may potentially be contrary to the Treaty in so far as they restrict imports.

Further cases adopted the approach in *Terminal Equipment* but did not cast any light on the rationale for the change in the Court's reasoning since *Sacchi* or attempt to distinguish the apparent contradiction with Articles 90(1) and 222 which appear to recognize the validity of State monopolies as such. The Court simply gave preference to the competition rules and the fundamental freedoms (free circulation of goods and freedom to provide services) over the rules permitting the existence of monopolies.

In *ERT*[5] (1991), the Court began to provide an explanation, stating that Member States must not adopt or maintain in force any measure which could deprive Treaty rules of their effectiveness. Article 90(1) prohibits the grant of exclusive rights where those rights are liable to create a situation in which the undertaking is *led* to infringe Article 86.

It was only in *Corbeau*[6] (1993), later confirmed in *Almelo*[7] (1994), that a further explanation was given and a limit was drawn to the permitted scope of monopolies.

Those cases provide that in order for a monopoly to be compatible with Treaty rules, it must be *necessary* in order to enable the performance of a task of general public interest by the monopoly holder. Such a 'task' can consist, for example, of collecting, transporting and delivering mail throughout a defined territory to all users at affordable prices as in *Corbeau*, or in supplying electricity throughout a defined territory to all users at affordable prices as in *Almelo.*

This book examines how the necessity test has been applied by the Court and the limits beyond which the holder of monopolies must accept competition.

The change in the Court's approach from *Sacchi* to *Terminal Equipment/ Corbeau/ Almelo* may be summarized as follows:

- The *Sacchi* approach: the grant of exclusive rights by the State and the existence of monopolies granted for reasons of general public interest is

---

[4] *Publico Ministerio* v. *Manghera.* Case 59/75 [1976] ECR 91. The Court held that the exclusive right over imports from other Member States held by commercial monopolies was contrary to Article 37.

[5] *Elliniki Radiophonia Tileorassi Anonimi Etairia (ERT AE) and Dimotiki Etairia Pliroforissis (DEP)* (the *ERT* case) C-260/89, 18 June 1991, [1991] ECR I-2925.

[6] *Paul Corbeau* C-320/91, 19 May 1993, [1993] ECR I-2533.

[7] *Gemeente Almelo and others and Energiebedrijf IJsselmig NV*, C-393/92, 27 April 1994, [1994] ECR 1477.

not incompatible with the Treaty and is not contrary to the competition rules. The competition rules apply but only to the relationship between the monopoly undertaking and third parties. In other words, undertakings benefiting from a monopoly which have been given a task of general interest must *behave* towards third parties in a manner consistent with the competition rules *unless* this would prevent them from carrying out their task. Only the *behaviour* of the monopoly towards third parties, i.e. the way in which the monopoly is *exercised*, may be contrary to Treaty rules.

- The *Terminal Equipment, Corbeau* and *Almelo* approach: The competition rules apply not only, as in *Sacchi,* to the way in which the monopoly is *exercised*, but also to the *existence* of the monopoly itself. The existence of monopolies granted by the State may be contrary to the Treaty. Monopolies are only compatible with the Treaty if they are necessary to perform a task of general economic interest.

Structure of the book

Part I of the book explains how Treaty rules apply to state monopolies, pursuant to Article 90, in the light of these changes in the Court approach. It first analyses in detail the provisions of Article 90 and in particular the scope of the derogation contained in Article 90(2). It then examines the application to state monopolies of the principal Treaty rules – including the competition rules, the rules on free movement of goods, on the freedom to provide services and the state aid rules.

Part II of the book examines the application of Treaty rules and the principal regulatory measures complementing these rules in the utility sectors – energy, telecommunications, broadcasting, post and transport. It describes in broad terms the liberalization measures taken by the Commission and the Council to open up these sectors to competition and the harmonization measures in order to create one single market in each sector. It also reviews some of the key Commission and Court cases, in each of the sectors.

The EC regulatory framework changes at a rapid pace in the utility sector. This book describes the principal regulatory measures and the case-law as at the *end of June 1997.*

Part I

ARTICLE 90

Chapter 1

ARTICLE 90(1): MEMBER STATES' OBLIGATIONS

> In the case of public undertakings and undertakings to which Member States grant special or exclusive rights, Member States shall neither enact nor maintain in force any measure contrary to the rules contained in this Treaty, in particular to those rules provided for in Article 6 and Articles 85 to 94.

This chapter discusses first the general meaning of Article 90(1) and then considers the meaning given to the various phrases contained in the Article.

General meaning

Article 90(1) is addressed to Member States. It requires them not to take any measure contrary to Treaty rules in favour of certain undertakings, namely public undertakings and undertakings to which they grant special and exclusive rights. These undertakings, the precise definition of which is examined below and which are referred to throughout the first part of this book as 'state-related undertakings', are singled out because they have a close relationship with the State. They are either owned or controlled by it or have been granted certain rights by it. The State will, in all cases, have a particular interest in them and will normally have the power to influence them.

The purpose of Article 90(1) is to ensure that this close relationship with the State does not create imbalances and distortions of the market to the detriment of other undertakings[1]. Article 90(1) refers to distortions due to two types of infringement – namely (i) breach of the competition rules

---

[1] The *Transparency Directive* case (1982), *France, Italy, UK* v. *Commission*, Cases 188 to 190/80 [1982] ECR 2545: '. . . the reason for the inclusion in the Treaty of the provisions of Article 90 is precisely the influence which the public authorities are able to exert over the commercial decisions of public undertakings' (ground 26); see also the European Commission XXIVth Report on Competition Policy 1994, point 215, second paragraph: 'Member States have close links with State enterprises and with enterprises to which Member States grant special or exclusive rights. Member States are likely to be tempted to use them to infringe Community law'.

including the State Aid rules and (ii) discrimination on grounds of nationality contrary to Article 6.

Article 90(1) thus specifically requires Member States not to discriminate in favour of state-related undertakings and against other EC undertakings. Member States must also not allow such undertakings to act in a manner contrary to the competition rules.

This is consistent with one of the general principles set out in Article 5 of the Treaty namely that Member States must 'abstain from any measure which could jeopardize the attainment of the objective of [the] Treaty'.

Meaning of specific phrases

The following phrases are examined below: Member States; public undertaking; special and exclusive rights; measures.

'Member States'

In the context of Article 90(1) the term 'Member States' includes public authorities at regional, provincial or communal level[2]. The term 'public authorities' has been defined, for the purpose of the Transparency Directive[3], to include the State, regional and local authorities.

'public undertaking'

The Treaty does not provide a definition of public undertaking. A possible reason is the great diversity of public undertakings across the EC as the Court itself recognized in the *Transparency Directive* case and the consequent difficulty of providing an all-embracing definition.

To understand the phrase 'public undertaking', it is useful to refer to the basic definition of the term 'public'. This term contrasts with 'private' and refers to one of the two basic types of property ownership. In that sense, the term 'public undertakings' would cover undertakings owned by the State, whilst 'private undertakings' would cover undertakings owned by individual interests, whether personal or corporate. The question arises whether the term 'public' must be limited to this definition of state ownership and, if so, is it necessary for the State to be sole owner or can a public undertaking be jointly owned by the State and third parties? The Commission has answered the question by taking a pragmatic approach to the issue which has subsequently been endorsed by the Court.

---

[2] *Corinne Bodson* v. *Pompes Funèbres des régions libérées SA* (the *Bodson* case), C-30/87, 4 May 1988, [1988] ECR 2479, para. 33.

[3] Commission Directive of 25 June 1980 on the transparency of financial relations between Member States and public undertakings, OJ 1980 L 195/35.

The Commission defines public undertakings in the Transparency Directive as:

> Any undertaking over which the public authorities may exercise directly or indirectly a dominant influence. Such influence is to be presumed when the public authorities (i) directly or indirectly hold the major part of the undertakings subscribed capital; (ii) control the majority of the votes, or (iii) can appoint more than half of the members of its administrative, managerial or supervisory body.[4]

The Court endorsed this definition in the *France, Italy and UK* v. *Commission* cases[5].

Thus both the Commission and the Court have been prepared to accept a wide definition of 'public undertaking' which is not limited to state ownership. They have defined the words in the context of the Transparency Directive by reference to the *influence* exercised by the State over certain undertakings. The Court recognized that the influence of public authorities may be exerted not only on the basis of financial participation but also on the basis of rules governing the management of the undertaking.

The absence elsewhere of a clear definition of the term 'public undertaking' suggests that the Transparency Directive's definition can be used in a general context but there is as yet no comprehensive definition. It follows that other criteria may later be used to establish State influence and thus bring an undertaking within the scope of 'public undertaking'.

'special and exclusive rights'

Certain undertakings may be under the close influence of the State even though public authorities neither participate in them nor control their management but have simply granted them special or exclusive rights; for example, the exclusive right to provide telecommunications services[6] or to sell

---

[4] Article 2 of the Commission's Directive. See also the definition of 'public undertaking' in the public procurement Directive of 1993 relating to the excluded sectors, Directive 93/38/EEC, OJ 1993 L199/84, Article 1(2).

[5] C-188/190, 6 July 1982, [1982] ECR 2545, paras 25 and 26. A similar definition was already recommended by the Advocate General in the *Müller (Hein)* case of 1971. The Advocate General suggested that for a company to be a public undertaking under Article 90(1) the following factors are relevant:

(i) the State participates in the capital of a company (this, however, is not, of itself, sufficient in the AG's view);

(ii) the establishment of the company must arise from a unilateral act of the public authorities e.g. a law;

(iii) the State must participate in the management of the company. This, in the AG's view is the key factor. State participation can be evidenced by the power to choose the number of members representing the public authorities on the managing bodies e.g. where the State can designate at least half of the board of directors and of the supervisory board.

[6] *Spain, Belgium and Italy* v. *Commission* (the *Telecom services* case), joined cases C-271/90, C-281/90 and C-289/90, 17 November 1992, [1992] ECR I-5833.

telecommunications[7] equipment over a national territory; the exclusive right granted to an employment bureau to place employees including senior employees[8]; or the exclusive right to collect, transport and distribute mail granted to a national post office[9]. The grant of special or exclusive rights may extend only to a region e.g. the grant to French *communes* (or local authorities) of an exclusive concession to organize the provision of funeral services[10]; or the exclusive right to collect and dispose of waste oil in a region[11].

The precise limit of what may constitute a special and exclusive right is not yet entirely clear. The terms have often been bundled together without any attempt by the Community authorities to define them clearly. In certain cases (*Waste Oils*[12]; *Ahmed Saeed*[13]) the Court has used the term 'exclusive' even though the rights in question have been granted to more than one undertaking. In *Waste Oils*, the Court referred to the grant of exclusive rights to one *or more* undertakings per geographical zone. In *Ahmed Saeed* the term was applied to a right to operate on an air route alone or with one or two undertakings.

The Court has criticized the Commission for not properly defining the words 'special rights' in two cases, *Terminal Equipment* and *Telecom Services*. In *Telecom Services* the Commission slightly clarified the words over its previous definition but the Court held that it was insufficient[14]. According to Advocate General Jacobs, at the oral hearing, the Commission further explained that:

> special rights are the rights held by a limited number of telecommunications organizations chosen in a discretionary and subjective manner by the State concerned[15].

This additional explanation which might have provided an adequate definition of 'special rights' was, however, missing from the original wording of the Directive which led the Court to hold that the definition in the Directive was incomplete and that the Directive's provisions relating to special rights must be annulled.

It certainly seems possible to consider that an exclusive right exists where a single undertaking has been granted the right to carry out a certain activity throughout a territory. This territory need not extend to a national territory and may comprise smaller administrative units or part of them (*Bodson*). The

---

7 *French Republic* v. *Commission* (the *Terminal Equipment* case) C-202/88, 19 March 1991, [1991] ECR 1223.

8 *Klaus Höfner Fritz Elser and Macrotron GmbH*, C-41/90, 23 April 1991, [1991] ECR I-1979.

9 *Paul Corbeau*, C-320/91, 19 May 1993, [1993] ECR I-2533.

10 See n. 2 above.

11 *Syndicat National des Fabricants Raffineurs d'Huile de Graissage and others* v. *Groupement d'Intérêt Economique 'Inter-Huiles' and others* (the *Waste Oils* case), C-172/82, 10 March 1983, [1983] ECR 555.

12 See n. 11 above.

13 *Ahmed Saeed Flugreisen, Silver Line Reisebüro GmbH and Zentrale zur Bekämpfung unlauteren Wettbewerbs e.V.*, case C-66/86, 11 April 1986, [1989] ECR 803.

14 *Telecom Services*, see n. 6 above, paragraphs 31 and 32.

15 *Telecom Services*, see n. 6 above, AG's opinion, paragraph 50.

right may still be regarded as special or exclusive according to the Court if it is shared with a small and defined number of undertakings, for example, as in *Ahmed Saeed* where the exclusive right to operate on an air route may be shared with one or two undertakings[16].

The question, however, remains as to whether the grant of rights may be regarded as special or exclusive if the rights can be given to anyone within a given class. The question was raised but not answered in *INNO* v. *ATAB*[17]. The Belgian State had granted all manufacturers and importers of tobacco in Belgium the right to fix the retail price of the tobacco. The Court merely queried whether this could qualify as an exclusive or special right since there was an unlimited number of undertakings which could qualify for the right – they merely had to be importers or manufacturers. The Court did not provide an answer[18]. The Advocate General, for his part, considered (at paragraph 5 of his opinion) that it could not be denied that Belgian importers and producers of tobacco had been granted special rights since the decisive factor was that 'the economic entities in question are placed in the unusually strong position of having the retail traders obliged to comply with the prices fixed by them'.

Contrary to the view expressed by the Advocate General in *INNO* v. *ATAB*, but nevertheless in line with the reasoning suggested by the Court, strong arguments can be made in support of the view that if rights can be granted to an unlimited number of people within a class on the basis of transparent objective and non-discriminatory criteria, then the rights will be neither exclusive nor special.

This is indeed the approach taken by the Commission in the *Satellite Directive*[19] which defines special rights as follows:

> the rights that are granted by a Member State to *a limited number of undertakings* through any legislative, regulatory or administrative instrument which, within a given geographical area,
> - limits to two or more the number of such undertakings authorized to provide a service or undertake an activity, *otherwise than according to objective, proportional and non-discriminatory criteria*, or
> - designates, otherwise than according to such criteria, several competing undertakings as being authorized to provide a service or undertake an activity, or
> - confers on any undertaking or undertakings, otherwise than according to such criteria, legal or regulatory advantages which substantially affect the ability of any other undertaking to provide the same telecommunications service or to undertake the same activity in the same geographical area under substantially equivalent conditions. (*emphasis added*)

---

[16] *Ahmed Saeed*, see n. 13 above, paragraph 50.

[17] *NV GB Inno-BM* v. *Vereniging van de Kleinhandelaars in Tabak (INNO* v. *ATAB)* C-13/77, 16 November 1977, [1977] ECR 2115.

[18] *INNO* v. *ATAB*, (see n. 17 above) paragraph 42.

[19] Directive 94/46/EC (OJ 1994 L 268/15) amending Directives 88/301 and 90/388 on satellite communications.

The definition of special and exclusive rights in the Satellite Directive appears to meet the criticism of the Court in the *Telecom Services* case and it is likely that the criteria set out in this definition can generally be applied to exclusive and special rights covered by Article 90.

'Measures'

Measures are acts of public authorities which may include legal, regulatory or administrative rulings or decisions which give a right to a state-related undertaking, impose an obligation on it or simply encourage it to act in a certain way.

Measures are contrary to Article 90(1) where they cause or lead the state-related undertaking to infringe Treaty rules.

There are two principal types of measures: first, measures which govern the relationship between the State or a state-related undertaking and third parties; second, measures which consist in granting an undertaking special or exclusive rights.

Measures which govern the relationship between the State or a state-related undertaking and third parties

These include laws or rulings permitting public undertakings or undertakings benefiting from special or exclusive rights (state-related undertakings) to impose unfair charges or conditions on their customers, or to discriminate between their suppliers[20]. The approval by public authorities of unfair charges set by state-related undertakings has repeatedly been held to constitute a measure contrary to Article 90(1). Such measures have included the approval by aeronautical authorities of tariff agreements on air routes contrary to Article 85(1)[21]; the setting by *communes* (local authorities) of unfair prices to be charged by exclusive providers of funeral services[22]; the approval by the relevant Minister of excessive and discriminatory piloting tariffs in maritime ports[23]; and the application of discriminatory landing fees by an airways authority pursuant to a Royal Decree[24].

Both the setting of unfair prices and the encouragement of unfair prices through a system of grant of approval by regulatory authorities constitute measures within Article 90(1).

---

[20] In *Sacchi* the Court held that an undertaking with a monopoly over television advertising which imposed unfair charges or conditions on users of its services or discriminated between commercial operators or national products on the one hand, and those of other Member States, on the other, as regards television advertising, would commit an abuse contrary to Article 86.

[21] *Ahmed Saeed*, (see n. 13 above) ground 50.

[22] *Bodson*, (see n. 2 above) ground 33.

[23] *Corsica Ferries Italia SRL and Corpo dei Piloti del Porto di Genova*, C-18/93, 17 May 1994, [1994] ECR I-1783, grounds 43–45.

[24] *Landing fees at Brussels airport*, Commission Decision, 28 June 1995, OJ 1995, L216/8.

Relevant measures include the act of a government which consists of imposing a burden only on a new entrant in the market so as to give an advantage to a state-related undertaking. An example of this is the obligation imposed by the Italian government on a new operator of GSM radiotelephony services in Italy to make a significant and discriminatory initial payment as a condition of access to the GSM market[25].

A measure also includes a national law which permits service providers benefiting from exclusive rights to restrict the provision of *ancillary* services by third parties. In *Van Ameyde*[26] the Court held that a national law which prohibits a national car insurance bureau from relying on third party loss adjusters would constitute a measure contrary to Article 90(1).

Arguably measures which limit the import or export of products contrary to Articles 30 to 34 EC constitute measures falling within Article 90(1) if they are taken in respect of public undertakings or undertakings benefiting from special or exclusive rights[27].

Measures which confer on undertakings special or exclusive rights

In certain cases a state measure which merely creates special or exclusive rights can itself constitute a measure contrary to Article 90(1). This is the case for measures which deprive Treaty provisions of their effectiveness *(effet utile)*[28]. Particularly relevant Treaty provisions are Articles 30, 59, 85 and 86. For example, in *ERT*[29], the Court held that the way in which the monopoly over TV and radio broadcasting is *organized* or *exercised* can lead to infringements of Treaty rules on free movement of goods, freedom to provide services and the competition rules. This was because the combination in one enterprise of the right to broadcast national programmes and retransmit foreign programmes led to the possibility of discrimination and could compromise the right of programmers of other Member States to have their programmes broadcast on a non-discriminatory basis to viewers. A national rule governing the way in which a monopoly is organized or exercised is contrary to Article 59 and Article 90(1) if it leads to discrimination in the way in which the services are provided.

This must be contrasted with the *Sacchi* approach where the Court held that the removal of radio and TV broadcasting, for non-economic reasons,

---

[25] *Conditions imposed on the second operator of GSM radiotelephony services in Italy*, Commission Decision, 4 October 1995, OJ 1995, L280/49.

[26] *S.r.l. Ufficio Henry van Ameyde* v. *S.r.l. Ufficio Centrale Italiano di Assistenza Assicurativa Automobilisti in Circolazione Internazionale (UCI)*, C-90/76, 9 June 1977, [1977] ECR 1091.

[27] See for example *Terminal Equipment* (see n. 7 above) (limits on imports); and *Waste Oils* (see n. 11 above) (limits on exports). Against this view, see *Campus Oil Limited and others* v. *Minister for Industry and Energy and others*, C-72/83, 10 July 1984, [1984] ECR 2727, (the *Campus Oil* case): the Court's ruling was based solely on Articles 30–36 and the Court rejected the application of Article 90. But the judgment was given in 1974; it is possible that the Court would take a different approach to Article 90 today and consider that the restriction on imports also constitutes a *measure* under Article 90(1).

[28] *INNO* v. *ATAB*, (see n. 17 above) paragraph 31.

[29] *ERT*, C-260/89, 18 June 1991, [1991] ECR I-2925 (see in particular paragraph 11).

from the field of competition was not contrary to Article 90. Even the extension of the monopoly, provided it took place pursuant to a state measure, was not contrary to Article 90. The Court effectively considered that the right of a monopoly to broadcast solely or primarily national programmes as opposed to programmes from other Member States was the essence of the monopoly and was compatible with Treaty rules.

The fact that measures creating monopolies can of themselves be contrary to Article 90(1) can be regarded as constituting a new approach by the Court. The first clear example is *Terminal Equipment*[30]. In this case the grant to certain undertakings of an exclusive right over the importation, marketing, connection and bringing into service of telecommunications terminal equipment was held to constitute a measure contrary to Articles 90(1) and 30 as this limited the sale of similar products from other Member States.

Other examples of Article 90(1) measures include the following:

- the grant to an undertaking of exclusive rights over the supply of certain telecommunications services[31], as this limits the freedom of private operators in other Member States to offer the same services;
- the grant to an undertaking of exclusive rights to act as an employment agency and, in particular, to place senior employees with prospective employers when that undertaking is manifestly unable to meet the demand on the market[32];
- the grant to an undertaking of exclusive rights to set norms, to verify compatibility with these norms and to grant approval for the placing on the market of products meeting the norms when the undertaking is itself active in the market for the sale of the products alongside other companies[33];
- the grant to an undertaking of exclusive rights to organize port operations together with the obligation to use only national workers when this hinders the use of modern technology and the skills of employees from other Member States[34];
- the extension of the dominant position of an undertaking which has already been granted special or exclusive rights[35].

These measures all share a common feature, namely that they resulted in the undertaking benefiting from exclusive rights infringing the fundamental

[30] *Terminal Equipment* (see n. 7 above).
[31] *Telecom Services* (see n. 6 above).
[32] *Höfner* (see n. 8 above).
[33] *Régie des Télégraphes et des Téléphones (RTT)* v. *GB-Inno-BM SA* (the *RTT* v. *GB* case), C-18/88, 13 December 1991, [1991] ECR I-5941.
[34] *Merci Convenzionali Porto di Genova SpA* v. *Siderurgica Gabrielli* SpA (the *Port of Genoa* case), C-179/90, 10 December 1991, [1991] ECR I-5923.
[35] *RTT* v. *GB*, (see n. 33 above) ground 21. Note that in *Sacchi* the Court had held that the extension of exclusive rights by the State was not as such contrary to the Treaty (ground 14). *RTT* v. *GB* reverses this approach.

freedoms and/or the competition rules set out in the Treaty without any corresponding justification.

Nature and form of measures

A measure is generally *an act of the public authorities* which permits or forces a public undertaking or undertaking benefiting from special or exclusive rights to act in a certain way.

The question arises as to whether *an act of the public undertaking or undertaking benefiting from special or exclusive rights* can itself constitute a measure for the purpose of Article 90(1). It appears from the *Van Ameyde* case[36] that an *agreement* between such undertakings which limits the right of third parties to supply services could fall within Article 90(1). It would appear that, to the extent that certain undertakings can issue regulations, an agreement or decision of such undertakings which resulted in the laying down of regulations could constitute a measure contrary to Article 90(1).

A measure generally takes the form of a positive act. It is not yet clear whether failure to act may also constitute a 'measure'. The German government submitted that this was a possibility in its pleadings in the *Transparency Directive* case[37]. It is also arguable that a Member State which maintained in force a law which failed to provide safeguards such as the right to appeal against certain administrative decisions could be held to have omitted to act. The omission to act in this case could conceivably constitute a measure contrary to Article 90(1).

[36] See n. 26 above.
[37] See n. 1 above.

Chapter 2

ARTICLE 90(2): PERMITTED DEROGATION FROM TREATY RULES

> Undertakings entrusted with the operation of services of general economic interest or having the character of a revenue-producing monopoly shall be subject to the rules contained in this Treaty, in particular to the rules on competition, in so far as the application of such rules does not obstruct the performance, in law or in fact, of the particular tasks assigned to them. The development of trade must not be affected to such an extent as would be contrary to the interests of the Community.

This chapter examines first the general meaning of Article 90(2), how it was applied in the past and its new scope of application. It then considers the various phrases and concepts contained in the article as well as the following specific issues: where does the burden of proof lie in Article 90(2) matters, does the article have direct effect and what are the possible defences based on the article?

General meaning

Article 90(2) provides for a *derogation* from the application of Treaty rules in the case of undertakings entrusted with a service of general economic interest or undertakings having the character of a revenue producing monopoly. As for all derogations, that contained in Article 90(2) must be interpreted narrowly. The derogation applies to all EC rules but in particular to the competition rules, the rules on freedom to provide services and the rules on free movement of goods.

Undertakings covered by the article *may* behave in a way that restricts competition, the free movement of goods, the free provision of services but *only if* they could not otherwise carry out the special task (i.e. the task of general economic interest or the task to produce revenue) with which they have been entrusted.

Article 90(2) thus provides a defence against claims that undertakings of general economic interest have infringed Treaty rules, but, as the case-law demonstrates, it has always been difficult to rely on it.

Original scope of application

Under the original interpretation of Article 90, as set out in the judgment in *Sacchi*, the circumstances when undertakings of general economic interest could rely on the Article 90(2) derogation were rare. The infringement of Treaty rules would only be allowed if it would otherwise be *impossible*, not just difficult, for these undertakings to carry out their special tasks[1].

The case-law involved state-related undertakings benefiting from a statutory monopoly. The infringement of Treaty rules was not due to the existence of the monopoly but to the way in which it was exercised, in other words, to the way in which the undertaking behaved towards third parties. The derogation was therefore not relied upon to justify the monopoly (the existence of the monopoly was taken for granted and was not open to attack) but to justify behaviour of the monopoly towards third parties, for example, discriminatory behaviour towards customers or suppliers, which had a restrictive effect on competition between these third parties.

In *Sacchi*, the Court stated, by way of example, that discriminatory behaviour by the holder of the monopoly towards third parties, including the purchase of equipment from national suppliers, as opposed to foreign suppliers, would be contrary to Article 86. The Court judgment implies that such conduct would not normally be exempted by the derogation in Article 90(2).

In *Anseau Navewa*, the Court held that 'a possible limitation of the application of the rules on competition can be envisaged only in the event that the undertaking concerned has no other technically and economically feasible means of performing its task'.

Recent scope of application

Following the new interpretation of Article 90, and, in particular, following the judgment in the *Terminal Equipment* case, there are greater opportunities for undertakings of general economic interest to rely on the Article 90(2) derogation. Why this change? Does it mean that undertakings benefiting from a monopoly are more likely than in the past to be allowed to behave in an anti-competitive manner towards third parties?

The answer to this question is clearly 'no'. Undertakings benefiting from a monopoly have no more right than before to behave in an anti-competitive

[1] See *Anseau Navewa*, OJ 1982 L 167/39.

manner towards third parties. The original application of Article 90(2) which is based on the *Sacchi* interpretation (referred to in this book as '*the first interpretation*') continues to apply with the same rigour to the anti-competitive exercise of a monopoly.

Application of Article 90(2) to the mere existence of a monopoly

It is under a new application of Article 90 by the Court (referred to hereafter as '*the new interpretation*') that Article 90(2) has a greater chance to apply successfully. Under this interpretation based on *Terminal Equipment*, *Corbeau* and *Almelo*, the Court applies Article 90 to the mere *existence* of a monopoly (as opposed to its exercise or behaviour) and examines whether the existence of a monopoly *per se* is valid under the Treaty. Previously, the existence of monopolies and their *per se* validity were taken for granted. Article 90(2) is now used to determine whether the existence of a monopoly which normally *per se* restricts potential competition may nevertheless be legal. This will be the case if the monopoly is *necessary* to perform a task of general economic interest. The Commission[2] and the Court[3] have shown that this necessity test could be proven and hence the Article 90(2) derogation successfully apply, in certain circumstances. The wider scope for the application of the derogation in Article 90(2) may be seen as a counterbalance to the new and far-reaching application of Article 90 by the Court.

Link between Article 90(2) and Article 90(1)

Under the new interpretation, Article 90(2) is used to assess the validity of measures within the meaning of Article 90(1). This was clearly stated by the Court in *Corbeau* (at paragraph 13):

> [Article 90(1)] must be read in conjunction with Article 90(2) which provides that undertakings entrusted with the operation of services of general economic interest are to be subject to the rules on competition in so far as the application of such rules does not obstruct the performance, in law or in fact, of the particular tasks assigned to them.

An Article 90(1) measure will thus only be compatible with the Treaty if it satisfies the conditions set out in Article 90(2) and, in particular, if it is necessary to perform a task of general economic interest.

For the first time, in *Corbeau*, the Court establishes a clear *link* between Article 90(1) and Article 90(2) and states that the grant of exclusive rights to an enterprise – for example, a monopoly – constitutes a 'measure' within Article 90(1), the validity of which must be checked against Article 90(2).

---

[2] See *Telecommunications Services Directive* of 28 June 1990, OJ 1990 L 192/10.
[3] See *Corbeau*, C-320/91, [1993] ECR I-2533.

The fact that the grant of exclusive rights could constitute a 'measure' was known since *Terminal Equipment* and *ERT*. But, there was still uncertainty as to the way in which the compatibility of the measure with the Treaty could be assessed. The Court clearly says, for the first time, that this must be done in the light of Article 90(2). This means that a special or exclusive right, including a monopoly, will only be legal if it is *necessary for the performance of a task of general economic interest* – in other words if it meets a 'necessity test'.

Corbeau and the necessity test

Corbeau is a turning point in that it clarifies the stance taken by the Court since *Terminal Equipment* whereby certain monopolies or part monopolies could *per se* be invalid. It leaves many questions unanswered; in particular, when is a monopoly necessary? But at least it sets a test and in doing so provides some *a posteriori* logic to some of the Commission and the Court's decisions: an absolute State monopoly over terminal equipment was held to be contrary to Treaty rules and not exempted by Article 90(2) because it was not necessary for the performance of its task of general economic interest, namely the maintenance of an adequate telecommunications network throughout the Community (*Terminal Equipment*). A monopoly over all telecommunications services, including value-added services, was again held to be contrary to Treaty rules and not exempted by Article 90(2) because it was not necessary for the performance of its task of general economic interest, namely the provision of adequate telecommunications services throughout the Community (*Telecom Services*).

On the other hand, the maintenance of a monopoly over voice telephony for a number of years has been held to be necessary to ensure the provision of adequate telecommunications services (*Telecommunications Services*). Similarly, a monopoly over basic postal services has been regarded as necessary to ensure the provision of adequate postal services throughout the Community (*Corbeau*). The reason for these decisions is set out on page 30, paragraph 3 and page 150 below.

The Article 90(2) necessity test is thus now used to *define the permitted scope of monopolies*, or rather *to determine their maximum limit*. Beyond the limit – determined by the necessity test – monopolies are likely to be incompatible with the Treaty.

Corbeau further confirms the 'twist' given by the Court to the precise words of the Treaty. The 'twist' consists in equating the grant of exclusive rights with a 'measure' under Article 90(1) and in stating that such a measure will only be compatible with the Treaty if it falls within the necessity test of Article 90(2); however, a 'normal' reading of Article 90(1) would not lead one to say that a 'measure' can consist in the grant of exclusive rights to an undertaking. On the contrary, the words:

> In the case of undertakings to which Member States grant special or exclusive rights, Member States shall neither enact nor maintain in force any measure contrary to the rules contained in this Treaty . . .

suggest that a 'measure' must be something different from the grant of an exclusive right.

It is regrettable that the Court should not have said clearly that it was giving Articles 90(1) and 90(2) a new interpretation.

It must be stressed that both the original and the more recent application of Article 90(2) remain relevant depending on the circumstances:

- the original application may apply where there is no question of the monopoly itself being illegal but the monopoly enterprise has behaved towards third parties in a way which breached Treaty rules, for example, the competition rules or the rules prohibiting export bans;
- the recent application will apply where the existence of the monopoly itself or its extension is challenged.

The case-law of the Commission and the Court on Article 90(2) relates principally to state-related undertakings which provide a service of general economic interest rather than to revenue producing monopolies. The specific phrases of Article 90(2) are examined in detail below.

Specific phrases

'services of general economic interest'

Following the Amsterdam Treaty signed in October 1997, the EC Treaty now includes a specific reference to 'services of general economic interest' in a new Article 7d but it does not define the concept. The Article provides:

> Without prejudice to Articles 77, 90 and 92, and given the place occupied by *services of general economic interest* in the shared values of the Union as well as their role in promoting social and territorial cohesion, the Community and the Member States, each within their respective powers and within the scope of this Treaty, shall take care that such services operate on the basis of principles and conditions which enable them to fulfil their missions (*emphasis added*).

It is not clear what the effect of this new article will be. It seems to be purely declaratory and merely to emphasize the importance of services of general economic interest in the Community. It does not impose any new obligation on the Member States, and is made expressly subject *inter alia* to Article 90.

The Declaration to the Final Act which forms part of the agreement reached by the signatories to the Amsterdam Treaty further provides that:

> The provisions of Article 7d on public services shall be implemented with full respect for the jurisprudence of the Court of Justice, *inter alia* as regards the principles of equality of treatment, quality and continuity of such services.

This declaration stresses three principles inherent in the concept of service of general economic interest but it does not call in question the jurisprudence of the Court on the application of the competition rules to those services.

An undertaking must establish that it is carrying out a service of general economic interest before seeking to rely on the Article 90(2) derogation unless it can, alternatively, show that it is a revenue producing monopoly[4]. The term 'service of general economic interest', however, is not clearly defined in the Treaty or in the case-law of the Court. Some national laws use the concepts of 'utilities'; of 'economic public service' or of 'public service of an industrial and commercial nature' but the concept of service of general economic interest is probably broader[5].

Member States decide whether to give certain undertakings a task consisting of carrying out a service of general economic interest. Member States have used this power in different ways with the result that the same service may be regarded as a service of general economic interest in the national law of one Member State and not in that of others. For example, the task given to an undertaking to provide funeral services is regarded as a service of general economic interest in France but not in several other Member States.

The qualification of a task as a service of general economic interest, for the purpose of Article 90(2), is subject to supervision by the Commission and final review by the Court[6]. It appears from the case-law of the Court that in order for an undertaking to qualify as a service of general economic interest a number of conditions *must* be fulfilled[7]:

- the undertaking must have been *given a specific task*[8];
- the task must have been granted by the *State*, pursuant to a *measure* of *the public authorities*[9]; an undertaking which manages the interest of artists such as a performing right society is not necessarily a service of general economic interest, even if its activity is regulated by law. This will depend on whether the State has granted the undertaking in question a *specific* task;
- a service of general economic interest may be publicly or privately owned[10]. The *form* of ownership is irrelevant.

---

[4] See page 127, paragraph 1 below.

[5] See *Ministère Public Luxembourg* v. *Madeleine Hein, née Müller and others* (the Müller case), C-10/71, 14 July 1971 [1971] ECR 723, AG opinion. The concept is defined in the Communication from the Commission (COM(96) 443 final) on Services of General Economic Interest: 'Services of general economic interest is the term used in Article 90 of the Treaty and refers to *market* services which the Member States subject to specific public service obligations by virtue of a general interest criterion. This would tend to cover such things as transport networks, energy and communications'.

[6] See *Italian Republic* v. *Commission* (the *BT* case), C-41/83, 20 March 1985, [1985] ECR 873, ground 30.

[7] *Belgishe Radio en Télévisie and Société belge des auteurs, compositeurs et éditeurs* v. *SV SABAM and NV FONIOR* (the *BRT* v. *SABAM* case), C-127/73, 27 March 1974, [1974] ECR 313.

[8] See note 7 above.

[9] *Züchner* v. *Bayerische Vereinsbank AG*, C-172/80, 14 July 1981, [1981] ECR 2021.

[10] *BRT* v. *SABAM*, see note 7 above.

The fulfilment of the following criteria[11] will normally be regarded by the Court as characteristics of a service of general economic interest, namely:

- the obligation to provide a service fulfilling certain *essential needs* of the population (e.g. distribution of electricity; transport of letters);
- the provision of an essential service throughout a defined *territory*;
- the provision of an essential service to *all consumers* within the territory;
- the provision of an essential service under *affordable conditions*, e.g. *uniform tariffs and conditions*, for example, the supply at all times of the quantity of electricity required.

If an undertaking fulfils some or all of these criteria, it is likely to satisfy the Commission (and ultimately the Court) that it is a service of general economic interest. The above list, however, is not exclusive and there might be other relevant criteria.

The Court has held that the following activities were services of general economic interest where they were specifically entrusted by the State to certain undertakings: the provision of telecommunication services[12] and, in particular, the operation of a public telephone network[13]; the operation by airlines of certain air routes[14]; the activity of an employment agency[15]; television broadcasting[16]; the provision of postal services[17]; and the distribution of electricity[18].

The Court has held that the obligation to ensure the navigability of a State's most important waterway[19] could be a service of general economic interest and the Commission has held that the distribution of water[20] was a service of general economic interest. The Court, however, has remained silent in respect of certain activities which its Advocate General considered to be capable of falling within Article 90(2) and of constituting a service of general economic interest, for example, the provision of funeral services[21]; and the collection and disposal of waste oils[22].

Certain services have been held *not* to constitute services of general economic interest, for instance the activities of performing rights societies[23]

---

11 *Corbeau*, C-320/91, [1993] ECR I-2533; *Almelo*, C-393/92, [1994] ECR 1477.

12 *BT*, see n. 6 above.

13 *RTT* v. *GB*, C-18/88, 13 December 1991, [1991] ECR I-5941.

14 Routes which are not commercially viable but which it is necessary to operate for reasons of the general interest; see *Ahmed Saeed*, case C-66/86, [1989] ECR 803.

15 *Höfner*, case C-41/90, [1991] ECR I-1979.

16 *Sacchi*, case C-155/73, [1974] ECR 409.

17 *Corbeau*, see n. 11 above.

18 *Almelo*, see n. 11 above.

19 *Müller*, see n. 5 above.

20 *Anseau Naweva*, see n. 1 above.

21 *Bodson*, case C-30/87, [1988] ECR 2479.

22 *Waste Oils*, C-172/82, 10 March 1983, [1983] ECR 555.

23 *BRT* v. *SABAM* case, see n. 7 above and *Gesellschaft zur Verwertung von Leistungsschutz rechten GmbH (GVL)* v. *Commission*, C-7/82, 2 March 1983, [1983] ECR 483.

where these activities were not specifically entrusted by the State to certain undertakings.

Where an undertaking – as a whole – is regarded as performing a service of general economic interest it seems that this feature applies to all its activities including those which have a purely commercial object such as advertising[24]. Conversely it seems that certain undertakings which do not qualify as a service of general economic interest for their overall activities may nevertheless qualify as such in respect of *part* of their activities. This can be the case for airlines which are forced to operate non-viable routes[25]. It could also be the case for certain activities of banks: the Court held in *Züchner/Bayerische Vereinsbank*[26] that the mere transfer of funds of one Member State to another is not a task which can be regarded as a service of general economic interest *unless* in carrying out such transfers they operate a service of general economic interest which has been entrusted to them by a measure adopted by the public authorities. It follows that the provision of a service for *certain transfers of funds* could be a service of general economic interest.

In *Banco de Credito*[27] the Court also leaves the question open as to whether the grant of an aid to a bank could be 'saved' from the prohibition of state aids on the ground that the specific bank could be an Article 90(2) undertaking[28].

It is possible that certain banking or insurance activities specifically entrusted to private or public undertakings by the State could qualify as services of general economic interest.

The qualification of a service as a service of general economic interest is subject to review by the Court. It appears from recent case-law, however, that the Court recognizes that, in the absence of harmonizing Community law, Member States have a large discretionary power to determine the scope of their services of general economic interest. In *Ladbroke Racing Limited* v. *Commission*[29], an Article 90 case, the Court held that Member States have a large discretionary power to regulate certain activities such as games and loteries pursuant to socio-cultural reasons including the protection of gamblers. They also have a large discretionary power to fix the costs that must be borne by the undertaking providing the service and the resources needed to finance the service[30].

---

[24] *Sacchi* (advertising by a television company), see n. 16 above.
[25] *Ahmed Saeed*, see n. 14 above.
[26] See n. 9 above.
[27] *Banco de Crédito Industrial SA, now Banco Exterior de Espana SA and Ayuntamiento de Valencia*, C-387/92, 15 March 1994, [1994] ECR I-877.
[28] See p. 157 and 158 below.
[29] Case T-32/93, 27 October 1994, [1994] ECR II-1015.
[30] *Fédération Française des Sociétés d'Assurance and others* v. *Commission*, 27 February 1997, case T-106/95, [1997] ECR II-229 (grounds 100, 108 and 192); but note that any State aid must not exceed what is necessary to carry out the task of general economic interest (ground 178).

'does not obstruct in law or in fact the particular tasks assigned to them'

The general meaning of this phrase is that the competition rules apply fully to undertakings which have been given a task of general economic interest or which are revenue-producing monopolies *unless* the application of these rules would prevent them from carrying out their tasks including the protection of revenue.

In many cases, these undertakings are also undertakings which have been given exclusive or special rights or which are public undertakings within the meaning of Article 90(1). The grant by the State of exclusive or special rights will often be counterbalanced by an obligation to perform certain tasks in the interest of the State – either a task of general economic interest in the direct interest of the population or the task of producing revenue to help finance the State. Public undertakings *controlled* by the State are also likely to be given specific tasks in the interest of the State as a whole.

So far, the Article 90(2) derogation has not been applied to revenue-producing monopolies but only to undertakings which have been given a task of general economic interest. For this reason and because, in addition, revenue-producing monopolies are becoming less common in the Community, the analysis which follows will deal solely with undertakings which have been given a task of general economic interest. The analysis which follows would nevertheless also apply to revenue-producing monopolies.

Article 90(2) has been applied first to the behaviour of state-related undertakings (i.e. public undertakings or undertakings which have been granted special or exclusive rights) and later to the existence of monopolies. Only in the latter case has the derogation been successfully applied.

Application of the Article 90(2) derogation to the behaviour of state-related undertakings

As mentioned above, it is difficult for state-related undertakings to rely on the Article 90(2) derogation to justify conduct which has a restrictive effect on competition. This difficulty may be illustrated by *Anseau Navewa*[31]. This case involved an agreement entered into between various parties including the National Association of Water Suppliers in Belgium and an association of manufacturers and traders in electrical appliances such as washing-machines. Members of the National Association of Water Suppliers had been given the task to maintain the purity of water. The purpose of the agreement was, *inter alia*, to monitor the conformity of washing-machines and dishwashers with the technical requirements prescribed for the preservation of drinking water by Belgian law. Conformity would be evidenced by a conformity label affixed to appliances. In practice, the labels were only awarded to

[31] See n. 1 above.

Belgian manufacturers or to individuals or undertakings designated by foreign manufacturers as their *sole* importers in Belgium. This had the effect of restricting parallel imports which, the Commission found, was also an essential objective of the agreement.

The Commission held that the agreement was contrary to Article 85(1) and did not qualify for an exemption because of its effect on parallel imports. It stated that the water companies involved in the agreement were undertakings entrusted with the operation of a task of general economic interest consisting in the preservation of the purity of water. But such undertakings remained subject to the competition rules unless the application of these rules would prevent the performance of this task:

> It is not sufficient in this regard that compliance with the provisions of the Treaty makes the performance of the particular task more complicated. A possible limitation of the application of the rules on competition can be envisaged *only* in the event that the undertaking concerned has *no other technically and economically feasible means of performing its task*. (*emphasis added*)

The Commission found that, on the facts of the case, the water supply companies had other means available to carry out their tasks which would be less restrictive of competition. Article 90(2), therefore, did not apply. The Court upheld the Commission decision on the application of Article 85(1). It did not even consider whether Article 90(2) could apply.

This case emphasizes that the Article 90(2) derogation will only apply in circumstances where there is absolutely no other way for the undertaking to carry out its task of general economic interest. If other less restrictive means exist, they should be applied even if they render the task of the undertaking more difficult.

In *Ahmed Saeed*, the Court mentioned that the Article 90(2) derogation might apply to certain behaviour such as the application by airlines (and subsequent authorization by aeronautical authorities) of air tariffs which restrict competition. But the derogation will be applied restrictively, namely only where it is demonstrated that the airline must operate on routes which are not commercially viable and where it is possible to determine the exact scope of the resources needed by the airline to carry out its task of general economic interest. The Court makes it clear that the burden of proving that the Article 90(2) derogation is applicable will be hard to discharge:

> Article 5 and 90 of the Treaty must be interpreted as:
> [. . .] not precluding a limitation of the effects of the competition rules in so far as it is *indispensable* for the performance of a task of general economic interest which air carriers are required to carry out but, *provided that the nature of that task and its impact on the tariff structure are clearly established*. (*emphasis added*)

In *Italian Republic* v. *Commission* (the *BT* case) the Court rejected Italy's arguments that the Commission was placing in jeopardy the performance by

British Telecom of its tasks of general economic interest by applying Article 86 to it and refusing to apply the Article 90(2) derogation. The Court held that Italy failed to demonstrate that the full application of the competition rules to British Telecom placed the company in jeopardy from an economic point of view.

The new generation of cases – involving cases decided after *Terminal Equipment* – shows that the Article 90(2) derogation remains as difficult to apply to the conduct of state-related undertakings which restricts competition.

In *ERT*, the Court held that preference given by a monopoly undertaking to the broadcasting of national TV programmes over the retransmission of foreign programmes constitutes discrimination which, in principle, is contrary to Article 86; it is for the national judge to examine whether the restriction may be justified under Article 90(2).

In *RTT* v. *GB*, the Court went even further and found that the *mere possibility* of discrimination by a telecommunications network operator regarding the connection of its own telecommunications equipment to the network in preference to third party equipment constituted an abuse contrary to Article 86[32]. The Court held that Article 90(2) could not justify the exclusion or restriction of competition in the market for telephone equipment[33].

ERT and *RTT* v. *GB* related to the choice of national services or equipment in preference to services or equipment from other Member States. Other forms of abuse have involved excessive prices and price discrimination. In these cases, the Court has held that price abuses fell within Article 86 and it did not even examine whether the abuse might be justified pursuant to Article 90(2). Thus in *Crespelle*[34], the Court held that:

> . . . approved [insemination] centres would be abusing their dominant position if they were to charge to users costs in excess of the additional costs actually incurred in obtaining and conserving until insemination semen imported at the request of a user from another Member State.

In *Bodson*, the Court held that Article 90(1) prohibits public authorities imposing excessive prices on undertakings to which they have granted exclusive rights. The undertakings involved in *Bodson* were undertakings in charge of providing funeral services. The Court found that these undertakings may be in a dominant position in France as a result of the exclusive right granted to them by local authorities. If they charged excessive prices, this would constitute an abuse contrary to Article 86. Again, in *Bodson*, the Court did not even mention the possibility of Article 90(2) applying.

---

32 *RTT* v. *GB* (see n. 13 above) paras. 23–24.

33 *RTT* v. *GB* (see n. 13 above) para. 22.

34 *Société Civile Agricole du Centre d'Insémination de la Crespelle and Coopérative d'Elevage et d'Insémination Artificielle du Département de la Mayenne*, C-323/93, 5 October 1994, [1994] ECR 5077.

Price discrimination is treated in a similar way to excessive prices and will not normally entail an examination of Article 90(2). *Corsica Ferries*[35] illustrates this. The Court held that the charging of different tariffs for similar transactions was discriminatory and contrary to Article 86(c) of the Treaty. It did not examine Article 90(2).

In none of these cases was Article 90(2) successfully applied to justify the conduct of the state-related undertaking. In some cases, it was not even mentioned. Yet the possibility that the Article 90(2) derogation might apply to the conduct of state-related undertakings remains, albeit in exceptional circumstances. The Court gives an example of a possible application of the Article to discriminatory prices in *Ahmed Saeed* – where this is necessary to operate routes which are not commercially viable. Such instances are bound to be rare and the burden of proving the application of the Article 90(2) derogation difficult to discharge. Some abusive conduct may even lead the Court to call into question the validity of the underlying monopoly; this was the case in *ERT* and *RTT* v. *GB*.

Application of the Article 90(2) derogation to the existence of monopolies

In *ERT*, the Court suggested that the grant of exclusive transmission and re-transmission rights to an undertaking *could* lead to discrimination and therefore to an abuse of Treaty rules. In *RTT* v. *GB*, the Court held that the combination of several exclusive rights within a telecommunications undertaking *could* lead to discrimination and, therefore, *did* constitute an abuse contrary to Articles 90(1) and 86. In these two cases, but even more clearly in the second case, the grant of exclusive rights was regarded as being the direct cause of a possible abuse and for this reason as *capable of constituting*, in the first case, and as *constituting*, in the second case, an infringement of the Treaty. The abusive conduct cannot be distinguished from the exclusive right which is at the origin of the conduct and it is the exclusive right itself which is called into question.

Where the exclusive right is bound to lead to an abusive conduct, as in *ERT* or *RTT*, it will only be valid under the Treaty if it is necessary to carry out a task of general economic interest. *ERT* and *RTT* illustrate the gradual shift of the Court from an abusive *conduct* to the abusive *existence* of a monopoly *per se*.

In *ERT* and *RTT*, the abuse consisted in discrimination and the Court considered that in so far as the discriminatory behaviour was the inevitable result of a monopoly, the monopoly itself *could be* (in *ERT*) and *was* (in *RTT* v. *GB*) contrary to EC law. In *Corbeau*, the Court went one step further: the abuse did not consist in discrimination in favour of national products to the detriment of foreign products which were on offer in the territory of the monopoly undertaking, but in the mere provision of a service by the monopoly undertaking in the absence of any discrimination or excessive pricing policy.

[35] See case C-18/93, [1994] ECR I-1783.

Corbeau related to the postal monopoly held by the Belgian postal undertaking, the Régie des Postes. In defiance of the postal monopoly, Mr Corbeau and his family collected mail in the city of Liège and its surroundings and distributed it to addressees situated within the same area by 12.00 noon the next day. Where the mail was addressed to people outside the area, Mr Corbeau would simply send it in the post so that distribution would be carried out by the Belgian post office. Mr Corbeau was prosecuted for infringing the Belgian postal monopoly by collecting and distributing mail in the city of Liège.

The Court of Justice was asked by the Criminal Court of Liège whether a monopoly such as the Belgian monopoly was compatible with the Treaty.

The Court did not consider that the postal monopoly was exercised in an abusive manner towards third party clients or suppliers in the sense, for example, that it was discriminatory. But it considered that the monopoly of the Belgian post office could be regarded as restricting or excluding competition because, by definition, it prevented third parties such as Mr Corbeau from offering their services in Belgium. This restriction or exclusion of competition which was a direct result of the monopoly was contrary to the competition rules but the Court held (in paragraphs 14 and 16) that it could be permitted if there was no other way of performing a task of general economic interest:

> [Article 90(2)] permits Member States to confer on undertakings to which they entrust the operation of services of general economic interest, exclusive rights which may hinder the application of the rules of the Treaty on competition, in so far as *restrictions* on competition or even the *exclusion of all competition*, by other economic operators are *necessary* to ensure the performance of the particular tasks assigned to the undertaking possessed of the exclusive rights. . . .
>
> The question which falls to be considered is therefore *the extent to which* a restriction on competition or even the exclusion of all competition from other economic operators is **necessary** in order to allow the holder of the exclusive right to *perform its task of general economic interest and in particular to have the benefit of economically acceptable conditions.* (*emphasis added*)

These two paragraphs set out a *necessity test*. Restrictions on competition and even the exclusion of all competition resulting from a monopoly may be permitted *to the extent that they are necessary to perform a task of general economic interest under economically acceptable conditions.*

The Court had no difficulty in holding that an undertaking such as the Belgian post office carried out a *task of general economic interest*. This consisted in 'the obligation to collect, carry and distribute mail on behalf of all users throughout the territory of the Member State concerned, at uniform tariffs and on similar quality conditions, irrespective of the specific situations or the degree of economic profitability of each individual operation' (see paragraph 15). In other words, the task of general economic interest consists in this case in an *obligation* to provide certain services or supply certain products[36] in the interest of the population as a whole over a defined territory.

The Court then considered how this obligation could be performed. In other words, what did the undertaking need to be able to discharge its obligation? The Court laid down a necessity test based on economic considerations: the undertaking entrusted with a task of general economic interest must be able to carry it out under acceptable conditions, which means that it must have *sufficient revenue* to carry out its task. The court recognized that sufficient revenue might not be guaranteed if the most profitable sectors are subject to competition since they may be creamed off by competitors. In order to guarantee sufficient revenue the Court considered that it may be necessary to provide that a sector of activity will be reserved or subject to a monopoly so as to avoid competitors cherry picking the most profitable clients and creaming off the sector leaving to the national undertaking the less profitable tasks. Reserving a sector of activity to the national undertaking may be necessary to enable it to secure sufficient revenue from its most profitable activities in order to cross-subsidize its loss making activities.

This necessity test based on economic considerations is a novelty. For the first time, the Court introduces the concept of *necessary economic equilibrium* for the undertaking in charge of a task of general economic interest. Maintaining or securing economic equilibrium may thus entail the grant of exclusive rights in favour of an undertaking and the correlated exclusion of competition over a sector of activity so as to prevent creaming off by competitors.

Once the scope of the task of general economic interest has been defined, the question which arises is how far competition needs to be excluded. Does it have to be excluded over the entire sector of activity covered by the task? The Court judgment suggests that this may be the case for postal services. Having defined the task of general economic interest in paragraph 15 of the judgment, the Court appears to take the view that undertakings entrusted with the task could not perform it if other operators were entitled to compete by providing a similar service but to people of their choice. They would cream off the most profitable sectors and postal administrations could not discharge their duty[37].

The Court then lays down the scope of the exclusion of competition by Member States in the postal sector where the need to balance revenues between profitable and non profitable sectors may justify an absolute ban on competition in certain postal services. This ban may cover what one may call an 'inner circle' of activities, namely those defined in paragraph 15 as falling within the task of general economic interest. It may not normally extend beyond this inner circle to outside services except in exceptional circumstances, namely, where they would compromise the economic equilibrium of

[36] See *Almelo*, n. 11 above.
[37] *Corbeau* (see n. 11 above), paragraphs 17 and 18.

the service of general economic interest. This could be the case where the services provided in the 'outer circle' attracted customers from the inner circle; in other words, where services in the outer circle encroach upon or 'eat into' the customers of the inner circle. This could be illustrated by figure 2.1.

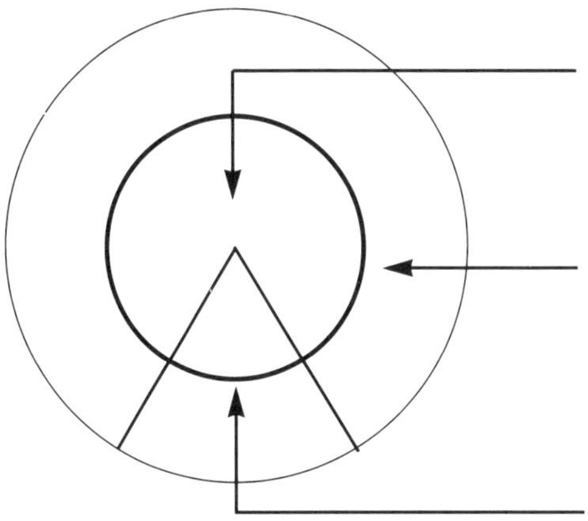

Figure 2.1

For example, an express courier service normally constitutes a specific service dissociable from the service of general interest, because it meets the needs of part – as opposed to all – of the population and principally those of companies and business people. If an express courier service is offered at the same price as ordinary letters, the service may attract customers from the inner circle covering the collection and delivery of ordinary mail and could thus jeopardize the provision of the post office task to deliver ordinary mail throughout a country. In this case, it may be necessary to provide that the express courier service must only be offered at prices above the tariffs for ordinary mail[38].

It may be argued that the exclusion of competition may not be permitted in cases where the outer circle does not 'eat into' the inner circle. Thus a Member State could not restrict competition in the outer circle simply because revenues from this sector would help finance the provision of services of general economic interest in the inner circle. For example, it might be argued that a Member State whose national post office also engages in financial services may not grant its post office specific advantages such as tax rebates in respect of its financial activity making it more difficult for private finance companies to compete with it. The fact that revenues from the

[38] See suggestion of AG Tesauro in his opinion in the *Corbeau* case (see n. 11 above) at ground 18, final paragraph.

financial sector could assist the performance by the post office of its task of general economic interest in the postal sector may not be a valid reason. In so far as financial services are totally distinct from postal services the provision by private operators of financial services cannot be held to 'eat into' the provision of postal services and cannot, therefore, be held to jeopardize the economic equilibrium of the post office. This may be illustrated by figure 2.2.

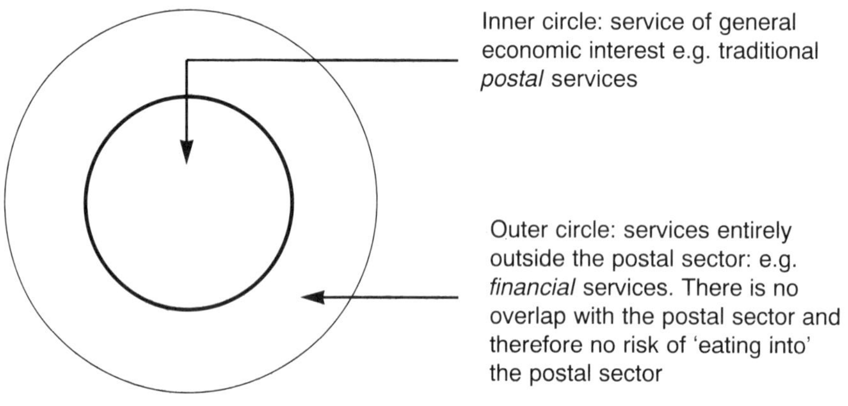

Figure 2.2

The need to provide postal services under acceptable economic conditions would not normally justify giving advantages to the national post office in relation to financial services activity. One may, however, ask whether there could be isolated and well defined cases, for example post offices situated in remote geographical areas where the State might give the post offices concerned specific advantages so that they may provide necessary financial services to the local population and where the revenue from these financial services might then be used to cross-subsidize essential postal activities[39].

Application of the Corbeau test

The *Corbeau* test has only been applied once since the *Corbeau* judgment, namely in *Almelo.* This case concerned the distribution of electricity in the Netherlands. The Court was asked whether an exclusive purchasing obligation imposed on a local distributor of electricity requiring it to source all its

[39] The issues of the link between postal services and financial services was examined by the French competition authorities in relation to advantages given by the French State to La Poste in respect of its financial services activities (see *Conseil de la Concurrence,* Avis n° 96-A-10 of 25 June 1996).

requirements in electricity from a regional supplier and prohibiting it from importing was compatible with the competition rules.

The Court expressly referred to the *Corbeau* judgment and, in particular, to paragraph 14 setting out what has been referred to above as the 'necessity test'. The Court first examined whether the undertaking which imposed the restriction on competition in the form of an exclusive purchasing obligation was in charge of a task of general economic interest. It concluded positively to the first question, finding that the undertaking in question (IJM, a regional distributor of electricity) had a task of general economic interest consisting in distributing electricity throughout a territory (a Dutch region), to all consumers, in the quantities requested by them, at uniform tariffs and on similar conditions. The Court then asked (ground 49) whether a restriction on competition was necessary:

> Restrictions on competition from other economic operators must be allowed in so far as they are necessary to enable the undertaking entrusted with such a task of general economic interest to perform it. In that regard, it is necessary to take into consideration the economic conditions in which the undertaking operates, in particular the costs which it has to bear and the legislation, particularly the environment, to which it is subject.

The Court thus applied the *Corbeau* test and slightly expanded on it by setting out some of the factors to take into account for the necessity test. These factors relate to the constraints applying to the undertaking in charge of a task of general economic interest and include:

- the cost which it must bear;
- the legislation applying to it including environmental legislation.

The Court then left it to the national judge to determine whether the necessity test was met, by assessing whether the restriction on competition was necessary to enable the regional electricity supplier to carry out its task of general economic interest. As in *Corbeau,* the responsibility given to the national judge is extremely heavy and the question may be asked whether a national judge will have the means to determine whether the test is satisfied. One should stress the importance of the word 'necessary' in the *Corbeau* test as reiterated in *Almelo*: there should be no other means available to carry the task of general economic interest[40].

The *Corbeau* test has not been applied since *Almelo.* This may seem surprising. Does this mean that the Court has abandoned the *Corbeau* approach on its assessment of monopolies? A doubt certainly arises following the *Crespelle* judgment of October 1994. This case concerned the validity of monopolies given to bovine insemination centres in French regions.

---

[40] See *Almelo* (n. 11 above), AG opinion, paras. 168 and 69. The monopoly must be *'indispensable'*; the prohibition on imports and the exclusive purchasing obligation must be the '*only means* available' to discharge the task of general economic interest.

French law provides that there can only be one insemination centre per region. A competing centre, known as Crespelle, was nevertheless maintained in the Mayenne region in defiance of the monopoly held by the regional centre. The Court was asked by the referring Court whether the monopoly system provided by the French law was compatible with Articles 5, 86 and 90(1).

The Court had the opportunity to give guidance on the compatibility of the monopoly in question in the light of the *Corbeau* test. Yet it failed to do so and only examined whether the regional centre could be regarded as having abused its monopoly by applying excessive prices. It reasserted the principle that the mere creation of a dominant position by the granting of an exclusive right within the meaning of Article 90(1) is not as such incompatible with Article 86 of the Treaty and referred (at paragraph 18) to *Höfner* and *Port of Genoa*:

> A Member State contravenes the prohibitions contained in those two provisions only if, in merely exercising the exclusive right granted to it, the undertaking in question cannot avoid abusing its dominant position.
>
> The alleged abuse in the present case consists in the charging of exorbitant prices by the insemination centres.

The Court then only examined the alleged price abuse even though it was invited by the referring Court and the regional centre, Crespelle, to examine the broader question of the validity of the monopoly. The precise reasons why the Court thus narrowed the question are not apparent from the case. Was it because the Commission itself did not question the validity *per se* of the regional monopolies in this particular sector? The answer is uncertain. It seems on the face of it that the Court could have examined whether a regional monopoly was absolutely necessary, in accordance with *Corbeau* and *Almelo*, and whether the French law might not instead have provided for a licensing system allowing for more than one undertaking to operate, thus introducing a degree of competition.

Does *Crespelle* reveal a new 'trend' in the examination of monopolies? Although the answer to this question is not certain, the preferred view is that *Crespelle* does not go that far. *Corbeau* and *Almelo* were not mentioned although they could have been and the result might well have been different. *Crespelle* certainly does not overrule *Corbeau* and *Almelo* which are left intact and are therefore likely to be relied upon in the future for the assessment of other monopolies. It is suggested that *Crespelle* simply shows a 'pause' in the Court's examination of monopolies. Judgments after *Crespelle* and up to the time of writing have not dealt with the existence of monopolies *per se* but new opportunities are likely to arise in the near future, in particular with the new *Electricity and Gas* cases[41].

---

[41] *Commission* v. *The Netherlands, Italy, France and Spain*, Cases C-157 to 160/94 (judgment pending at the time of writing; see stop press section).

The *Electricity and Gas* cases deal with the compatibility of exclusive rights to import and export electricity and gas with the rules on the free movement of goods contained in Article 30 of the Treaty. The judgment of the Court will be important for all state-related undertakings and not only for undertakings in the energy sector. At the time of writing the Advocate General opinion has been published. It suggests a widening of the *Corbeau* and *Almelo* approach in relation to the interpretation of Article 90(2): the derogation to restrictions on the free movement of goods might be permitted not only where this is necessary to enable an undertaking to carry out its task of general economic interest under acceptable economic conditions – in other words under conditions of financial equilibrium – but also where this is necessary to enable the undertaking to meet all the obligations imposed upon it by the State, provided these obligations do not go beyond what may be reasonably regarded as forming part of a task of general economic interest[42]. Advocate General Cosmas thus suggests that the exclusive right over the import of electricity granted to the Dutch company SEP is justified in so far as it is necessary to enable SEP to meet its statutory obligation to keep the costs of production as low as possible and to allocate the costs of production and distribution between all consumers in the fairest possible manner[43]. It remains to be seen whether the Court will follow Advocate General Cosmas's interpretation of Article 90(2).

Even if the necessity test of Article 90(2) has been discharged one must still show that the last requirement of the Article relating to the effect of the restrictions on trade between Member States is also fulfilled.

'The development of trade must not be affected to such an extent as would be contrary to the interest of the Community'

In practice, the Court has not, so far, given much importance to the last phrase of Article 90(2). No mention of it is made in *Corbeau* and *Almelo*. One possible interpretation is that this last requirement will be met if the 'necessity test' is met: once it has been proven that a restriction on the free movement of goods or the free provision of services is necessary to carry out a task of general economic interest, a restriction on trade between Member States will be regarded as an inevitable and acceptable consequence; the need to preserve the development of trade between Member States will be regarded as secondary to the need to carry out the task of general economic interest and trade will be deemed not to be unduly affected.

Although it might be dangerous to overlook the last requirement of Article 90(2) entirely, it may be fairly assumed, on the basis of the case-law to date, that it is secondary compared with the necessity test which forms the core part of Article 90(2). Advocate General Cosmas briefly examined the phrase

---

[42] C-157 to 160/94 opinion of AG Cosmas, 26 November 1996, not yet reported, para. 96; see Stop Press section in preliminary pages.
[43] See n. 42, paragraph 97.

in the *Electricity and Gas* cases. He found that the exclusive right over import of electricity and gas would not affect trade to an excessive extent because, as the facts demonstrate, despite these exclusive rights, a significant amount of imports has continued to take place between Member States.

The last phrase in Article 90(2) might thus be used in practice to support an analysis of the necessity test and to show that within a sector of activity, which might be broadly defined, exchanges between Member States in the form of imports and exports of goods or cross-border services will continue to take place.

Burden of proof

Where does the burden of proof lie? Do state-related undertakings have the burden of proving that their monopoly, exclusive or special rights are *necessary* to enable them to carry out their task of general economic interest? Or do undertakings wishing to offer products or services in competition with the state-related undertaking have to prove that the latter's monopoly, special or exclusive rights are *not necessary*?

The question has important consequences. If the burden of proof lies on the competing new entrant rather than on the state-related undertaking, the chances of the competing new entrant proving his case are slim. The reason for this is that national judges may be inclined to adopt a favourable attitude to national monopolies and may therefore find that special or exclusive rights are valid pursuant to domestic laws. If the burden of proof lies on the state-related undertaking the chances of success for the new entrant are greater. The presumption will be that the monopoly, exclusive or special right might not be necessary and it will be for the state-related undertaking to prove that it is; in other words, the state-related undertaking will have to prove that it could not, without its privilege, carry out its task of general economic interest. It will have to prove that no other organization of the market less restrictive of competition could enable it to carry out its specific task.

The answer to the question will depend on whether one considers that the principle is that competition is the rule and that monopolies holding special and exclusive rights are the exception or, whether this principle does not apply to state-related undertakings in charge of a task of general economic interest. In the latter case, a different rule would apply: the special or exclusive rights of monopolies would be presumed valid; the presumption could be rebutted if it was proven that they were not necessary.

Which is the correct starting point? It is essential to be able to answer this question since it is the key to the issue of where the burden of proof lies.

The Court keeps repeating that 'the mere creation of a dominant position by the granting of an exclusive right within the meaning of Article 90(1) is not as such incompatible with Article 86 of the Treaty'[44], which suggests that

---

[44] See, for example, *Crespelle*, (n. 34 above) para. 18.

the Treaty raises a presumption of legality in favour of monopolies, exclusive or special rights. This presumption is, however, neutralized by statements which immediately follow this first statement:

> The Treaty nonetheless requires the Member States not to adopt or maintain in force any measure which might deprive those provisions of their effectivens[45].

The link between the two statements is further emphasized in *Corbeau*, where the Court held (at paragraph 13) that:

> [Article 90(1)] must be read in conjunction with Article 90(2) which provides that undertakings entrusted with the operation of services of general economic interest are to be subject to the rules on competition in so far as the application of such rules does not obstruct the performance, in law or in fact, of the particular tasks assigned to them.

The consequence of the Court's ruling is that state-related undertakings are subject to the rules of competition *unless* the undertakings in question could not carry out their specific tasks. Since the Court clearly said that this principle applies to the grant of special or exclusive rights, the logical consequence of this appears to be that competition is the rule; it is only where there is *no* other way for the state-related enterprise to carry out its specific task that special or exclusive rights will be compatible with the Treaty. *Corbeau* thus seems to confirm that the old presumption (prior to *Terminal Equipment*), namely that monopolies with special and exclusive rights were in principle compatible with the Treaty, is no longer true and has now been reversed. Special and exclusive rights are not presumed legal. Competition is the rule but it is open to state-related undertakings to *prove* that special and exclusive rights are necessary and, therefore, legal.

According to this interpretation (which stems from *Corbeau*) the burden of proof clearly lies on the state-related undertaking. This interpretation is further supported by the opinion of Advocate General Darmon in *Almelo* which has not been rejected by the Court in its judgment. In paragraph 169 of his opinion, Advocate General Darmon said: 'It is for the operator who wishes to rely on the exception of Article 90, paragraph 2, to justify its application'. The Advocate General continues by saying that it would accordingly be for IJM (the electricity operator benefiting from exclusive rights) to demonstrate that the prohibition on import and the exclusive purchasing obligation are the only means available to ensure the performance of the task of general economic interest.

The Court generally followed the Advocate General although it did not spell out that the operator benefiting from exclusive rights had to prove that they were necessary. It may only be inferred from the judgment that the Court approved this statement of its Advocate General. Although the law is

---

[45] *ERT*, C-260/89, [1991] ECR I-2925, para. 35; *Corbeau* (see n. 11 above), para. 11.

not yet settled on this, it is submitted that the better view is that, given the current state of the law, the burden of proof lies on the state-related undertaking seeking to rely on the Article 90(2) derogation.

Direct effect of Article 90(2)

The question of whether Article 90(2) has direct effect has met with different answers from the Court over the years but has now been definitely resolved with a positive answer. It is worth looking at the evolution.

The question was whether Article 90(2) gave undertakings clear and unconditional rights which national courts must enforce. In other words, could undertakings entrusted with the operation of services of general economic interest rely on the Article 90(2) derogation before their national Courts to claim that the competition rules did not apply to them; and were national courts under a duty to apply the derogation if they were satisfied that its conditions were met?

The Court's answer to the question was first that Article 90(2) did not have direct effect. Thus, in *Müller*, in 1971[46], the Court held that Article 90(2) did not have direct effect because it was not unconditional; it required an appraisal of two issues: (i) whether the undertaking in question had been entrusted with a particular task; (ii) whether the non-application of Treaty rules and, in particular, the competition rules would be contrary to the interest of the Community. The Court considered that the appraisal of the two issues depended on 'the objectives of general economic policy pursued by the States under the supervision of the Commission'. So long as these objectives have not been laid down and the Commission has not exercised its power of supervision, possibly under Article 90(3), Article 90(2) could not, according to the Court, create individual rights which the national courts must protect (paragraph 16 of the judgment). This approach was later reversed, first in *Ahmed Saeed*, in 1989 and then in *ERT*, in 1991.

In *Ahmed Saeed*[47], the Court held that Article 90(3) places a duty on the Commission to address directives or decisions to Member States to ensure the application of Article 90. It does not, however, preclude the application of Articles 90(1) and 90(2) where the Commission fails to act. In other words, and contrary to the Court's ruling in *Müller*, Article 90(2) is independent of Article 90(3) and of a ruling of the Commission on what constitutes the Community interest.

In *ERT*, the Court clearly said that it was for the national judge to examine if the behaviour of the state-related enterprise in question may be justified by Article 90(2)[48].

Finally, the direct effect of Article 90(2) was further confirmed in *Rendo I*. The Court reasserted the fact that the Commission does not have sole power

---

[46] See n. 5 above.

[47] *Ahmed Saeed* (see n. 14 above), paras. 53 and 57.

[48] *ERT* (see n. 45 above), para. 34.

to decide on the application of the Article 90(2) derogation. National courts have the power to decide whether the Article 90(2) derogation applies in any given case[49].

Possible defences

This chapter has shown that the Article 90(2) derogation will rarely apply to justify an anti-competitive *conduct*, for example a discriminatory conduct towards third parties. The derogation has greater chances to apply successfully to justify the *existence* of an exclusive or special right, including a monopoly. The burden of proving that the special or exclusive right is necessary probably lies on the undertaking seeking to rely on the Article 90(2) derogation. Circumstances where the derogation has been applied successfully so far include:

- telecoms services: in the Directive on telecoms services[50], the Commission held that voice telephony could be reserved until 1997 in order not to jeopardize the economic equilibrium of national telecommunications operators;
- postal services: the Court recognized in *Corbeau* that national postal administrations should be entitled to reserve – and therefore exclude from competition – certain basic activities where this was necessary to perform their task of general economic interest;
- electricity: the Court recognized in *Almelo* that the grant of exclusive rights restricting competition would be compatible with the Treaty if they were necessary for its beneficiary to provide a service of general economic interest under acceptable economic conditions and taking into account the regulatory constraints such as environmental regulations applying to it.

The economic defence and the need to avoid creaming off the most profitable sectors as well as the need to comply with certain regulations such as environmental regulations are the most likely defences pursuant to Article 90(2).

Other defences are examined in the following chapters. It should be remembered that none of the defences examined are cast in stone. Article 90 involves economic sectors which are subject to rapid economic changes. Because the Article 90(2) defence is primarily based on economic grounds the availability of the defence may change as economic conditions evolve. A clear example of this is telecommunications. Article 90(2) provided a valid

[49] *Rendo NV, Centraal Overijsselse Nutsbedrijven NV, NV Regionall Energiebedrijf Salland* v. *Commission*, case T-16/91, 18 November 1992, [1992] ECR II-2417 (Rendo I), para. 99.

[50] *Commission Directive of 28 June 1990 on competition in the markets for telecommunications services*, OJ 1990 L192/10.

derogation from the application of the competition rules until the end of 1997 only. After that date, the Commission considers that Member States no longer need to grant national telecommunications undertakings exclusive or special rights to carry out their task of general economic interest under acceptable economic considerations.

Chapter 3

ARTICLE 90(3): COMMISSION POWER TO ISSUE DIRECTIVES OR DECISIONS

> The Commission shall ensure the application of the provisions of this Article and shall, where necessary, address appropriate directives or decisions to Member States.

This chapter first examines the general meaning of Article 90(3). It then analyses the Commission's power to issue directives and decisions.

General meaning

Article 90(3) covers two important points:

- first, it imposes on the Commission a supervisory duty in relation to Article 90(1) and (2); it provides that the Commission must check that Member States do not enact or maintain in force any measure benefiting state-related undertakings contrary to Treaty rules; it must also check that state-related undertakings do not act in a manner contrary to the Treaty and, in particular, the competition rules;
- second, it gives the Commission the power to issue Directives and Decisions addressed to Member States to ensure that Member States do not implement any measure falling within the prohibition contained in Article 90(1) and contrary to other Treaty rules.

Article 90(3) thus sets out the Commission's supervisory duty in relation to Article 90 and gives the Commission the power to enforce this article.

Commission's supervisory power

The Commission has been granted a general power to supervise the application of Treaty rules under Article 155 of the Treaty. The exercise of this power is one of the key roles of the Commission which is known as the guardian of the Treaty.

Article 90(3) reinforces this role in respect of measures taken by Member States and acts of state-related undertakings covered by Articles 90(1) and (2). The power of the Commission to act as a supervisory authority is thus strengthened by this specific power of enforcement.

Commission's enforcement power

Article 90(3) gives the Commission powers of enforcement in relation to Article 90 which expressly mentions the competition rules. But Article 90 is much wider and refers to all other Treaty Articles. It follows that the Commission's power of enforcement pursuant to Article 90(3) is potentially extremely broad and covers not just infringements of the competition rules but of any other principles of the Treaty and, in particular, those relating to the freedom to provide services and the free circulation of goods.

As far as competition law is concerned, the Commission already has enforcement powers based on Articles 87, 89 and Regulation 17/62 all of which define the Commission's means of action. These, however, only relate to infringements by *undertakings*. The Commission may issue Decisions against undertakings asking them to bring their infringements to an end, and may take sanctions against them. This power applies in relation to all undertakings covered by Articles 85 and 86 including state-related undertakings. It does not, however, extend to Member States.

Article 90(3) gives the Commission the express power to enforce Treaty rules against infringements by *Member States* of their obligations under Article 90(1). If Member States take measures in favour of state-related undertakings contrary to Articles 85 and 86, Article 90(3) gives the Commission the power to act against the infringing Member States and to ask them to remedy the situation. Article 90(3) thus extends the power of the Commission to act against undertakings in the field of competition law by permitting it also to act against Member States.

Article 90(3) also enables the Commission to act by making Decisions or Directives to enforce other rules, for example, the rules on freedom to provide services or the rules on free movement of goods. The Commission has, for example, issued a Decision against Spain for infringing Article 90(1) in conjunction with Article 86 by reserving the activity of express couriers to its national postal administration. Such a measure was held to restrict competition in the field of express couriers and to lead to an abuse of a dominant position by the Spanish postal administrations contrary to Article 86[1].

Further examples of Article 90(3) Decisions are examined in more detail later in this chapter.

---

[1] *Commission Decision of 1 August 1990 concerning the provision in Spain of international express courier services*, OJ 1990 L233/19.

The Commission has issued a number of Directives based on Article 90(3) which have been controversial. The question causing this controversy was whether the Commission was entitled to use this legislative instrument in circumstances where:

- a Member State infringed the Treaty: here it is argued that the more appropriate procedure would be to take a decision against that Member State which could then be appealed before the Court or to start proceedings pursuant to Article 169 which could then also lead to an action before the Court;
- no specific infringements have been committed: here it is argued that the issue of a Directive amounts to legislating – a power which is best exercised by the Council following a proposal from the Commission and debate in the European Parliament.

These questions on the power of the Commission to issue Directives were examined in detail by the Court particularly in the *Terminal Equipment* case and are explained below.

Differences between Decisions and Directives under Article 90(3)

Decisions are generally taken in relation to a specific Member State although the Court accepts that they can also be taken in relation to several Member States[2]. They relate to an infringement of Treaty rules and require the Member State(s) concerned to put an end to the infringement. Decisions, therefore, require a full assessment of the infringement in the light of EC law. The Court explained in paragraph 30 of the *Dutch Courier* case that the Commission's power to take such Decisions against Member States is the corollary of its power to take Decisions against undertakings pursuant to Article 87 of the Treaty.

Directives are addressed to all the Member States. In contrast with Decisions they do not relate to a specific infringement of EC law by one or more Member States[3]. The Court stated in *Terminal Equipment* that Article 90(3) 'empowers the Commission to specify in *general terms* the obligations arising under Article 90(1) by adopting Directives' (*emphasis added*).

A Decision defines the infringement committed by one or several Member States and the consequences for those Member States of their infringement.

A Directive sets out in general terms, without referring to specific infringements, the obligations arising from the Treaty which Member States must observe.

---

[2] *The Netherlands* v. *Commission*, cases C-48/90 and C-66/90, [1992] ECR I-565; and see *Commission Decision of 20 December 1989 concerning the provision in The Netherlands of express delivery services*, OJ 1990 L10/47.

[3] See *Terminal Equipment*, case C-202/88, [1991] ECR I-1223 (paragraph 17).

Whilst Decisions may be regarded – referring to the Commission's own view[4] as the expression of the *repressive* power of the Commission, Directives are the expression of a *preventive* power, which the Commission also has. The question which remains is, what is the scope of the Commission's power to issue Directives and, in particular, do Commission Directives constitute legislative acts or administrative rulings?

The issue was first raised in relation to the *First Transparency Case* of 1982 and again in relation to the *Terminal Equipment* and the *Telecom Services* cases.

In the *First Transparency Case*, Member States had argued that the Commission did not have a general law-making power as this belonged to the Council alone. The UK government argued that all the Commission could do by way of Directives was to issue specific provisions addressed to one or more Member States and which should be analogous to Decisions[5]. If the Commission's power was not limited in this way, this would mean the Commission had a legislative power which might be exercised without any of the usual procedural safeguards, including consultation of the European Parliament, of the Economic and Social Committee and a vote by the Council. Moreover, since Article 90(1) potentially relates to all the rules in the Treaty, this would give the Commission virtually unlimited legislative competence.

The Court disagreed with the United Kingdom. It explained that the Treaty defines only one form of Directive, described in Article 189. It follows that there is no distinction between the nature of Directives adopted by the Commission and by the Council. The Commission is not limited to addressing Directives laying down specific measures; it can issue Directives of general application. The Commission, therefore, has a clear legislative power[6].

This power, however, is limited in scope by Article 90. This article 'concerns only undertakings for whose actions states must take special responsibility by reason of the influence which they may exert over such actions' and the Commission can only take Directives which are necessary to perform effectively the duty of surveillance imposed upon it by this Article (see grounds 12 and 13).

How far does this power to issue Directives go? In the *First Transparency Case*, the Commission had issued a Directive which required Member States to disclose information on the public funds which they made available to public undertakings. This information, the Court held, was necessary to enable the Commission to exercise its duty of surveillance. The Directive was thus limited to obliging Member States to provide certain information so that the Commission could properly carry out its supervisory duties. The Directive

[4] See *France, Italy, UK, Ireland* v. *Commission* (the *First Transparency Case*) 1982, report for the hearing (p. 2554).

[5] This was the view put forward by the UK in the report for the hearing, p. 2562.

[6] This is the direct result from the Court judgment; yet the Commission does not regard the right to issue a Directive as constituting a right to legislate; see n. 10 below.

merely gave the Commission a tool to carry out its task. Could the Commission, however, go beyond this and issue a Directive, the object of which would not be to help it carry out its supervisory tasks but rather to impose obligations on Member States to give effect to other rules of the Treaty, for example, the rules on the free movement of goods? If this were the case, the legislative power of the Commission would largely overlap with that of the Council, but would be without equivalent procedural safeguards.

This problem was raised in the *Terminal Equipment* case which involved the issue by the Commission of a Directive relating, not to the effectiveness of its supervisory power as in the Transparency Directive, but to state-related undertakings' compliance with the rules on free movement of goods. It imposed on Member States an obligation to abolish their monopolies over telecommunications terminal equipment. This was an extremely far-reaching piece of legislation. France and other Member States contested the Commission's arguments that the Treaty gave it these extensive powers. They raised two principal arguments:

(i) the Commission had misused Article 90(3); it should have relied on Article 169 instead;

(ii) the Commission had encroached on the Council's power to issue Directives pursuant to Articles 87 and/or 100a.

Article 90(3) versus Article 169

The French government referred to a recital in the Directive, to the effect that state monopolies over terminal equipment were not compatible with the rules on free movement of goods, in particular Article 37[7]; and infringed the competition rules laid down by the Treaty[8].

The French government argued that since the Commission believed that there were infringements of the Treaty by a number of Member States it should have initiated proceedings based on Article 169. Article 169 gives express power to the Commission to deliver a reasoned opinion to Member States which, in its view, have failed to fulfil their obligations under the Treaty. The Commission must give Member States a period of time within which to comply with its opinion. If Member States do not comply within the time given, the Commission can bring the matter before the Court. This procedure gives Member States the power to defend themselves fully first, by commenting on the Commission's reasoned opinion and second, if the Commission still disagrees and brings the matter to Court, by pleading their case before the judges

Article 169 expressly covers failures by Member States to comply with their obligations under the Treaty. It was, therefore, in the view of the French government this procedure which the Commission should have followed.

---

7 *Terminal Equipment*, 88/301/EEC of 16 May 1988 OJ 1988 L 131/73, recital 5.

8 See n. 7 above, recital 9.

Article 90(3) Directives should only be issued, in its opinion, 'where it is unclear how compliance with the Treaty is to be achieved' and so as to inform Member States 'of the means which must be used to ensure such compliance'.

The Court did not comment on this interpretation of Article 169. Article 169 proceedings remain one of the options available to the Commission where it considers that Member States have failed to fulfil their obligations under the Treaty. But the Court expressly disagreed with the narrow interpretation of Article 90(3) given by the French government. The Court held as follows (in ground 17):

> Article 90(3) of the Treaty empowers the Commission to specify in general terms the obligations arising under Article 90(1) by adopting Directives. The Commission exercises that power where, without taking into consideration the particular situation existing in the various Member States, it defines in concrete terms the obligations imposed on them under the Treaty. In view of its very nature, such a power cannot be used to make a finding that a Member State has failed to fulfil a particular obligation under the Treaty.

The Court held that Commission Directives are *not* limited to stating precisely what infringements one or more Member States have committed. On the contrary, a Directive cannot be used to make a finding that a specific Member State has infringed the Treaty. The Court also rejected the argument that Commission Directives are limited to making clear how compliance with Treaty rules should take place. It states that the Commission has a power to legislate consisting in setting out in *general terms* the obligations imposed by the Treaty on Member States. The word 'general' must be construed by contrast with the consequences resulting from a *specific* infringement of the Treaty by one or more specific Member States. In addition, the role of the Commission under Article 90(3) is to express in *concrete terms* the obligations arising from the Treaty. The Terminal Equipment Directive provides a good example: the Commission stated that observance of Treaty rules on free movement of goods and on competition matters meant in concrete terms that Member States must abolish their monopolies over the sale of terminal equipment. This Commission statement was upheld by the Court.

It follows from the judgment of the Court in *Terminal Equipment* that the Commission has wide legislative powers. The judgment confirms that the Commission may issue a Directive requiring Member States to abolish certain monopolies.

Another question which arose was whether the Commission may issue a Directive in areas where the Council is also empowered to do so, that is, whether the Treaty allows for the dual competence of both the Commission and the Council in certain cases. France had argued that the Commission could not legislate in areas where the Treaty specifically empowered the Council to act.

Dual competence of the Commission and the Council to issue Directives

France and Belgium stressed that the Terminal Equipment Directive provided for the abolition of a substantial part of the telecoms monopoly and required the establishment of an independent body responsible for drawing up specifications relating to equipment, monitoring their application and granting type-approval. They considered that the Directive thus imposed an obligation on Member States to restructure their telecommunications terminal equipment which, in their view, could only be imposed by the Council acting under Article 100a. This article gives power to the Council, following a proposal from the Commission, the opinion of the European Parliament and consultation of the Economic and Social Committee to issue a Directive aimed at harmonizing Member States' laws to ensure the effective operation of the Common Market.

In addition, Belgium and Italy considered that the Commission Directive was contrary to Article 87 of the Treaty which gives sole power to the Council to issue Directives applying the principal competition policy provisions of the Treaty, namely Articles 85 and 86.

The Court rejected the argument that the Commission Directive wrongly encroached upon the Council's sole competence under Articles 100a and 87. The Court explained that the scope of these two Articles overlapped to some extent with Article 90(3) – but not entirely – and that they were, in fact, broader than this article. The Court confirmed that Article 100a was concerned with the harmonization of measures for the proper functioning of the Common Market. Article 87 was only concerned with measures regulating undertakings in the field of competition. Article 90(3) was concerned only with measures adopted by the Member States in relation to state-related undertakings. The Court stressed, in ground 24, that it is only with regard to the latter measures that 'Article 90 imposes on the Commission a duty of supervision which may, where necessary, be exercised through the adoption of Directives and Decisions addressed to the Member States'.

The Commission may thus only issue Directives to Member States on matters affecting their dealings with state-related undertakings. These Directives may have a harmonizing role and/or deal with competition issues. The Council may also adopt such Directives but it has a wider power not limited to state-related undertakings. The Court further explained that the existence of an overlap between the Commission and the Council's powers is not precluded by the Treaty. The possibility of an overlap and of dual competence had already been mentioned by the Court as a possibility permitted by the Treaty in the *Transparency Directive* case.

In summary, the Commission may, pursuant to Article 90(3), exercise its supervisory and implementing power by way of *Decisions* against Member States which have taken measures contrary to Article 90(1) of the Treaty and by way of *Directives* comparable with Council Directives but which are limited in scope to measures taken by Member States in relation to state-related undertakings. Commission Directives may, nevertheless, be far-reaching since they may require Member States to abolish certain monopolies.

Article 90(3) Directives

Nature of Commission Directives

The Treaty defines Directives in Article 189 as Community instruments which are binding as to the results to be achieved, upon each Member State to which they are addressed, but which leave the national authorities the choice of form and methods.

The Treaty contains only one definition of Directive. The Court confirmed in the first *Transparency Directive* case that this definition applies to both Commission and Council Directives. There are, nevertheless, certain differences between Commission and Council Directives. First, the Council has a wider power to legislate. The Commission's power is limited to dealings with Member States' measures under Article 90(1) relating to state-related undertakings[9]. The Commission denies that it has a power to legislate[10] but this view cannot be accepted following the Court's judgment in *Terminal Equipment*. Directives constitute legislation albeit secondary legislation. The Commission states that it is 'limited to specifying the implications of existing Treaty rules' but it may be argued that this is precisely one of the roles of secondary legislation. The second difference relates to the applicable procedure.

Applicable procedure

Whilst all Council Directives are subject to a detailed procedure involving lengthy discussions within the Parliament and the Commission, both of which may propose amendments and which may also sometimes block a proposed text[11], a Commission Directive requires only the involvement of the Commission, at least in theory. The Treaty contains no specific procedure for the adoption of a Directive by the Commission. In practice, however, the Commission will discuss its proposed Article 90(3) Directive with both the Parliament and the Council of Ministers but it will not submit its proposed Directive to a vote of Member States' representatives. The

---

[9] See the first *Transparency Directive* case, n. 4 above; and see *Terminal Equipment*, n. 3 above.

[10] See the Commission's XXIVth Report on Competition Policy (1994), point 215, paragraph 4: 'Article 90(3) does not give legislative or quasi-legislative powers. The Commission has no power to create new substantive obligations for Member States, or for State enterprises, under Article 90. The Commission can only specify the implications of existing Treaty rules, and set up procedures for making sure that existing obligations on Member States are complied with. "Horizontal" measures can be adopted, but as they cannot involve new obligations, they are not "legislative".'

[11] See, for example, text of Article 100a.

Commission will hold substantial consultations with interested parties such as Member States' representatives, state-related undertakings, users and consumers. It may issue a draft in the Official Journal and leave time for these parties to comment. But the Commission is not *obliged* to follow any procedure or to carry out such consultations.

The procedure which the Commission generally (but voluntarily) follows is, therefore, considerably simpler than that which applies to Council Directives. The Commission is free but not obliged to take into account the comments which it receives. In practice, a Commission Directive is issued much more rapidly than a Council Directive. This has two effects:

- first, Article 90(3) Directives constitute a formidable weapon in the hands of the Commission which it may use to 'threaten' Member States, which are reluctant to agree to changes to their laws pursuant to the normal procedure involving a Council Directive. The threat of an Article 90(3) Directive which would precipitate such changes (see for example the Telecoms Directive) often pushes Member States towards a consensus. It also appears that such pressure was applied in the railway sector where Council Directive 91/440 was finally accepted by Member States following threats by the Commission to issue an even harder Directive based on Article 90(3);
- the second effect (and this flows from the first) is that the issue of Commission Directives lacks the essential procedural safeguards which apply to Council Directives. Given the far-reaching scope of Commission Directives which may oblige Member States to abolish certain monopolies, the lack of procedural safeguards raises a fundamental question as to the compatibility of such Directives with a democratic system based on the balance of powers between institutions. The Commission is aware of this problem which is no doubt why it has endeavoured to minimize its potential adverse effects by introducing consultation procedures before issuing such Directives.

This lack of procedural safeguards is not always helpful to the Commission. It may, paradoxically, impede the issue of Directives: because of their far-reaching nature, there is a risk that the threat of a new Directive will cause major political wrangle throughout the Community; there is also a risk that the issue of a new Directive will not be accepted and will not, therefore, be properly implemented by the Member States even if it is made in accordance with the Treaty, because it lacks the legitimacy of a democratic process.

The conduct of extensive consultations may not cure this problem entirely. The Commission is aware of this and in the postal sector, for example, it has held back from issuing an Article 90(3) Directive, in the face of Member States' opposition, and has instead proposed a Directive based on Article 100a. In the energy sector it has proposed Council Directives.

Article 90(3) Directives remain an extremely effective weapon but one which the Commission may be reluctant to use in the face of determined political pressure from the Member States.

Commission's power to take normative measures by way of Directives only

The Commission can enact laws providing for new norms to be applied by Member States but only by way of Directives.

It may not take normative measures by way of a 'Communication' or a 'Notice'. In *France* v. *Commission*, the Court held that the Commission had wrongly attempted to impose a new obligation on Member States, namely to disclose certain information on a systematic basis to the Commission, by way of a mere Communication which did not provide any legal basis. The Court held that the additional obligation could only be imposed by way of a measure based on Article 90(3)[12]; in practice, it appears that a measure imposing a new obligation and setting out a new norm may only be taken in the form of a Directive and not of a Decision. It can never be taken by way of a Notice or of a Communication.

In December 1995, the Commission published a draft Notice setting out competition guidelines in the postal sector. A Notice like a Communication is not binding and should, therefore, not contain any new binding norms. Whilst the draft Notice in question aimed at setting out the Commission's interpretation of the competition rules laid down in the Treaty, it provided *inter alia* that Member States should further limit their postal monopolies and set up an independent regulatory body responsible for supervising postal organizations; such requirements may be regarded as going beyond a mere interpretation of Treaty rules and as imposing new obligations on Member States. It may be argued that the types of obligations imposed should not be set out in a Notice but only in a Directive.

Commission's power to legislate in respect of Article 90(1) measures

The Commission's power to legislate is limited to issuing Directives addressed to *Member States* and to dealing with *measures* adopted by them in relation to state-related undertakings. Commission Directives may not impose obligations directly on state-related undertakings but only on Member States. This is clear from both the *Terminal Equipment* case[13] and the *Telecom Services* case[14]. The Court held that the prohibition of long-term contracts entered into by telecoms administrations with their customers could not

---

[12] *France* v. *Commission*, case C-325/91, [1993] ECR I-3283, paragraph 27.

[13] See n. 3 above.

[14] Joined cases C-271/90, C-281/90 and C-289/90, [1992] ECR I-5833.

form the subject of an Article 90(3) Directive. The reason was that such contracts were entered into by state-related undertakings on their own initiative and were not Member State measures.

The Commission cannot issue a Directive prohibiting a measure taken by a state-related undertaking as opposed to a Member State. On the other hand, the Commission can take a *Decision* against the acts of a state-related undertaking (for example, the signing of an agreement), if it considers that the act infringes Article 85 or 86.

Examples of Commission Directives pursuant to Article 90(3)

The Commission has issued the following Directives based on Article 90(3) of the Treaty:

- Transparency of financial relations between Member States and public undertakings[15] 1980; amended in 1985[16]; and in 1993[17];
- Competition in the markets for telecommunications terminal equipment[18], 1988;
- Competition in the markets for telecommunications services[19], 1990;
- Satellite communications[20], 1994;
- Cable television[21], 1995;
- Mobile telephony[22], 1996;
- Full competition in telecommunications markets[23], 1996.

The above Directives relate to two principal fields: (i) the transparency of public undertakings' accounts and (ii) communications. The first field includes three Directives which aimed at improving the weapons at the Commission's disposal to carry out its supervisory task. The second field includes Directives which are aimed at reducing Member States' monopolies in various communications sectors.

Principal features of Article 90(3) Directives leading to the abolition of monopolies

These Directives contain detailed recitals which set out Article 90(3) as the legal basis and explain why the exclusive and special rights in question are

[15] Commission Directive 80/723/EEC, OJ 1980 L 195/35.
[16] Commission Directive 85/413/EEC, OJ 1985 L229/20.
[17] Commission Directive 93/84/EEC, OJ 1993 L 254/16.
[18] Commission Directive 88/301/EEC, OJ 1988 L131/73.
[19] Commission Directive 90/388/EEC, OJ 1990 L192/10.
[20] Commission Directive 94/46/EC, OJ 1994 L268/15.
[21] Commission Directive 95/51/EC, OJ 1995 L 256/49.
[22] Commission Directive 96/2/EC, OJ 1996 L20/59.
[23] Commission Directive 96/19/EC, OJ 1996 L 74/13.

contrary to Articles 86 and 59 or, as the case may be, Article 30. They indicate whether any derogation from the application of Treaty rules may be permitted either on the basis of the exceptions expressly provided for in the Treaty, e.g. Articles 55 and 56, or on the basis of the case-law of the Court on the doctrine of mandatory requirements, or on the basis of Article 90(2).

In the Terminal Equipment Directive, the Commission states that the derogation based on Article 90(2) does not apply because the conditions provided in the Article were not fulfilled[24]. In the Satellite Directive, the Commission states that no derogation under Article 90(2) is justified except in respect of voice telephony[25]. In neither Directive does the Commission explain in any detail why Article 90(2) may not apply to most activities under consideration. The Telecommunications Services Directive, however, indicated in detail why the Article 90(2) derogation *may* apply to voice telephony; the reasons were, first, the need to maintain the financial stability of the telecommunications organizations and, second, the need to preserve emergency services related to public safety. The Commission indicated that it would review, at a later date, the scope of special or exclusive rights maintained by Member States and authorised on the basis of Article 90(2) 'to take account of technological developments and the evolution towards a digital infrastructure'. The Telecoms Directive of 1996 now provides for the full liberalization of telecommunications services by 1998 subject only to transitional periods for certain Member States. This shows that the extent of the derogation permitted under Article 90(2) may fluctuate with technological advance. It follows that the maintenance of special rights which may be covered by the derogation of Article 90(2) and thus be legal at a particular time may no longer be covered by the derogation and will become *illegal* at a later date. This evolutionary nature of the application of Article 90(2) is an important feature of this Article.

In addition to an obligation to reduce or abolish monopolies these Directives contain other important rules such as:

- the obligation to draw up and publish the technical specifications to be met by telecommunications equipment so as to ensure maximum transparency for interested operators; (see Terminal Equipment Directive, recital 15);
- the obligation to draw up objective and non-discriminatory conditions governing access to the network (Telecom Services Directive, Article 4);
- the obligation to arrange for a body independent of the telecoms organizations and operators of the network to be responsible for monitoring type-approval specifications, the grant of operating licences, the

[24] *Terminal Equipment*, see n. 3 above, recital 11.
[25] Satellite Directive, 94/46/EC (OJ 1994 L 268/15), recital 16.

allocation of frequencies (Terminal Equipment Directive, recital 17; Telecom Services Directive, Article 7).

Article 90(3) decisions

These decisions are taken against Member States in respect of measures connected with state-related undertakings which infringe Treaty rules. The principal Article 90(3) decisions are the following:

(i) Decision of 1985 concerning the insurance in Greece of public property and loans granted by Greek state-owned banks[26];

(ii) Decision of 1987 concerning reductions in air and sea transport fares available only to Spanish nationals resident in the Canary Islands and the Balearic Islands[27];

(iii) Decision of 1989 concerning the provision in the Netherlands of express delivery services[28];

(iv) Decision of 1990 concerning the provision in Spain of international express courier services[29];

(v) Decision of 1993 concerning a refusal to grant access to the facilities of the port of Rodby in Denmark[30];

(vi) Decision of 1995 relating to discriminatory landing fees at Brussels airport[31];

(vii) Decision of 1995 concerning the conditions imposed on the second operator of GSM radiotelephony services[32] in Italy.

Decision (i) above relates to infringements of Article 90(1) and Article 52 on establishment; Decision (ii) above relates to infringements of Article 90(1) and Article 7 (now Article 6). Decisions (iii) to (vii) above relate to infringements of Article 90(1) and Article 86. The majority of Decisions thus related to infringements of the competition rules. The more recent Decisions and the Commission statement in the draft Notice on postal services[33] show that the Commission is determined to use its powers under Article 90(3) to fight State measures which infringe the competition rules.

In certain cases, the Commission may also be empowered to take Decisions against Member States and state-related undertakings on a different basis, which does not involve Article 90(3), pursuant to specific regulations. Where

[26] Commission Decision of 24 April 1985, OJ 1985 L152/25.
[27] Commission Decision of 22 June 1987, OJ 1987 L194/28.
[28] See n. 2 above.
[29] See n. 1 above.
[30] *Commission Directive of 21 December 1993 (the Port of Rødby – Denmark)* OJ 1994 L55/52.
[31] OJ 1995 L216/8.
[32] OJ 1995 L280/49.
[33] Draft Notice on the application of the competition rules to postal services 1995 OJ C-322/3.

a Member State's measure falls within a sector governed by a Community regulation which gives the Commission power of enforcement by way of Decision, the Commission Decision should be taken under that regulation rather than under Article 90(3). An example of this is the Commission Decision of 27 April 1994 relating to *TAT-Paris (Orly)-Marseille and Paris (Orly)-Toulouse*[34] which was taken pursuant to Council Regulation 2408/92 on access for Community air carriers to intra Community air routes even though TAT, the complainant, had based its complaint both on Article 90 and on Regulation 2408/92. It is, therefore, important in any particular case to check whether a specific regulation governing the sector concerned applies, rather than Article 90(3).

Article 90(3) decisions are similar in form to Article 85 and 86 decisions. The Commission starts by describing the facts in detail and, in particular, the State measure at issue. It then proceeds with the legal analysis setting out which Treaty Articles besides Article 90(1) have been infringed and examining, where relevant, whether the Article 90(2) derogation may or may not apply. Of the five decisions mentioned above, the Commission examined Article 90(2) only in the three cases involving Articles 90(1) and 86 i.e. those involving pure competition issues only. In all these cases, it held that the derogation should not apply.

Two of the decisions were appealed to the Court: these were the *Greek insurance and banking case* and the *Dutch express courier* case. The latter was successfully appealed on procedural grounds. Both cases provide interesting lessons relating to the right of the defence in Article 90(3) cases.

Right of defence in Article 90(3) Decisions

Article 90(3) Decisions may be appealed before the Court under Article 173 of the Treaty. Member States and state-related undertakings may bring an action before the Court against a Decision of the Commission which is of direct and individual concern to them. They have two months to appeal from the date of publication of the Decision. The Decision is binding upon its addressee(s) unless and until the Court issues an order suspending or annulling it.

The *Dutch express courier* case made it clear that a state-related undertaking may bring an action against a Commission Decision addressed to a Member State if the Decision directly affects it. Both the Member State and the state-related undertaking concerned could seek a ruling under Articles 173 and 174 that the Commission Decision was void.

Even though Article 90(3) contains no specific procedure setting out the rights of defence, the Court confirmed that these constitute fundamental rights which Community institutions must observe.

---

[34] Commission Decision of 27 April 1994, OJ 1994 L 127/32.

The consequences of this are outlined below.

The rights of Member States

- the Member State concerned should receive an exact and complete statement of all the objections raised against it;
- it should be informed of the observations submitted by third parties;
- it should be given the opportunity to be heard and give its opinion on these observations.

The rights of state-related undertakings

A state-related undertaking which, as in the *Dutch express courier* case, is a direct beneficiary of the State measure attacked by the Commission and is expressly named in the measure in question, is directly affected by the contested Decision and bears the economic consequences of that Decision. It must, therefore, have a right to be heard. In particular:

- it must be informed in precise terms of the Commission's specific objections to the state measure at issue (informal discussions with the Commission are not sufficient);
- it must be given the opportunity to be heard on the Commission's objections.

The Court found that the right of defence of the Netherlands and the state-related undertaking concerned, the Dutch PTT, had not been observed in the Commission's Decision. It accordingly annulled the Decision.

It is important to stress that proceedings may only be instituted within two months of publication of the Decision. After that date, a Member State or a state-related undertaking directly affected may no longer seek the annulment of the Decision unless it can be shown that the Decision infringed a fundamental requirement of the Community legal order. This is an exceptional defence which will rarely apply. The Court held in the *Greek insurance and banking* case that it would only examine it 'if the measure at issue contained such particularly serious and manifest defects that it could be deemed non-existent'. The Court held that this was not so in the case before it.

Limited rights of complainants in Article 90(3) cases

Besides the question relating to the rights of the defence, the question also arises as to what are the rights of complainants in Article 90(3) cases. To what extent can a potential competitor whose entry into the market is hampered by the exclusive rights of a state-related undertaking require the Commission to bring proceedings?

It follows from the judgments in the first *Rendo* case[35] of 1992, in *Ladbroke Racing Limited* v. *Commission*[36] and in *Bundesverband der Bilanzbuchhalter eV* v. *Commission*[37] that the Commission has a discretionary power whether to act pursuant to Article 90(3). It is not obliged to act on a complaint by an individual or a company relating to a breach of Treaty rules by a state-related undertaking. This is particularly so in cases involving the potential application of the derogation contained in Article 90(2). But as the Court stressed in *Rendo*, the Commission does not have exclusive competence on such matters and individuals may seek redress before their national courts which also have the power to decide whether Article 90(2) may or may not apply.

The Commission's discretionary power also exists in cases involving a choice between starting proceedings pursuant to Article 90(3) and starting an action based on Article 169 before the Court of Justice. In the *Rendo* judgment[38] of 1995, the Court held that the Commission was entitled to take the view that the most appropriate procedure was the initiation of proceedings against the Netherlands, pursuant to Article 169 for failure to amend its law in conformity with the Treaty rather than acting against a state-related undertaking which had relied on the Dutch law. Before starting action against a state-related undertaking, the Commission was entitled to decide that it was more logical to examine first whether the national law on which that undertaking had relied was compatible with Article 90 and, if it was not, to start proceedings against the Member State pursuant to Article 169. This is what the Commission did in relation to the Dutch law on electricity, which gave rise to the *Rendo* cases. It started proceedings against the Netherlands and other Member States pursuant to Article 169 for allegedly infringing Articles 30 and 37[39].

The rights of a claimant may be summarized as follows:

- it may bring a complaint to the Commission under Regulation 17/62 for infringement of the competition rules by a state-related undertaking. The Commission must seriously consider the complaint, but it is not obliged to adopt a Decision. If it terminates an investigation without taking action, it must state its reasons for doing so in order to enable the Court of First Instance to verify that it has not committed an error of fact or law or is not guilty of a misuse of powers[40] (*Rendo* 1995);
- it may bring a complaint to the Commission in relation to a Member State's infringement of Treaty rules (e.g. the competition rules or the rules on free movement of goods or on the freedom to provide services);

---

[35] *Rendo NV, Centraal Overijsselse Nutsbedrijven NV, NV Regional Energiebedrijf Salland* v. *Commission*, case T-16/91, 18 November 1992 (ground 100).
[36] Case T-32/93, 27 October 1994, [1994] ECR II-1015 (grounds 38, 39).
[37] Case T-84/94, 23 January 1995, [1995] ECR II-101 (grounds 23, 24).
[38] Case C-19/93 P, 19 October 1995, [1995] ECR I-3319 (ground 23).
[39] The *Electricity and Gas* cases, C-157 to 160/94 (not yet reported).
[40] Case C-19/93 P, (ground 27). Note that the Court quashed part of the *Rendo* judgment of 1992 relating to the alleged absence of legal effects of a Commission Decision to terminate proceedings.

- it may not force the Commission to start proceedings under Article 90(3) against a state-related undertaking or a Member State pursuant to an action before the Court under Article 175 on the ground that it has failed to act (*Ladbroke* 1994);
- it may not force the Commission to start proceedings under Article 169; in *Bundesverband der Bilanzbuchhalter* the claimants started an action pursuant to article 173 asking the Court to annul the Decision of the Commission not to bring proceedings against Germany for infringement of Treaty rules. The Court of First Instance dismissed the claimants' application;
- it may start proceedings before its national Court for breach of Article 90. The Court held in *Van Schijndel*[41] that where in proceedings before a national Court the parties failed to rely on Article 90 it is for the national Court to apply the Article of its own initiative where domestic law allows it to do so.

[41] *Jeroen van Schinjdel and Stichting Pensioenfonds voor Fysiotherapeuten*, case C-430/93 joined to case C-431/93, 14 December 1995, [1995] ECR I-4705.

Chapter 4

STATE-RELATED UNDERTAKINGS AND THE COMPETITION RULES: ARTICLES 85 AND 86 AND ARTICLE 90

Articles 85 and 86 are the principal competition rules applying to undertakings. This chapter focuses primarily on Article 86 and examines cases where a state-related enterprise may be infringing Article 86 by abusing its dominant position. It also examines cases where state-related undertakings may, between themselves or together with other undertakings, infringe Article 85 by entering into an anti-competitive agreement or concerted practice.

Article 86

Summary

This section examines the conditions under which an infringement of Article 86 may be caused by Member States or state-related enterprises. It reviews the concepts of undertaking, effect on trade between Member States, dominance and abuse. It concentrates on the forms of abuses which Member States or state-related enterprises may commit. There are two principal forms:

- first, the abuse committed by a state-related enterprise in its dealings with third parties e.g. through unfair or discriminatory practices (the 'traditional' form of abuse);
- second, the abuse consisting in preventing all competition by third parties in the undertaking's principal field or ancillary field of activity, as a result of the exercise of the exclusive right held by the undertaking in question (the 'new' form of abuse).

The second form of abuse results from the recent case-law.

Article 86 was first applied in the 'new' sense – and dramatically – in *Höfner*[1]. There the Court found that a national law could give rise to a monopoly which was *incompatible with Treaty rules on competition if the monopoly was manifestly unable to meet the demand on the market.* This dramatic approach was followed in *Port of Genoa*[2] where the Court held that a statutory grant of exclusive rights liable in themselves to lead to an abuse of a dominant position, was contrary to the Treaty.

The logic of these two cases was that the monopolies concerned were illegal *per se* and should be abolished. In other important cases Article 86 has been applied to limit the scope of the monopoly but not to abolish it entirely. In *Telemarketing*, for example, the Court held that a monopoly enterprise committed an abuse by reserving to itself an ancillary activity; this became the leading case on the concept of abusive extension of a dominant position to ancillary activities and was followed in *RTT* v. *GB*. Another example of partial limitation of a monopoly is *Corbeau* where the Court dealt with the extension of a monopoly to new services, contrary to Treaty rules.

Article 86 criteria

In order to fall within Article 86 the following conditions must be met:

(i) there must be an *undertaking* subject to the competition rules;

(ii) the undertaking must act in a way that *affects trade between Member States*;

(iii) the undertaking must be *dominant in a substantial part of the common market*;

(iv) the undertaking's action must constitute an *abuse* of the kind described in Article 86 paragraphs (a) to (d).

Undertakings which are subject to the competition rules

It is important to determine first whether the undertaking in question is an 'undertaking' for the purpose of Articles 85 and 86. For this it must be carrying on an *economic activity*. In *Höfner*, the Court held, in ground 21, that 'the concept of an undertaking encompasses every entity engaged in an economic activity, regardless of the legal status of the entity and the way in which it is financed'[3].

It is not always easy, however, to determine whether the State or a state-related enterprise carries out an economic activity. In its Communication on Services of General Interest, the Commission states that Article 90 does not apply to 'non-economic activities (such as compulsory education and social

---

[1] Case C-41/90, 23 April 1991, [1991] ECR I-1979.

[2] Case C-179/90, [1991] ECR I-5923.

[3] See n. 1 above, ground 21.

security) or matters of vital national interest, which are the prerogative of the State (such as security, justice, diplomacy or the registry of births, deaths and marriages)'[4].

In *Commission* v. *Italy*[5] (the *transparency* case) the Court established a distinction between those cases where the State merely exercises 'public powers', i.e. conducts an activity of a governmental and non-economic nature and those cases where it offers goods and services on the market and thus carries on economic activities of an industrial or commercial nature.

In two important cases the Court held that the State or a state-related enterprise acted by exercising 'public powers': *Eurocontrol* and *Poucet.* Eurocontrol is an enterprise which carries on certain activities concerned with air control. The Court noted that the undertaking has been entrusted with the task of strengthening cooperation between contracting states in the field of air navigation and was requested to ensure maximum freedom for all users of air space consistent with the safety requirements. To discharge its duty, it has been vested with certain powers of enforcement outside the scope of ordinary law. Its functions include establishing and collecting route charges levied on users of air space. The charges are based on a rate which is determined by each contracting State. The Court held that in fixing and collecting the charges Eurocontrol must be regarded as a *public authority acting in the exercise of its powers.*[6] Its activities are 'connected with the exercise of powers relating to the *control and supervision of air space* which are *typically those of a public authority.* They are not of an economic nature justifying the application of the Treaty rules of competition'. The levying of a charge was merely ancillary to the task of control and supervision; it was meant only to help that task; in addition, Eurocontrol had no independence or freedom in setting the charge.

Poucet[7] was a further case where the Court held that the activity in question was of a non-economic nature. The Court held that 'sickness funds and the organizations involved in the management of the public social security system fulfil an exclusively social function'. Their activity is based on 'the principle of national solidarity and is entirely non-profit making. The benefits paid are statutory benefits bearing no relation to the amount of the contributions'. The Court concluded that the funds or organizations in question did not perform an economic activity and accordingly were not enterprises within Articles 85 and 86.

In contrast with *Poucet,* the Court held in *Fédération Française des Sociétés d'Assurances and others* v. *Ministry of Ágriculture and Fisheries*[8] that a national

---

[4] COM(96) 443 final, p. 5.
[5] *Commission* v. *Italy* C-118/85 [1987] ECR 2599, ground 7.
[6] See also *LTU* v. *Eurocontrol* [1976] ECR 1541 and *Bavaria Flugesellshaft and Germanair* v. *Eurocontrol* [1977] ECR 1517.
[7] *Poucet et Pistre,* cases C-159/91 and C-160/91 [1993] ECR 637, ground 18.
[8] Case C-244/94, 16 November 1995, [1995] ECR I-4013.

mutual insurance fund dealing with pension funds for farmers was an undertaking even though it was a non-profit making organization which had been established principally for social reasons; its rights and obligations were based on regulations adopted under public law; its pension schemes were based on the principle of solidarity so that members who were unable to pay contributions were exempted from payment (responsibility for the contributions would be assumed by a social fund financed by a levy on contributions made by other members). The Court explained that *Poucet* was different because it involved 'bodies entrusted with the management of certain *compulsory* social security schemes based on the principle of solidarity'; in contrast, the national mutual pension fund for farmers was an *optional* scheme based on the principle of capitalization; benefits depended solely on the amount of contributions paid by the members and the financial results depended on the financial results of the investments made by the managing organization. The Court concluded that the fund therefore carried on an economic activity in competition with life assurance companies[9].

The difficult question of whether a government agency which fulfils a social purpose should qualify as an 'undertaking' for the purpose of Articles 85 and 86 was raised in the *Höfner* case. The Court held in *Höfner* that the placement of employees is an economic activity and the German employment agency, Bundesanstalt, whose task consisted in finding jobs for unemployed people, was accordingly an undertaking within the meaning of Articles 85 and 86[10]. The Court did not, however, give any convincing explanation for its ruling. The placement of employees raises difficult social and economic concerns which, at first sight, could legitimately fall within the reserved activity of Member States.

To understand the Court's ruling, it is helpful to refer to the opinion of Advocate General Jacobs. He made the following points:

(i) the service in question fell within the normal definition of services in Article 60;

(ii) the service in question was a service which is normally provided for remuneration;

(iii) the fact that the German employment agency provided its service free of charge did not change the situation.

The Advocate General then examined whether the activity of employment procurement could be regarded as bound up with the exercise of official authority pursuant to Articles 55 and 66 of the Treaty as the defendants claimed. He found it doubtful that the agency exercised official authority when performing its statutory functions and rejected the argument that Article 55 could apply in this case and thus limit the freedom to provide

---

[9] Ground 17. The Court held that the fund was an undertaking subject to the competition rules but left open the possibility that the Article 90(2) derogation might apply (ground 20).

[10] *Höfner*, (see n. 1 above) ground 23.

services. Article 55 which derogates from the normal application of Treaty rules must be interpreted narrowly and could not apply to this case. The Court readily followed the Advocate General.

A question now arises as to whether there remain circumstances where a Member State may still freely decide to take a sector of activity outside the field of competition and provide a service free of charge through an undertaking which will not be subject to the rules contained in Articles 85 and 86.

The Court has evidently asserted its power to review decisions by Member States as to what constitutes an economic activity on the basis of the general Community interest. There is thus little doubt that Member States no longer have the power to decide entirely for themselves – without taking into account the interests of the Community as a whole – that certain of their activities will be of a non-economic nature and that they may thus be provided by undertakings which do not fall within Articles 85 and 86. It is clear from *Höfner* that it makes no difference that the service is provided free of charge with a view to meeting social needs of a non-economic nature. The service may still be deemed to constitute an economic activity. It is uncertain, however, how far Member States have lost this power of decision[11]. Certainly the fact that other Member States generally treat a given activity, such as the placement of employees, as an economic activity subject to competition from undertakings in a liberalized market, will play an important part in determining whether Member States do or do not maintain their power of decision.

The undertaking must be dominant in a substantial part of the common market

An undertaking is dominant if it enjoys a position of economic strength which enables it to hinder the maintenance of effective competition on the relevant market by allowing it to behave to an appreciable extent independently of its competitors and customers and ultimately of consumers[12]. The traditional case-law of the Commission and the Court applies to Article 86 within the context of Article 90. The fact that the situation of dominance is created by law makes no difference.

Undertakings vested with a monopoly may be regarded as having a dominant position[13] according to the constant case-law of the court; furthermore if the monopoly extends to the territory of a Member State the dominance may be regarded as taking place over a substantial part of the common market[14].

---

[11] In the judgment issued by the French Cour de Cassation (Chambre Commerciale), 12 December 1995 (*Météorologie Nationale*), the French Supreme Court held that there was no actual market for the supply of information based on aeronautic weather forecast. But it is at least arguable that if the case had been referred to the Court of Justice pursuant to Article 177 of the Treaty, the Court would have ruled that the supply of such information could constitute a market and could be subject to competition.

[12] See *Hoffman la Roche* v. *Commission* [1979] ECR 461 subsequently cited in many cases, in particular, *Michelin* v. *Commission* [1983] ECR 3461; *CBEM* v. *CLT* and *CPB* [1986] ECR 3261.

[13] See, for example, *Port of Genoa*, n. 2 above; *RTT* v. *GB*, C-18/88, [1991] ECR I-5941.

[14] *Michelin* [1983] ECR 3461.

What is the position of a monopoly which does not extend to the entire territory of a Member State? In that case it is necessary to examine whether the territory in question is sufficiently large or important in terms of the trade which it generates to constitute a substantial part of the Common Market. The Court held that the Port of Genoa constituted a substantial part of the Common Market because of the significant volume of traffic in the port and its importance for maritime import and export operations for Italy as a whole[15].

Where an undertaking holds exclusive rights which it exercises over a number of towns within a Member State it may be relevant to examine the overall percentage of the population concerned rather than the percentage of towns and cities actually covered. This was the analysis followed by the Court in *Bodson*[16] which concerned the funeral services market. Whilst the group of undertakings concerned (Pompes Funèbres) only held an exclusive right to provide funeral services in less than ten per cent of French local authorities *('communes')*, these represented more than one-third of the total population of France. This was a significant factor in determining whether the group was dominant. The Court held that it was also relevant to examine the financial strength of the group and assess its influence in local authorities in which it did not hold an exclusive right for the provision of its services[17].

The undertaking acts in a way that may affect trade between Member States

For this condition to be fulfilled it is sufficient that the situation of dominance or monopoly *should be capable of* affecting trade between Member States. In other words, a potential effect on trade between Member States is sufficient and it is not necessary to prove an actual effect[18]. This was made clear in the *Höfner* case which was concerned with a purely internal situation involving a German defendant, a German plaintiff and a request for employment of a German national. The Court, nevertheless, considered that the monopoly at issue could have an effect on trade between Member States. It held[19] that a 'potential effect on trade between Member States arises in particular where executive recruitment by private companies *may* extend to the nationals or to the territory of other Member States' (emphasis added). In other words it was sufficient that the monopoly situation should make it more difficult for companies in Germany to recruit executives from other Member States or for recruitment firms outside Germany to provide their services in Germany.

---

15 See n. 2 above.
16 See case C-30/87, [1988] ECR 2479.
17 See n. 16 above, ground 29.
18 *Michelin* [1983] ECR 3461, paragraph 104.
19 *Höfner*, (see n. 1 above) paragraph 33.

An effect on trade between Member States may result from barriers to entry and a partitioning of the Common Market created by a monopoly. In *Bodson* the Court held that it was for the national judge to determine whether the monopoly situation over a significant part of the Member State concerned (in this case France) had the effect of making more difficult the imports of products into France or the provision of services in France by companies situated outside[20].

In certain cases a monopoly over the entire territory of a Member State may nevertheless not affect trade between Member States to any significant extent and therefore fall outside the prohibition in Articles 85 and 86.

This can be the case with monopolies over territories situated on the fringe of the EC for example Ireland, Greece, Denmark, where, in certain sectors of activities and because of geographical reasons, trade between Member States is unlikely. It is a question of fact in each case. An example may be found in the electricity decisions concerning Scotland where the Commission held that 'the geographical location of Scotland on the fringe of the Community and the resultant physical constraints of the system does not make electricity exchanges with other Member States very likely'[21].

Abuses

There are various forms of abuses. First, there is the traditional Article 86 abuse of the type mentioned in *Sacchi* which may also be committed by state-related undertakings if they are in a dominant position in a substantial part of the common market. Such abuses include discrimination between customers, excessive prices and excessively long-term purchasing agreements. Second, there are certain types of abuse which are characteristic of state-related undertakings (although they may also be committed by other undertakings) and which are of more recent application by the Commission and the Court. These include the unjustified extension of a monopoly, manifest inability to meet market demand, denial of access to an essential facility, preventing the supply of a new service or product by a competitor and cross-subsidization of non-monopoly activities from revenue generated by a statutory monopoly.

Traditional Article 86 abuses

A distinction must be made between (i) abuses committed by undertakings by themselves without any State or regulatory intervention, (ii) abuses committed with the support of, or prompted by, the State or through other regulatory intervention and (iii) abuses committed by the State alone.

---

[20] *Bodson* (see n. 1 above) ground 25.

[21] Article 19(3) Notice concerning the reorganization of the electricity industry in Scotland, OJ 1990 C-245/9.

Abuses committed by the undertaking itself without any State intervention

State-related undertakings may infringe Article 86 by abusing their dominant position following actions taken on their own initiative and without any intervention or prompting from the State[22].

In *Sacchi* the Court held that certain acts of state-related undertakings can constitute abuses contrary to Article 86. It gave the example of an undertaking with a monopoly over television advertising which imposes unfair charges or conditions on its customers or which discriminates between commercial operators seeking to advertise in favour of national operators. The Article 86 prohibition applies to such discrimination if the undertakings concerned are unable to prove that the Article 90(2) derogation must apply; in other words, the prohibition applies unless they can prove that without committing such discrimination they could not perform the specific task which has been entrusted to them.

An example of a state-related undertaking infringing Article 86 by itself is the *BT* Decision[23]. The Commission held that BT used its rule making power in a way that contravened Article 86: it limited competition by third parties by restricting the use of its network by telex forwarding agencies. It did this by imposing charges not related to the nature of the service provided in order to protect the revenue of other telecommunications administrations and restricted technical progress.

The Court upheld the Commission's decision. It ruled that the activity of telex forwarding fell outside the scope of BT's monopoly and was not regulated by law. The decisions taken by BT accordingly fell to be examined under Article 86, as they were taken by BT on its own initiative in a field that was not within its monopoly[24]. The Court held that an undertaking having a monopoly over a telecommunications network infringes Article 86 when it prohibits the activities of private message forwarding agencies handling international telecommunications traffic, unless such agencies harm the public network. It found that private agencies do not harm the network by using new technology in the interest of the public. The prohibition imposed by BT, therefore, limited the introduction of new technology contrary to Article 86. The Court examined the argument put forward by the plaintiff (the Italian government) that the Article 90(2) derogation applied so as to exempt BT's conduct from the Article 86 prohibition. It held that Italy had not proved that the activity of the telex forwarding agencies were detrimental to BT or that they placed the company in jeopardy in economic terms[25].

---

[22] Commission Decision on BT, Decision 82/861/EEC, OJ 1982, L 360/36.
[23] Commission Decision 82/861/EEC, OJ 1982 L360/36.
[24] *BT*, case C-41/83, [1985] ECR 873, paragraph 22.
[25] See n. 24 above, paragraph 33.

Excessive prices

It is contrary to Article 86 for a state-related undertaking to charge excessive prices on its own initiative and without any encouragement from the State.

In *Ahmed Saeed*[26] the Court held that a tariff agreement between airlines over certain routes could infringe Article 85 and also Article 86 if the airlines held a dominant position in the market. The Article 90(2) derogation might apply in certain circumstances but only if it is established that the airlines have been granted specific tasks of general economic interest and that the state has approved the tariffs (see below).

The abuse can consist in applying excessively high or predatory prices. Predatory tariffs will normally be applied with a view to excluding competitors from the market. In order to avoid abusive pricing on an air route on which an airline occupies a dominant position, the Court recommends referring to Article 3 of Directive 87/601 which provides guidance on prices:

> tariffs must be reasonably related to the long term fully allocated costs of the air carrier, while taking into account the needs of consumers, the need for a satisfactory return on capital, the competitive market situation, including the fares of the other air carriers operating on the route, and the need to prevent dumping.

The Court accepts that if an airline has been entrusted with a task of general economic interest it may be constrained in its pricing policy. But, if the constraints have the effect of restricting competition, they should be strictly limited to what is necessary to enable the airline to carry out its specific tasks. Airlines which have been given the task of operating uneconomic routes for social reasons may fall partly within Article 90(2). For their other activities, for instance, for operations on 'normal' routes, they may not rely on the Article 90(2) derogation and the competition rules should apply to them even though they may be undertakings with special or exclusive rights by virtue of the licences that they have been granted.

In *Ahmed Saeed* the Court examined the alleged breach of the competition rules by the airline by treating them, first, as if they were ordinary undertakings subject to the full rigour of the competition rules. It referred to the fact that they might come within Article 90(1) and benefit from the derogation in Article 90(2) only at a second stage, and as an exception; the Court clearly stated that the derogation could only apply to part of their tasks. As far as the majority of their tasks was concerned – operation on 'normal' routes as opposed to clearly uneconomic routes – it did not even consider the possible application of Article 90(2).

In *Bodson* the Court also examined Article 86 on its own and verified whether the Article 86 criteria (see page 60 above) were met. It held that the prices charged by Pompes Funèbres should be compared with the prices charged by enterprises which had not been granted exclusive rights. The

[26] Case C-66/86, [1989] ECR 803.

comparison could serve as a basis to determine whether Pompes Funèbres had committed an abuse contrary to Article 86. The Court said that it appeared from the documents before it that Pompes Funèbres was entirely responsible for the prices charged and that the prices were not dictated by the local authorities (*communes*).

The Court, however, explained that had the *communes* in fact been responsible for the prices, in other words, had they imposed the prices on Pompes Funèbres as a condition for granting them the exclusive right, then the pricing constraint would constitute a measure under Article 90(1) which could be contrary to Article 86. Interestingly, in *Bodson*, the Court did not even examine the possibility that Article 90(2) might apply. This suggests that if prices are clearly excessive, in particular as a result of a comparison with the prices charged by other enterprises, this may constitute evidence of an Article 86 infringement and there is then no need to examine whether the Article 90(2) derogation might apply.

If unfair prices are imposed, authorized or encouraged by a public authority then they will be treated as a measure under Article 90(1). In that case the unfair prices are open to challenge under Article 90(1) in conjunction with 86.

Long-term contracts

The use of long-term contracts by enterprises holding a monopoly or exclusive rights may constitute an abuse contrary to Articles 85 and 86. In its original Terminal Equipment Directive 88/301/EEC the Commission required Member States to take the necessary steps to terminate long-term leasing or maintenance contracts between their national telecoms undertakings and customers. The Court, however, stated that it was inappropriate for the Commission to make such a ruling against Member States, pursuant to Article 90, in so far as the anti-competitive conduct had not been compelled or encouraged by the State but had been adopted by the telecoms undertakings on their own initiative[27]. The Commission should instead have pursued the alleged infringement against the telecoms undertakings directly pursuant to Article 86. This is a further example of cases where abuses committed by state-related undertakings on their own initiative will fall to be considered only under Article 86 rather than also under Article 90.

Abuses committed by an undertaking with the support of the State or other public authority

If a public authority, for example a local authority, imposes unfair prices as a condition for granting and operating an exclusive right, this will constitute a measure contrary to Article 90(1) : 'public authorities may not *assist* undertakings holding concessions to charge unfair prices by imposing such prices

[27] *Terminal Equipment*, C-202/88, [1991] ECR 1223, paragraphs 55 and 56.

as a condition for concluding a contract for a concession' (emphasis added)[28].

In *Ahmed Saeed*, the Court held that aeronautical authorities are precluded by Article 5 and by Article 90(1) from *approving* tariff agreements contrary to Article 85(1)[29]. They are also precluded from *encouraging* airlines to conclude such agreements.

It follows from *Ahmed Saeed* and *Bodson* that if the abuse is committed by a state-related undertaking alone with no intervention of a public authority, the abuse falls to be examined under Article 86 only. If a public authority is involved in any way it falls to be considered under Articles 90(1) and 86. In the latter case the Commission and the Court may be prepared (although not always) to examine whether the Article 90(2) derogation may apply. In the decision concerning *Landing Fees at Brussels Airport*[30], the Commission held that the Belgian airway authority had imposed discriminatory landing fees which benefited principally the national airline Sabena to the detriment of most other airlines. It found that the fees had been applied pursuant to a State measure in the form of a Royal Decree and that the abuse infringed Article 90(1) read in conjunction with Article 86. The Commission briefly examined whether the Article 90(2) derogation applied and held that it did not because the application of the competition rules would neither jeopardize the task of the airway authority nor that of the national airline.

In *Bodson*, however, the Court did not mention Article 90(2), presumably because it considered that there was not even a remote possibility of the derogation applying. The same approach was followed in *Corsica Ferries*[31]. A corporation of pilots in the Port of Genoa had been given the exclusive right by the public authorities to provide piloting services in the port. The corporation proposed piloting tariffs which were then approved and brought into force by ministerial decree. In accordance with a ministerial decree, the corporation had thus imposed tariffs which in fact discriminated mostly against vessels which did not fly an Italian flag[32]. The Court held that the discrimination constituted an abuse contrary to Article 90(1) and Article 86(c). It did not even examine the possibility of a derogation under Article 90(2).

In *Crespelle*, the Court held that the charging of excessive prices by a bovine insemination centre benefiting from exclusive rights could constitute an abuse of a dominant position contrary to Article 86. This would be the case if it charged fees disproportionate to the economic value of the services

---

[28] *Bodson*, (see n. 1 above) paragraph 34.
[29] *Ahmed Saeed*, (see n. 26 above) paragraphs 48 and 49. See also *Vereniging van Vlaamse Reisbureau* [1987] ECR 3801.
[30] Commission Decision of 28 June 1995, OJ 1995 L216/8. The decision is particularly interesting because it involved a system of discounts linked to increased usage of the airport facility which appeared to be objective. In reality, the discount system was not linked to economies of scale; its purpose was to favour the national airline Sabena.
[31] *Corsica Ferries*, C-18/93, [1994] ECR I-1783.
[32] See n. 31 above, paragraphs 32 and 33.

provided[33] or if the centre charged costs for extra services to users in excess of the additional cost actually incurred in procuring the extra services requested.

The Court rejected the possibility that the alleged abuse fell under Article 90(1) and resulted from the law authorizing the insemination centre to operate. The mere grant of an exclusive right to the centre did not of itself constitute an *inducement* to commit an abuse. This must be distinguished from the situations examined in *Bodson, Ahmed Saeed* and *Landing Fees at Brussels Airport* where the public authorities themselves would expressly impose or approve tariffs.

The situation is, therefore, clearer as a result of *Crespelle*: for a pricing abuse to be contrary to Article 90(1) in conjunction with Article 86, the abuse should be the result of a measure of the public authorities. Such measure will consist in an approval, authorization or obligation to price in a certain way. The mere grant of an exclusive right to an undertaking which then uses its dominant position on the market to impose excessive prices is not sufficient to trigger the application of Article 90(1).

The position taken by the Court in *Crespelle* is different from its previous position in *Port of Genoa*. *Crespelle* appears to constitute a drawing back from the position adopted by the Court in *Port of Genoa*. There the Court held that:

> a Member State is in breach of the provision contained in [Articles 90(1) and 86] if the undertaking in question merely by exercising the exclusive rights granted to it cannot avoid abusing its dominant position *or, when such rights are liable to create a situation in which that undertaking is induced to commit such abuses*[34] (*emphasis added*).

The Court took the view that 'the undertakings enjoying exclusive rights in accordance with the procedures laid down by the national rules in question are, *as a result, induced* to demand payment for services which have not been requested, to charge disproportionate prices[35] . . . (*emphasis added*).

The Court found that the unfair prices charged by the port authorities were directly linked to the exclusive right which they had been granted. It held that the exclusive right created a situation contrary to Article 86.

In *Crespelle*, the Court refused to establish a direct link between excessive charges and the exclusive right. It considered that the excessive price was not the direct consequence of the law on which the exclusive right was based. In other words, it was open to the insemination centre in question not to apply unfair prices. Interestingly, in *Crespelle*, the Court only quoted the first part of the above statement in *Port of Genoa* i.e. it examined whether the exclusive right created a situation where the undertaking cannot avoid abusing its

---

[33] *Crespelle*, C-323/93, [1994] ECR 5077, paragraph 25, see also *General Motors* v. *Commission*, case 26/75 [1975] ECR 1367; *British Leyland*, case 226/84 [1986] ECR 3263.

[34] *Port of Genoa*, (see n. 2 above) paragraph 17.

[35] *Port of Genoa*, (see n. 2 above) paragraph 19.

dominant position; it did not mention the second part of the ruling emphasized above, namely the likelihood that the law would induce an abuse, which constitutes a much broader test. This leaves it uncertain whether the broader test has been abandoned so that *Port of Genoa* has been reversed in part[36].

Had the Court in *Crespelle* adopted the same approach as in *Port of Genoa*, it might have asked whether the exclusive right given to the centres was likely to place them in a situation that would induce them to apply excessive prices[37]. If the answer was positive, the Court could then have examined whether the exclusive rights granted to the centres were nevertheless necessary to enable the centres to carry out their specific tasks, in accordance with its reasoning in *Corbeau* and *Almelo*. It might have examined whether alternatives less restrictive of competition existed. Could the Member State concerned not have authorized other undertakings besides *Crespelle*, provided of course that they met the various technical and health conditions required by community and national law?

The position following *Crespelle*, therefore, appears to be as follows:

- price abuse by a state-related undertaking constitutes an infringement of Article 86 which will rarely qualify for a derogation under Article 90(2) unless the infringement is prompted by the State. The following distinction must be made:
  - (i) if the abuse is the result of a law or a regulatory act of the public authorities (which approve, authorize or impose the prices in question) it will constitute an infringement of Article 90(1) in conjunction with Article 86 subject to Article 90(2);
  - (ii) if the law simply grants an exclusive right but does not interfere with prices (even though in practice, because the undertaking is in a dominant position, it is in a position to apply unfair prices), this in itself is not sufficient to call into question the validity of the monopoly/exclusive right. It follows that if the undertaking has committed a price abuse on its own initiative, Article 86 will normally apply on its own and will not be subject to Article 90(2) except in rare cases.

Abuses committed by the State alone

The two sections immediately above examined abuses committed by a state-related undertaking on its own initiative, or with the backing of the State or following instructions of the State.

---

[36] This is the analysis of Mr Buendia of DG IV in EC Competition Policy Newsletter Autumn/Winter 1994 p. 53.

[37] If one adopts an analogous approach to that of the Merger Regulation one could argue that the mere creation of a dominant position through the grant of an exclusive right creates a likelihood of abuse and that unless the exclusive right is necessary for public interest reasons and there are effective counterweights to prevent abuses, the exclusive right should be prohibited.

Abuses may also result from a measure taken by the State alone. An abuse will be committed if the measure strengthens the dominant position of a state-related undertaking by limiting competition by third parties. Examples of such measures have included:

- A Dutch law which prohibited undertakings other than the Dutch postal administration from providing certain express courier services unless they charged a minimum price fixed by Decree. The effect of the law was to reserve a section of the express courier market to the Dutch postal administration and thus reinforce the dominant position of the state-related company by sheltering it from competition. The Commission[38] held that the Dutch law constituted a measure contrary to Article 90(1) and Article 86 because *inter alia* it resulted in an extension of the dominant position of the Dutch postal administration and to restrictions of supply by third parties. The Commission Decision was reversed on appeal to the Court[39] but on procedural rather than substantive grounds.
- Another similar measure in the field of express postal services was a Spanish Decree which totally precluded private undertakings from providing such services by reserving the activity to the Spanish postal administration. The Commission[40] held that the Decree had the effect of extending the dominant position of the Spanish administration in the field of basic services into the area of express courier services. This was contrary to Articles 90(1) and 86. It could not be exempted by Article 90(2) because the introduction of competition in the express courier sectors would not, on the basis of existing figures, jeopardize the basic postal service provided by the Spanish postal administration.
- Letters from the Danish Ministry of Transport refusing to allow a company involved in ferry services from operating from the public port of Rødby or from building a private commercial port near that port. The Commission [41] held that the letters of refusal from the Danish Ministry constituted measures which infringed Articles 90(1) and 86. Their purpose and effect was to protect the dominant position of the Danish public company which owned the port of Rødby and also operated ferry services from the port. The refusals of the Ministry would, in particular, reinforce the monopoly of the Danish public company on a key maritime route to the detriment of competition.

---

[38] Commission Decision concerning the provision in the Netherlands of express delivery services, OJ 1990 L10/47.

[39] *The Netherlands* v. *Commission* (the *Dutch Express Courier* case), cases C-48/90 and C-66/90 [1992] ECR 565.

[40] Commission Decision concerning the provision in Spain of international express courier services, OJ 1990 L233/19.

[41] Commission Decision concerning a refusal to grant access to the facility of the port of Rødby [Denmark]; OJ 1994 L55/52.

- The obligation imposed by the Italian government on the second applicant for access to the GSM radiotelephony services market in Italy to pay a significant and discriminatory initial payment. The first applicant (the public operator, Telecom Italia) had not been subject to such an obligation. The Commission[42] held that the obligation which was one of the conditions imposed on potential bidders for the right to become the second operator of GSM radiotelephony services was therefore discriminatory. Furthermore, it would oblige the successful applicant (Omnitel Pronto Italia) to use part of its revenues to make the initial payment which would restrict its ability to charge low tariffs and to compete. Telecom Italia wold thus be protected from aggressive competition by the new entrant into the market.
  Another effect of the measure would be to reduce Telecom Italia's competitive spirit and to delay investment decisions which it might otherwise have made in order to maintain its position in the market. The measure was thus likely to limit the development of the market contrary to Article 86(b).

All these measures have in common that they were taken by the State alone, pursuant to a law, a decree or by mere decisions, with a view to protecting the dominant position of a public undertaking. The measures had the effect of limiting competition into lucrative markets which were also new markets in three cases – the Dutch and Spanish express courier cases and the application for access to the GSM market in Italy.

Article 86 abuses characteristic of 'state-related' undertakings

In principle, the creation of a dominant position by the grant of exclusive rights within the meaning of Article 90(1) is not, as such, incompatible with Article 86[43]. But there are cases where the grant of exclusive or special rights has nevertheless been held to be contrary to Article 86[44].

The following cases are examined below:

- cases where the mere grant of exclusive or special rights constitutes an abuse of a dominant position;
- cases where the exercise of part of a monopoly constitutes an abuse of a dominant position;
- specific forms of abuse which may or may not call the monopoly into question:

---

[42] Commission Decision concerning the conditions imposed on the second operator of GSM services in Italy; OJ 1995 L280/49.

[43] *Sacchi*, C-155/73, [1974] ECR 409, paragraph 8; *Höfner* (see n. 1 above), paragraph 29; *ERT*, C-260/89 [1991] ECR I-2925, paragraph 32; *Port of Genoa* (see n. 2 above), paragraph 16; *Corbeau*, C-320/91 [1991] ECR I-2533, paragraph 11; *Crespelle* (see n. 33 above), paragraph 18.

[44] *ERT* (see n. 43 above), paragraph 35, *INNO* v. *ATAB*, C-13/77, [1977] ECR 2115, paragraphs 31 and 32.

(i) extension of a dominant position;
(ii) manifest inability to meet the demand on the market;
(iii) denying access to an essential facility;
(iv) preventing the offer of a new service/product by a competitor;
(v) using monopoly revenue to cross-subsidize activities subject to competition.

Cases where the grant of exclusive or special rights constitutes an abuse of a dominant position

Nowhere does the Court clearly say that the *mere* grant of exclusive or special rights may constitute an abuse of a dominant position. Indeed it says the contrary quite expressly[45]. But the logical conclusion of a number of cases is precisely that the grant of certain exclusive or special rights by a public authority will constitute a measure which may be or may become an abuse under Article 86.

In *ERT* the Court starts by stating, quoting *Sacchi*, that the grant of an exclusive right is not contrary to the Treaty[46] and adds:

> Nevertheless, it follows from Article 90(1) and (2) of the Treaty that the manner in which the monopoly is *organized* or *exercised* may infringe the rules of the Treaty, in particular those relating to the free movement of goods, the freedom to provide services and the rules on competition[47]. (*emphasis added*).

The Court seems to distinguish between the mere *grant* of a monopoly which is not contrary to the Treaty provided this occurred for 'considerations of a non-economic nature'[48] and the *organization* and *exercise* of that monopoly which may be contrary to the Treaty. Yet this distinction may be artificial as is illustrated by *ERT* itself, *Terminal Equipment* and by *Höfner*.

In *ERT*, the Greek radio and television undertaking had been granted exclusive rights over broadcasting in Greece, including a right to broadcast its own programmes and the exclusive right to receive and retransmit programmes from other Member States. This aggregation of two exclusive rights enabled ERT to discriminate in favour of its own programmes against foreign programmes, contrary to Article 59 of the Treaty. The Court held that if it could be shown that discrimination actually took place then the *grant* of the exclusive rights over broadcasting and retransmission, which caused the discrimination, would of itself constitute a measure contrary to Article 59 on freedom to provide services (unless the derogation in Article 56 applied[49]).

In other words, the Court found that the scope of the monopoly granted to ERT was wide enough to make it possible for ERT to discriminate in favour

---

[45] *Sacchi*, see n. 43 above.
[46] *ERT* (see n. 43 above), ground 10.
[47] *ERT* (see n. 43 above), ground 11.
[48] *Sacchi* (see n. 43 above), ground 14.
[49] *ERT* (see n. 43 above), grounds 23 and 24. The Court found that ERT had not given any justification falling within Article 56 (ground 25).

of its own programmes, in particular because there were no safeguards to prevent it. If ERT actually discriminated, thus using the power to discriminate which was inherent in its monopoly, then the monopoly would be contrary to the Treaty. The Court's conclusion is that the monopoly itself can be contrary to the Treaty. One notices a shift in the Court's analysis from the *exercise* of the monopoly which can be discriminatory, to the *grant* of the monopoly which makes discrimination possible and which itself may constitute an infringement of Treaty rules.

The Court establishes an even clearer link between the grant of a monopoly and possible infringement of the Treaty through the concept of abuse under Article 86, in a different section of the *ERT* judgment. Having quoted *INNO* v. *ATAB* and the obligation of Member States not to deprive Treaty rules of their effectiveness, the Court stated (at paragraph 37):

> Article 90(1) of the Treaty prohibits the *granting of an exclusive right* to retransmit television broadcasts to an undertaking which has an exclusive right to transmit broadcasts, where those rights are *liable* to create a situation in which that undertaking is led to infringe Article 86 of the Treaty by virtue of a discriminatory broadcasting policy which favours its own programmers. (*emphasis added*)

It follows from this statement that the mere grant of an exclusive right or monopoly can constitute a measure contrary to Article 90(1) and Article 86.

The possibility of a monopoly being itself contrary to Treaty rules is already apparent from *Terminal Equipment*; the Court held (in paragraph 22):

> . . . even though that article [i.e. Article 90(3)] presupposes the existence of undertakings which have certain special or exclusive rights, it does not follow that all the special or exclusive rights are necessarily compatible with the Treaty. (*emphasis added*)

It follows from this statement that certain special or exclusive rights may be contrary to the Treaty. This conclusion is supported by the Court's ruling: it held that the exclusive right over the import and marketing of telecoms equipment deprived traders of the possibility of their products being purchased by consumers and thus restricted intra-community trade contrary to Article 30.

In this case, it is clearly the *existence* of the rights which may have the effect of restricting trade. The distinction between the grant and the existence of exclusive rights and the exercise of these rights becomes blurred and the Court attacks the existence of the right itself.

Höfner constitutes a logical development of this case-law which starts with *INNO* v. *ATAB* and continues with *ERT* and *Terminal Equipment*. The following three paragraphs of *Höfner* illustrate the points:

> paragraph 27: . . . any measure adopted by a Member State which maintains in force a statutory provision that *creates a situation in which a public employment agency cannot avoid infringing Article 86* is incompatible with the rules of the Treaty . . . (*emphasis added*)

> paragraph 29: A Member State is in breach of the prohibition contained in [Articles 90(1) and 86] only if the undertaking in question, *merely by exercising the exclusive right granted to it*, cannot avoid abusing its dominant position (*emphasis added*)
>
> paragraph 30: A Member State *creates* a situation in which the provision of a service is limited [contrary to Article 86(b)] when the *undertaking to which it grants an exclusive right* [ ] is manifestly not in a position to satisfy the demand prevailing on the market . . . (*emphasis added*).

Höfner makes it clear that the *grant* of exclusive rights is contrary to Article 86 if it creates a situation which automatically leads to an abuse of a dominant position by the undertaking benefiting from exclusive rights.

In *Port of Genoa*, the Court held that if the undertakings benefiting from exclusive rights were induced by these rights to abuse their dominant position then the adoption of the rules creating the exclusive rights, in other words the grant of the exclusive right, would be contrary to Article 86 of the Treaty[50].

Cases where exercise of part of the monopoly constitutes an abuse of a dominant position

The grant to a telecommunications network operator of exclusive rights to lay down standards for telephone equipment and to check compliance by third parties[51] constitutes an abuse of a dominant position if the operator also competes with third parties for the sale of equipment. Only part of the monopoly is abusive: the telecommunications network operator's monopoly over the telecommunications infrastructure is not called into question; it is the extension of the monopoly to the market for telecommunications equipment which constitutes an abuse. It is immaterial that the extension of a dominant position is the result of a law[52]. In the latter case, the law itself constitutes an abuse.

Similarly in *Corbeau*, what was at issue was not the entire monopoly but only its application to distinct services consisting in the accelerated collection and delivery of mail.

Specific forms of abuse

Extension of a dominant position

An undertaking which occupies a dominant position on a market and *which reserves to itself*, on a neighbouring but separate market, *an ancillary activity* which could be carried out by others, abuses its dominant position contrary to Article 86, unless the reservation of this activity corresponds to an objective necessity.

---

[50] *Port of Genoa* (see n. 2 above), paragraphs 19 and 20.

[51] *RTT* v. *GB* (see n. 13 above), paragraph 28.

[52] *RTT* v. *GB* (see n. 13 above).

The leading case on extension of a dominant position is *Telemarketing*[53]. Compagnie Luxembourgeoise, the defendant, had been granted a legal monopoly over broadcasting, including television advertising. It started using the services of a third party, the plaintiff, known as Centre Belge de Telemarketing, to carry out telemarketing services. It then decided to reserve telemarketing activities entirely to itself and refused to grant the plaintiff access to broadcasting time unless it used the Compagnie Luxembourgeoise telephone number. This, in effect, prevented Centre Belge de Telemarketing from providing its own telemarketing services. The Court held that the defendant's refusal to grant access to broadcasting time except on conditions which prevented the plaintiff from carrying out telemarketing amounted to a refusal to provide services analogous to the refusal to supply a raw material in *Commercial Solvents*[54].

Refusal to supply constitutes an abuse where this is done to enable the dominant company to reserve to itself an ancillary activity that could be carried out by others[55] with a view to eliminating competition on the ancillary market.

There are three key notions in this case, namely:

- ancillary activity;
- refusal to provide indispensable services;
- absence of valid justification.

(1) *Ancillary activity*: the Court held that telemarketing services constitute a separate market from that of television advertising although they are closely associated. It is a neighbouring market.

(2) *Refusal to provide indispensable services*: Compagnie Luxembourgeoise refused to provide services which were indispensable to the activities of another undertaking on another market. The obligation imposed on Centre Belge de Telemarketing to use Compagnie Luxembourgeoise's number as a condition for access to broadcasting time amounted to a refusal to provide services essential to another telemarketing company to enable it to carry on its activities in competition with Compagnie Luxembourgeoise. By refusing to supply broadcasting time save for the above condition, Compagnie Luxembourgeoise was able to reserve to itself the neighbouring, albeit separate, market of telemarketing and to eliminate all competition in that market. Such refusal of an indispensable service constituted an abuse of a dominant position unless it could be justified by an objective necessity.

---

[53] Telemarketing: centre belge d'études de marché – *Telemarketing (CBEM) SA* v. *Compagnie Luxembourgeoise de télédiffusion SA and information publicté Benelux SA*, case C-311/84, 3 October 1985, [1985] ECR 3261.

[54] *Commercial Solvents and others* v. *Commission* [1974] ECR 223, referred to in paragraph 23 of *Telemarketing*.

[55] *Telemarketing* (see n. 53 above), ground 26.

(3) *Absence of valid justification*: an abuse will be committed unless the undertaking in a dominant position can prove that its act was *objectively necessary*. The Court suggests that *technical* or *commercial requirements* relating to the nature of the television broadcasting in question might constitute proper justification. This must be contrasted with an intention to reserve to oneself an ancillary activity and to eliminate competition in a neighbouring market which, in effect, amounts to an extension of a dominant or monopoly position. The mere fact that the two activities are closely linked is no justification for a refusal to provide services.

The other justification put forward by the defendant, namely the preservation of the television station's image, is not a proper justification. This type of justification is often put forward by state-related undertakings which try to demonstrate that the importance of the task which they have been assigned in the public interest justifies the fact that they alone should provide the ancillary services in question. Because their image is closely linked to that of the State they argue that they should have control over these ancillary services. Preservation of image, however, is unlikely (in most cases) to constitute a valid justification.

Commercial Solvents and *Telemarketing* deal with the reservation by an undertaking in a dominant position of an ancillary activity in a neighbouring but separate market. In *RTT* v. *GB* the Court used the the terms 'extension of a dominant position'. The Court held that the fact that a monopoly reserves to itself a neighbouring but separate market amounts to the extension of a dominant position. Where this extension is not caused by an objective necessity, it constitutes an abuse contrary to Article 86. If the extension of dominant position was caused by a measure of the State the measure constitutes an infringement of Article 90 in conjunction with Article 86 of the Treaty[56].

RTT v. *GB* takes the concept of extension of a dominant position, which is derived from existing case-law, significantly further in three respects:

(1) *Sole power to control the quality of a competitors' products may constitute an extension of a dominant position:* RTT's monopoly was for the establishment and operation of the public telecommunications network in Belgium. In addition, it had sole power to grant and withdraw authorizations for connection to the network, to connect telephone equipment to the network, to lay down technical standards to be met by that equipment and to check whether any equipment, which it had not itself produced, was in conformity with the standards that it had laid down[57]. This additional power could have far-reaching consequences on the market for the sale of equipment where RTT was itself present as a vendor of equipment.

---

[56] *RTT* v. *GB* (see n. 13 above), paragraph 21.
[57] *RTT* v. *GB* (see n. 13 above), paragraph 15.

The Court inferred from RTT's power to vet its competitors' products, that it had reserved to itself another market where it was also present, namely 'the neighbouring but separate market for the importation, marketing, connection, commissioning and maintenance of equipment for connection to the network, thereby eliminating all competition from other undertakings'[58].

The power to control the products of its competitors in an ancillary market was regarded as constituting an extension of a dominant position. This differed from *Commercial Solvents* and *Telemarketing*. In *Commercial Solvents*, a company which was dominant in the market for raw material entirely reserved to itself the market for derivatives; in *Telemarketing*, a company which was dominant in the television market, entirely reserved to itself the market for telemarketing services. In contrast, RTT which had a monopoly over the public telecommunications network, did not entirely reserve to itself the market for telecommunications equipment – it simply sold telecommunications equipment alongside others *but* had sole power to control the quality of its competitors' equipment through the grant and withdrawal of authorizations. In other words it could *control* the presence of other operators as opposed to straightforwardly prohibiting their activities (as in *Commercial Solvents* and *Telemarketing*). This is nevertheless construed as constituting an extension of a dominant position.

An important feature common to *RTT* v. *GB*, *Commercial Solvents* and *Telemarketing* is the fact that the dominant undertakings were present in the separate neighbouring markets where they could make use of their dominance. This must be contrasted with the case where the dominant undertaking can exercise a control over the downstream market but is not present in that market. This was the case of *Banchero*[59] where the dominant undertaking, AAMS, which had an exclusive right to distribute Italian tobacco to retailers and had a *de facto* monopoly over the wholesale distribution of locally produced and imported tobacco in Italy also had the exclusive statutory right to grant authorizations to retail outlets to sell tobacco to final users. The dominant undertaking and holder of the exclusive rights was not, however, present at the retail level. The Court held that the system of retail sales through outlets authorized by AAMS only did not result in an abuse of a dominant position by AAMS. One of the significant factors relied upon by the Court was the fact that AAMS was not present in the retail downstream market. The Court held that the Italian law establishing the system was not contrary to Articles 90 and 86.

(2) *The restriction, as opposed to elimination of competition, may give rise to an extension of a dominant position:* It follows from the Court's analysis that

[58] *RTT* v. *GB* (see n. 13 above), paragraph 19.

[59] *Banchero*, case C-387/93, 14 December 1995, [1995] ECR I-4663 (see, in particular, ground 49).

control of an ancillary market may amount to the reservation to oneself of an ancillary market and, accordingly, to an extension of a dominant position. Contrary to what the ordinary meaning of the words might suggest, it was not necessary that *the entire market should be reserved* to RTT and that RTT should have *excluded* all other players, for the market to be regarded as reserved to it. It was sufficient that the market should be *restricted*[60]. The Court refers to 'the exclusion *or* the restriction of competition on the market in telephone equipment'[61] (emphasis added). In the report for the hearing[62], RTT argued that there were many other players in the relevant ancillary market, in particular in the field of PABX and fax. In relation to certain of the relevant products it was either not present or had only a minimum market share. Yet RTT's power to vet the products on the ancillary market, even if it did not have a controlling market share, was regarded as constituting an extension of a dominant position. This is a new application of the concept of dominance and abuse[63].

(3) *The mere possibility to discriminate against others may give rise to an abuse:* The mere power to restrict competition on the market by discriminating against others was held to constitute an abuse. It was not necessary to show that this power had actually been used[64].

The grant to RTT of a vetting power over all terminal equipment meant that RTT 'had the power to determine at will which terminal equipment may be connected to the public network' and 'placed RTT at an obvious advantage over its competitors'[65].

The Court uses the words 'extension of a dominant position' (paragraph 21) and 'extension of a monopoly' (paragraph 24) even though the Court neither states nor presumes that RTT held a monopoly over terminal equipment or had a dominant position in that market.

Although the Court's analysis of the concepts of dominance and reservation of an ancillary market is not rigorous in *RTT* v. *GB*, the lessons from the case are important:

- the extension of a monopoly or dominant position to an ancillary market without proper justification will constitute an abuse of Article 86 and Article 90 if this is due to a measure from the State (this confirms *Telemarketing*);

---

[60] There is no need to eliminate all competition as in *Telemarketing* (see n. 53 above), paragraph 26, and as suggested by paragraph 19 of *RTT* v. *GB* (see n. 13 above).

[61] *RTT* v. *GB* (see n. 13 above), paragraph 22.

[62] *RTT* v. *GB* (see n. 13 above), pages 25 and 26.

[63] Contrast with what the Advocate General said in the *Dutch express courier* case (see n. 39 above), the extension of a dominant position by the Dutch PTT in the market for express services was not proved because it had not been established that it had acquired a dominant position in that neighbouring market.

[64] *RTT* v. *GB* (see n. 13 above), paragraph 23.

[65] *RTT* v. *GB* (see n. 13 above), paragraph 25.

- the extension of a monopoly or dominant position may result from the accumulation of powers in an undertaking, for instance, the establishment and operation of a telecommunications network together with the control of equipment to be connected to that network, in circumstances where the undertaking also operates on the ancillary market for the sale of equipment;
- the undertaking may be extending its dominant position in an ancillary market even if it does not hold a monopoly or even acquire a dominant share of that market, provided it has the power to control the presence of third parties on that market;
- the abuse is committed even if the undertaking does not use its power to discriminate unfairly against its competitors. It is sufficient that it should have a vetting power over its competitors' products, which gives it the possibility to discriminate against them.

Manifest inability to meet market demand
The leading case is *Höfner*:

> A Member State creates a situation in which the provision of a service is limited when the undertaking to which it grants an exclusive right extending to executive recruitment activities is *manifestly not in a position to satisfy the demand prevailing on the market* for activities of that kind. (*emphasis added*)

In *Höfner*[66], the German employment agency, Bundesanstalt, controlled by the State, had a monopoly over the placement of employees. It followed that firms of 'head hunters' – an activity which was commercially widespread and profitable across Europe – were illegal in Germany and could only be exercised in breach of the monopoly. The German employment agency did not, however, cope with all the demand on the market and, in practice, tolerated certain abuses of its monopoly by not prosecuting infringements of which it was aware. The Advocate General's Opinion mentions that the Bundesanstalt only dealt with 28 per cent of vacancies[67]. Höfner and Elser, the defendants, reported during the hearing that there were between seven and eight hundred private consultants for the placement of senior employees in Germany. These figures demonstrated that there was a serious need for other operators and that the Bundesanstalt was manifestly not able to satisfy the demand on the market.

The Advocate General said that consumers did not receive the quality of service that they were entitled to expect and would receive if the sector of activity in question was open to free competition[68]. Referring to the *Volvo Veng* case, he added that the situation was comparable with that 'of the owner

---

[66] See n. 1 above.
[67] *Höfner* (see n. 1 above), AG opinion paragraph 44.
[68] *Höfner* (see n. 1 above), AG paragraph 45.

of a Volvo who cannot obtain a new body panel for his car because the owner of the registered design does not manufacture the part and refuses to allow anyone else to do so'[69].

According to the Advocate General, if the beneficiary of an exclusive right granted by law 'fails to produce the goods or services covered by the exclusive right, that failure may amount to an abuse of a dominant position'[70].

The Court's ruling supports the approach of the Advocate General.

The question arises as to how far the Court's ruling may apply to other cases. Are all 'incompetent' monopolies, i.e. monopolies which are manifestly unable to meet the demand on the market, illegal and bound to be held – sooner or later – incompatible with Article 86?

Postal services, for example, are of varying quality throughout the Community[71]. The quality of the service is poor in certain Member States where it by no means achieves the level which customers are entitled to expect. Does it follow that the postal monopoly in those Member States is illegal? Following the *Corbeau* case it might be argued, on the contrary, that the monopoly is essential in order to provide a postal service whatever its level of quality[72]. Without the monopoly the public postal service might entirely disappear and be replaced by private companies offering certain niche services to defined customers as opposed to the public as a whole. The Green Paper on postal services already recognized this problem and stated that the Article 90(2) derogation might continue to apply longer in favour of those Member States needing a longer period of time to reach a minimum level of quality of service[73].

Corbeau versus Höfner

Does this mean that, after *Corbeau*, the *Höfner* test is dead and that a monopoly may no longer be illegal merely on the ground that it is manifestly unable to meet the demand on the market? Does it mean that irrespective of the quality of the service or product offered, the monopoly must be maintained because it is necessary to provide a minimum level of service to the population as a whole[74]?

The preferred view is that *Höfner* is not dead[75] but that it cannot apply to all cases of inability to meet market demand. *Höfner* concerned services

---

[69] *Höfner* (see n. 1 above), AG paragraph 45.

[70] *Höfner* (see n. 1 above), AG paragraph 46.

[71] See the Green Paper on Postal Services COM(91) 476 final, p. 2 paragraph 3.2.

[72] See Advocate General in *Corbeau* (see n. 43 above), paragraph 16.

[73] In the Green Paper on Postal Services (see n. 71 above), the Commission suggests that Article 90(2) might be relied on by certain Member States whose postal services meet greater financial difficulties than others, in order to maintain wider postal monopolies (see page 188, paragraph 4 and page 207, paragraph 3 of the Green Paper).

[74] See the concept of 'universal service' (p. 175 below).

[75] The *Höfner* test relating to the manifest inalibilty to meet the demand on the market was mentioned in *Banchero*, case C-387/93, 14 December 1995, ground 53.

consisting in the placement of employees – an area already deregulated in most EU countries. It may, however, be difficult to apply to certain essential postal services or other utility services.

Circumstances in which Höfner is likely to apply

Höfner is likely to apply to less 'vital' monopolies i.e. monopolies not related to the basic services provided by the utilities, namely water, electricity, gas, other energy supplier, transport, post. But it could also apply to some of the essential utility services in certain circumstances, for example:

- to services offered in remote areas situated at the end of a national utility network, if the services offered in these areas are of a significantly lower quality than those provided in the rest of the territory. An example of this would be the poor supply of electricity in a remote region situated at the end of the electricity network near a border with another country. If the national electricity supplier across the border or private operators could provide a better service, this might justify a relaxation of the monopoly in favour of these third parties on the grounds that the monopoly enterprise is manifestly unable to meet the needs of customers in the remote region;
- to certain utility services where the monopoly provider tolerates infringement of its monopoly, i.e. where it accepts competition by others by failing to challenge them vigorously or at all.

The mere sending of a letter before action by the monopoly holder would not of itself constitute a vigorous challenge; unless this is followed by proceedings it may mean that the monopoly provider is effectively turning a blind eye to infringements. Even though certain national laws (e.g. French law) provide that failure to enforce a right does not amount to a waiver of the right, nevertheless such failure may make it more difficult to justify the monopoly at a later stage and to enforce it. It would be open to challengers of the monopoly to argue either that it is not necessary or that the presence of infringers shows that the monopoly is manifestly not able to meet the demand on the market.

The question which arises is: how strong must the presence of competitors be for a *Höfner*-like situation to exist; at which point can it be said that the monopoly provider is *manifestly unable* to meet market demand? There is no clear answer. The situation in *Höfner* was extreme if one refers to the figures quoted above (there were apparently over seven hundred private consultants alongside the so-called monopoly holder). One could, however, imagine cases where *Höfner* could apply in less extreme circumstances. Where similar services are open to competition in most other Member States, it may be difficult to justify the monopoly and economic arguments are likely to be unhelpful.

Denying access to an essential facility

Definition of access to an essential facility

The concept[76] was first used and defined by the European Commission in *Sea Containers* v. *Stena Sealink*, a decision of 21 December 1993[77].

> an undertaking which occupies a dominant position in the provision of an essential facility and itself uses that facility (i.e. a facility or infrastructure, without access to which competitors cannot provide services to their customers), and which refuses other companies access to that facility without objective justification or grants access to competitors only on terms less favourable than those which it gives its own services, infringes Article 86 if the other conditions of that Article are met.

In the *Stena Sealink* Decision, a competitor of Stena Sealink, Sea Containers, wished to introduce a new fast ferry service on the Holyhead–Dun Laoghaire route, using a wave-piercing catamaran. The route in question is the central corridor between the UK and Ireland and is the most popular route between these two Member States. Stena Sealink owned the port of Holyhead and also operated a fast ferry service on the route. Sea Containers alleged that Stena Sealink either delayed its decision to offer it slots at the port or asked for higher charges than for its own ferry services thus discriminating against it. In its Decision, the Commission made a preliminary finding that Stena Sealink 'in its capacity as port authority was in a dominant position in the market for the provision of port facilities for car and passenger ferry operations on the central corridor route between the UK and Ireland'[78]. It held that Stena Sealink had abused its dominant position by obstructing access to the port – an essential facility without which Sea Containers could not introduce what constituted a new service at the time. The Commission found that Stena Sealink had not behaved as an independent operator would, i.e. trying to maximize the traffic and accordingly examining how it would be possible to accommodate new entrants.

There are three principal characteristics of this specific form of abuse:

- a dominant position resulting from the management of an essential facility;
- a denial of access to the essential facility to competitors or application of discriminatory conditions to obtain access;
- an absence of objective justification.

(1) *Dominant position resulting from the management of an essential facility*: An essential facility is defined as 'a facility or infrastructure, without access to which competitors cannot provide services to their customers'[79]. It

[76] See further p. 165.
[77] *Stena Sealink*, OJ 1993 L15/8 (paragraph 66).
[78] *Stena Sealink* (n. 77 above), paragraph 65.
[79] *Stena Sealink* (n. 77 above), paragraph 66.

must be an indispensable means of providing a service. Examples of essential facilities may, in certain circumstances, apply to all the major utility networks including telecoms networks, railway tracks, electricity grids and water. The owner and/or operator of an essential facility will normally be regarded as occupying a dominant position.

(2) *Denial of access to the essential facility or making access subject to discriminatory conditions*: Outright denial of access may constitute an abuse. Certain types of obstruction in the form of discriminatory and unfair conditions may be tantamount to a refusal to grant access (see *Telemarketing*). In *Stena Sealink* the Commission found that delaying tactics amounted to a refusal to grant access. The dominant undertaking had failed to undertake proper consultations about its projected developments with all interested parties, even though these developments could affect their access to the port[80]. On the face of it, it had imposed a competitive disadvantage on some competitors. Absence of discrimination also means that conditions of access should be no more onerous for third parties than for the owner of the essential facility; for example, the oil companies owning a sole refuelling facility at an airport must not charge higher prices to third party oil companies than to themselves[81]; slots should be allocated on a fair basis at airports[82] and ports; schedules for ferry services should be arranged in a fair manner[83].

(3) *Absence of objective justification*: Refusal or obstruction of access should be objectively necessary. If the real reason for denying access is to reserve an advantage to oneself this is likely to constitute an abuse. Economic or technical necessity (such as disruption of the network and worsening of the service to the majority of customers) or environmental or security concerns, may constitute objective justifications provided they can be substantiated and the obstacle to access is proportionate to the alleged justification.

Failure to give proper access may also result from a failure to facilitate the entry of competitors. An airline operator holding a dominant position in the market and withdrawing the interlining facilities[84] which were necessary to a new entrant, at least during a start-up period, was held to abuse its dominant position[85]. The grant of access may thus mean that the dominant undertaking must assist positively in the entry of a competitor to its own market by

---

80 *Stena Sealink* (note 77 above), paragraph 73

81 *Malpensa Airport*, 30 July 1993, IP/93/084.

82 See Council Regulation (EEC) 95/93 of 18 January 1993 on common rules for the allocation of slots at Community airports, OJ 1993 L14/1; and Commission Regulation (EEC) 1617/93 OJ 1993, L155/18.

83 *B & I* v. *Sealink*, 11 June 1992, (1992) 5 CMLR 255; EC Bulletin, 6-1992, (1992), point 1.3.30.

84 'Interlining facilities' are facilities granted by one airline to another on the same route for the interchange of seats, reservation and tickets.

85 *British Midland* v. *Aer Lingus*, OJ 1993 L96/34.

providing interlining facilities or other services that it normally uses for itself in return for fair remuneration. In some cases, the Commission will intervene by way of decisions to deal with these issues. Thus in the *Night Services*[86] Decision (an Article 85 case) the Commission granted an exemption on condition that the national railways, which were parties to the joint venture agreement, would provide new entrants the services which they may require in order to compete, namely the provision of a locomotive, train crew and path on the national rail network and in the Channel tunnel. This decision has, however, been appealed and the decision of the Court of First Instance is pending at the time of writing.

Development of the concept of essential facility

The concept was first used in US antitrust law several decades ago, pursuant to sections 1 and 2 of the Sherman Act. The underlying principle has been applied in the European Community since *Commercial Solvents* in respect of goods, and since *Telemarketing* in respect of services, but the term 'access to an essential facility' was applied for the first time in *Stena Sealink*. In this Decision, the Commission recites a large number of cases and Commission Decisions[87] which, in its view, provide the legal basis for the concept.

It is in the transport sector that the concept as such has been the most widely applied to date[88]. But the concept can be applied to many other sectors[89] and in particular to state-related undertakings which have been given exclusive or special rights over key utility infrastructure.

Further possible developments

The concept is likely to have a wide application to Article 90 cases, particularly those involving undertakings benefiting from exclusive rights. These undertakings might be able to retain some exclusive rights provided they give third parties access to part of their services. Third parties are likely to include new entrants who can provide a new service or product, or similar services or products, but on improved conditions.

Examples of the access principles can be found in chapter 8 below.

---

[86] Night Services Commission Decision of 21 September 1994, OJ 1994 L259/20; on appeal, case T-374/94.

[87] *Stena Sealink* (n. 77 above): page 16, note 3: the Commission mentions the following Court judgments: cases 6 and 7/73, *Commercial Solvents* v. *Commission* [1974] ECR 223; Case 311/84, *Telemarketing* [1985] ECR 3261; Case C-18/88 *RTT* v. *GB-INNO* [1991] ECR I-5941; Case C-260/89, *Elliniki Radiophania Teleorassi* [1991] ECR I-2925; Cases T-69, T-70 and T-76/89, *RT, BBC and ITP* v. *Commission* [1991] ECR II-485, 535, 575; and the following Commission Decisions: 76/185/EEC, 1976 *National Carbonizing Company*, OJ L35/6; 88/589/EEC, 1988 *London European–Sabena*, OJ L317/47; 92/213/EEC, 1992 *British Midland* v. *Aer Lingus*, OJ L96/34; *B&I Sealink* [1992] 5 CMLR 255; EC Bulletin, 6 – 1992, point 1.3.30.

[88] For example, *Stena Sealink* (n. 77 above); *the Port of Rødby* (n. 41 above); *London European-Sabena* (n. 87 above); *British Midland* v. *Aer Lingus* (n. 87 above); *B & I* v. *Sealink* (n. 87 above).

[89] E.g. broadcasting, TV listings, computer software. See the extensive analysis of the concept of access to essentital facilities by John Temple Lang in *Twenty First Annual Fordham Corporate Law Institute International Antitrust Law and Policy*, October 1994.

Preventing the offer of a 'specific' or 'new' service or product by a competitor

In *Corbeau*, the Court defined a new type of abuse by state-related undertakings benefiting from a statutory monopoly, namely preventing third parties offering a new service or product to consumers in the area covered by the monopoly. The Court explained that a monopoly undertaking could not exclude competition from third parties resulting from 'specific services'[90]:

> The exclusion of competition is not justified as regards *specific services* dissociable from the service of general interest which meet special needs of economic operators and which call for certain additional services not offered by the traditional postal service, such as collection from the sender's address, greater speed or reliability of distribution or the possibility of changing the destination in the course of transit, in so far as such specific services, by their nature and the conditions in which they are offered, such as the geographical area in which they are provided, *do not compromise the economic equilibrium of the service of general economic interest performed by the holder of the exclusive right.* (*emphasis added*)

A monopoly undertaking cannot prevent the supply of certain specific services, which may be referred to as *new* services[91]. What are the characteristics of 'specific' or 'new' services? The Court, in *Corbeau*, laid down a test of general application which is not limited to the postal sector. It was used shortly after *Corbeau* in the electricity sector, in the *Almelo* case. The characteristics of 'specific' or 'new' services are as follows:

(1) *Specific services dissociable from the service of general interest:* The services must not be similar to the core service of general economic interest of the monopoly undertaking. In the postal sector this core service is the collection and delivery of ordinary mail. In order to be 'specific', services must:
   (a) Meet the special needs of economic operators: The services in question thus do not meet the needs of all the population but of certain economic operators only. In the postal sector they should not relate to the common needs of all users but, for example, to the special needs of the business sector.
   (b) Call for certain additional services not offered by the traditional monopoly operator: The Court gives examples of additional services in the postal sector namely collection from the sender's address, greater speed or reliability of distribution, or the possibility of changing the destination in the course of transit.

   Yet certain postal administrations may already be offering these services. Even if they did not do so in the past, they may now offer such services as a result of the pressure of their customers or of the

[90] *Corbeau* (see n. 43 above), paragraph 19.
[91] 'New services' are the words used in the Green Paper on postal services COM(91) 476 final, p. 194.

competition. In those circumstances, can the service nevertheless qualify as 'additional' and as falling outside the monopoly? The answer to this question may largely depend on timing. If a particular idea 'meeting the special needs of business people' is first devised by competitors of the monopoly or holder of exclusive rights, then it is likely to be regarded as a 'new' service since it could not be said to form part of the services offered by the traditional monopoly operator. An example of this would be the provision of express mail services.

(2) *The service must not, however, compromise the economic equilibrium of the service of general economic interest performed by the holder of the exclusive right:* This second condition introduces an economic test: it provides that the 'specific' or 'new' service must not jeopardize the economic equilibrium of the monopoly operator. It may be argued that in many cases, if the first condition is satisfied, then the second condition will also automatically be met. The reason for this is that a 'new' service will by definition not have been provided by the monopoly operator and will, therefore, not take away any of its business and revenue. This argument may not, however, apply where the 'new' service does not merely increase the range of existing services offered by the monopoly operator but replaces those services. In this case, it may be appropriate to limit the scope of the new service so that the economic test may be met.

The Court mentions that for this reason it may be appropriate for the new service to be provided only in certain geographic areas or on certain conditions[92]. In the case of *Corbeau*, the Advocate General suggested that the Member State concerned could require that the 'new' service be supplied only above the postal administration's tariffs so as to reflect the added value provided by the 'new' service over and above the postal administration's standard service and so as not to take away revenue from the postal administration[93].

When is a service specific or new within the Corbeau definition?

Having laid down this double test, the Court said that it was for the national judge to say whether the criteria were met. But this is an extremely difficult task with far-reaching consequences, since, if a service meets the *Corbeau* test, it must be allowed. The national monopoly operator may not then rely on its legal monopoly to stop the provision of the new service. It must accept competition from the provider of the new service. One of the dangers is that national judges may reach different solutions. Some of the difficult issues which are likely to arise are examined below.

---

[92] *Corbeau* judgment (n. 43 above), paragraph 19.

[93] *Corbeau* (n. 43 above), see point 18, paragraph 10 of the AG opinion.

One Corbeau or one hundred Corbeaux?

This question asks, how should the economic test be applied? The *Corbeau* case involved only one individual and his family carrying on mail services in one Belgian city, Liège. The national judge could well find that, on the facts, such operation will not have a significant influence on the nationwide revenue of the monopoly or holder of exclusive rights and would therefore not jeopardize its finances.

Assuming the first part of the test is met, should a *Corbeau*-type operation be allowed? The problem is that if a *Corbeau*-type operation is permitted this means that it constitutes a new service which does not jeopardize the economic equilibrium of the post office. It should, therefore, be allowed in all other similar cases. It would follow that the post office (or other monopoly) cannot rely on its monopoly to prevent this type of service. But what may be financially acceptable in respect of one *Corbeau*-type activity may not be so if it is multiplied one hundred or one thousand times. Should the national court limit itself to examining one *Corbeau*-type situation? The answer must surely be in the negative because allowing *Corbeau*-type activity amounts to limiting the scope of the monopoly in all *Corbeau*-like situations. It is, therefore, the effect of the general limitation of the monopoly that must be envisaged.

The question must be, whether by allowing this new type of service, and not just this particular service by this individual, the economic equilibrium of the post office will be jeopardized[94]. Evidence of the economic value of the service in question will have to be produced. In the express courier sector, the European Commission had shown that this represented only a small part of the postal administration's activity[95].

Who must have the idea of the new service first; and does it matter?

In theory, who had the idea first, should make no difference; if a service meets the *Corbeau* test it should be regarded as a new service and therefore fall within the competitive sector. In practice, however, the question is likely to make a difference. If the private sector comes up with the idea first, as was the case with express couriers, it will be more difficult for the monopoly undertaking to argue that it is not a new service. If the postal administrations have the idea first, they may be able to claim that the service in question is

---

[94] *Contra*: The issue has been raised before a few Belgian and French Tribunals, for example, *La Poste* v. *SPRL Delta Express and SA Hays Business Services and others*, case A.C./5.884/94 and A.C./7.829/94 before the Tribunal de Commerce de Bruxelles (22 May 1995); *SARL Pro-Mess and Direction Départementale de la Poste de Seine Maritime*, T.A. de Rouen, 22 February 1995. Interestingly, first instance judges have held in all these cases that the defendants (individuals or companies wishing to provide a *Corbeau*-type service) had met the double test in *Corbeau* and could, therefore, not be stopped on the ground that they infringed the legal monopoly of the postal administration. Some of these cases have been appealed.

[95] For example, in the Commission Decision concerning the provision in Spain of international express courier services (OJ 1990 L233/19) the Commission stated that express international courier had represented only 0.00039% of the postal items handled by the Post Office.

only an adaptation of their existing services, that it constitutes a normal evolution and, accordingly, forms part of their monopoly.

How new should a 'new' service be?

The fact that someone has a new idea and is able to offer a slight improvement over an existing service may not be enough to give rise to a 'new' service. *Corbeau* does not provide clear guidance. The question may be asked whether Mr Corbeau himself really provided a 'new' service or only an enhanced service. Whilst express courier clearly constitutes a distinct service which meets the double test laid down by the Court, it is not certain that Mr Corbeau's services constituted express courier and, therefore, met the test.

Can monopolies evolve to include new services?

This is a difficult issue. Any monopoly undertaking must evolve as technology changes and as its customers' needs change. In most Member States the speed of postal services has increased as a result of technological progress and changes in the transport infrastructure as well as in the means of transport. Postal administrations have thus constantly modernized their services. The same applies to all utilities, for example gas and electricity where, in most countries, performance has improved not only in quality but in the extension of the service to more customers.

Does the *Corbeau* judgment mean that one should be able to identify a cutoff time which would set the frontier between traditional postal services which would continue to be covered by the monopoly and new services introduced after this cutoff time which would be open to competition?

This would be an extreme construction of the *Corbeau* judgment which would probably be unrealistic. It would mean that the use of improved technology resulting in changes in the service provided would almost automatically result in the provision of a new service falling outside the monopoly. If this were the case, postal administrations could not improve and change their basic services to customers without risking the loss of their monopoly. It is unlikely that the Court in *Corbeau* could have intended such consequences for basic monopolies.

But, how can the *Corbeau* test of new services be applied in practice? Other criteria in the judgment must be taken into account in addition to the distinction between traditional and new services.

Certainly the distinction between traditional services and other services is important, but as explained above, it can probably not be reduced to establishing a cutoff point in time between monopoly and non-monopoly activities. The statement in *Corbeau* is nonetheless important because it shows *the tendency of the Court to limit the further expansion of monopolies and to introduce as much competition as possible into new services.*

The following paragraphs tentatively suggest some possible guidelines.

Improved services by the traditional monopoly operator can remain within its monopoly. To take the postal sector, for example, the following services would constitute an improvement over the existing service but would probably remain within the monopoly:

- further increases in the speed of delivery;
- seven-day collection and delivery service (i.e. to include Sunday and most bank holidays);
- additional early and/or late deliveries.

These improved services would probably fall within the monopoly provided they were offered to all users throughout the territory at uniform tariffs and on similar quality conditions. The consequence might be that a private operator could not offer a new early and/or late delivery service without infringing a postal administration's monopolies.

Improved services proposed by competitors are likely to fall within existing monopolies. For example a proposal by a private company to offer an improved quality of service on a train journey by offering newspapers, magazines and books; new forms of catering services; or a children's entertainment area with a child minder, would probably not constitute a new service. Even though this type of service is not currently offered on trains it is unlikely to constitute a new service in the *Corbeau* sense because it involves the same train journey, i.e. the same route and the same schedule; it does not involve a new *transport* service meeting special needs of economic operators.

New services with significant added value falling outside the monopoly

New services should provide more than an improved existing service offered to everyone. They should include significant added value[96] for which economic operators are prepared to pay, because the services meet their special needs.

New services are likely to aim at the following economic operators: business people, tourists, families (provided in the latter case the service is not introduced principally for social reasons). They will apply either across the country or in defined geographical areas and they will be offered at tariffs generally higher than the monopoly's standard tariffs.

The European Night Services (ENS) introduced by a number of national railways following construction of the Channel Tunnel may fall within the category of new service[97]: it aims at servicing principally the needs of business people; and provides a value added service: a new route, new schedules, new quality of service with shower rooms; it is offered at a price above national railways' standard price. Even though the idea stems from the national railways a new entrant may wish to offer a similar competing service provided it

[96] See the Advocate General's opinion in *Corbeau* (n. 43 above) point 18, paragraphs 5 and 6.

[97] The Commission recognized in the Night Services Decision (OJ 1994 L259/20) that ENS would introduce new high quality transport services (see paragraphs 59 and 61 of the Decision).

meets the conditions for entry into the market as set out in Directive 91/440[98].

Another example would be the offer of a weekend train service between well-known cities in Europe leaving on a Friday night, returning on a Monday morning, with bathroom and catering facilities, and including a hotel night on Sunday. The service would be aimed at tourists or weekenders; it would be a complete train service (not just an improved existing service) and would involve significant added value reflected in the tariffs.

The distinction is far from clear. However, to take the transport sector for example, the difference between an improved service and a new service may be that the new service relates to key features of transport namely the route, the schedule and speed; in contrast the improved service may be limited to non-transport facilities such as the provision of a children's entertainment area.

Using monopoly revenue to cross-subsidize activities subject to competition

Cross-subsidization means using the revenues of one sector to help another sector. It is customary business practice to use revenues from the most profitable activities to balance less profitable activities or to provide investment for the development of a new business sector, product or service. Such practices are normally compatible with the competition rules but raise competition law problems in two main circumstances:

(1) the undertaking is in a dominant position in one market and uses the revenues of that market to cross-subsidize activities in an ancillary market in a *predatory manner*[99];

(2) the undertaking enjoys a *statutory monopoly* or has been granted exclusive rights by the State in a market and uses the revenues generated from such monopoly or exclusive rights to cross-subsidize activities in another market where it is in competition with other undertakings.

Both circumstances can apply to monopoly undertakings. The first one, predatory pricing, is a classic Article 86 abuse applying to any undertaking in a dominant position. The second one, cross-subsidization of competitive services with revenue from monopoly services, applies only to state-related monopolies and is examined below.

There has been limited activity so far at the Court or the Commission on this; however, one example is the Commission decision concerning the alleged cross-subsidization by the French post office of its express mail subsidiary. This was followed by a judgment of the Court of First Instance[1] which confirmed the principle. The rule is clearly stated in the Commission

[98] Directive 91/440 OJ 1991 L237/25.

[99] *AKZO* v. *Commission* [1991] ECR 3359. *Tetra Pak II*, case T-83/91, 6 October 1994 ECR II-649.

[1] *Syndicat Français de l'Express International, DHL International Service CRIE, May Courier* v. *Commission*, T-77/95, 15 January 1997, para. 69, [1997] ECR II-1.

guidelines on competition in the telecommunications sector[2] and is also clearly mentioned in the green paper on postal services[3] and in the draft Notice on competition in the postal sector[4].

The principle applies to all undertakings which have been granted a monopoly by the State. The proposed Directives in the electricity and gas sector, and in the railway sector all contain rules on the separation of accounts. The purpose of these rules is *inter alia* to monitor the revenues of each activity and control their transfer from one activity to another. National regulators also tend to take the same approach.

The *rationale* behind this principle is that where a monopoly undertaking competes with others in certain sectors it is important that competition should be fair. If the monopoly undertaking uses revenues from an area where it is not subject to competition to boost its revenues in another area where it is subject to competition, thus enabling it to undercut its competitors in that other market, this is likely to distort competition contrary to Article 86. The monopoly undertaking has, so to speak, guaranteed revenues in the monopoly sector; it should not be allowed to use them to compete unfairly in the competitive sector.

It can be argued, however, that the monopoly undertaking should be able to use its revenues coming from the monopoly sector in two cases: (i) for exceptional, e.g. social reasons[5]; (ii) where the cross-subsidization does not distort competition and is necessary for commercial reasons. For example, cross-subsidization may be necessary in order to enable the monopoly undertaking to diversify its activities and enter into a new area. But the cross-subsidization should not extend beyond a commercially fair start- up period.

Article 85

There have been a few cases involving Article 85 and Article 90. The major cases which are examined below are: *Almelo*[6], *Uniform Eurocheque*[7], *IJsselcentrale*[8], *EBU*[9], *Metropole Television*[10], *Jahrhundertvertrag*[11]. They show that Article 90(2) applies to agreements between state-related undertakings which have been entrusted with a task of general economic interest. These agreements are subject to the competition rules – whether or not they have been

[2] *Guidelines on the application of the competition rules in the telecommunications sector*, OJ 1991 C-233/2.
[3] The Green Paper on Postal Services COM(91) 476 final, p. 250 point 6.4; an exception is provided in point 6.6.
[4] See paragraph 5.6 of the draft Notice (OJ 1995 C-322/11).
[5] See proposed Commission Notice in the postal sector of 30 January 1997, paragraph 5.76.
[6] *Almelo*, C-393/92, [1994] ECR 1477.
[7] *Uniform Eurocheques:* Commission Decision 10 December 1984, OJ 1985 L35/43.
[8] *IJsselcentrale:* Commission Decision 16 January 1991, OJ 1991 L28/32.
[9] *EBU: EBU/Eurovision System* Commission Decision, 11 June 1993, OJ 1993 L179/23.
[10] *Metropole Television SA and Reti Televisive Italiane* v. *Commission*, joint cases T-528/93, T-542/93, T-543/93 and T-546/93, 11 July 1996, ECR II-649.
[11] *Jahrhundertvertrag:* Commission Decision of 22 December 1992, OJ 1993 L50/14.

encouraged by public authorities – unless these rules would obstruct the performance of a task of general economic interest.

Article 90(2) may apply to exempt an agreement entirely from the application of Article 85, whether or not the agreement has been notified and therefore irrespective of Article 85(3) and the rules for exemption under Regulation 17/62. Thus in *Almelo*, the Court held that Article 90(2) might apply to an exclusive purchasing clause in agreements between a regional distributor of electricity and local distributors even though this significantly restricted trade between Member States and would normally be prohibited by Article 85. The Article 90(2) derogation would apply if the exclusivity clause was necessary to enable the regional electricity supplier to carry out its task of general economic interest.

Article 90(2) may also apply within the context of a request for exemption under Article 85(3)[12]. In that case, it may be used to reinforce arguments in favour of an exemption from the prohibition of Article 85(1). In those cases, the examination of Article 90(2) will take place prior to the assessment under Article 85(3).

The undertakings seeking to rely on Article 90(2) must prove that they have been entrusted with a task of general economic interest and that the application of Article 85(1) would prevent the fulfilment of that task.

The undertakings involved in the *Uniform Eurocheque* organization which had entered into an international agreement relating *inter alia* to the payment of banking commission claimed that Article 90(2) applied and that the agreement was therefore exempted from the application of the prohibition in Article 85(1). The Commission disagreed. It held that the double test for the application of Article 90(2) failed:

(1) the banking institutions involved in the Eurocheque agreement were not entrusted with the operation of a service of general economic interest by a measure adopted by the public authorities. It made no difference that the Eurocheque system had the express approval of the authorities concerned or that in some countries there had been an explicit legal act in favour of the system or part of it. An *ex post facto* approval even if it took the form of a legal act was not sufficient, in this case, to confer to the undertakings concerned a task of general economic interest under Article 90(2) (paragraph 29 of the Decision);

(2) the Commission said that in any event the application of the competition rules would not in any way obstruct the fulfilment of a 'hypothetical special assignment'.

The Commission, therefore, ignored Article 90(2) in its assessment of the validity of the agreement under Article 85(3). However, it found, on the facts, that the conditions of Article 85(3) were fulfilled and that the *Uniform Eurocheque* agreement could benefit from an exemption.

---

[12] See *Metropole Television*, n. 10 above.

In another case, *IJsselcentrale*, the Commission also examined whether Article 90(2) applied. It held that the first test of Article 90(2) regarding the grant of a task of general economic interest by public authorities was fulfilled. The undertakings concerned were electricity companies which had been entrusted by the State, pursuant to a law, with such a task[13]. The Commission, however, found that the second test of Article 90(2) was not fulfilled: the application of the competition rules would not obstruct the fulfilment of the task. The Commission thus found that an agreement between Dutch electricity generators which contained a total import and export ban on certain operators went beyond the restrictions imposed by national law and was not necessary in order to enable the operators to carry out their specific task of general economic interest[14].

The Commission noted that the agreement had not been notified in accordance with Regulation 17 and it could therefore not qualify for an exemption under Article 85(3). The Commission went further saying that even if the agreement had been notified it would not have qualified for an exemption under Article 85(3) presumably because the restrictions on competition were not indispensable. This suggests that the necessity test in Article 90(2) is close to the indispensability test in Article 85(3). Failure to meet the necessity test in Article 90(2) is likely to lead to a failure under Article 85(3). In *Uniform Eurocheque*, the Commission had not examined whether the necessity test of Article 90(2) had been met before concluding that the indispensability test of Article 85(3) was fulfilled and examined the case solely under Article 85(3). The *Uniform Eurocheque* Decision, therefore, does not contradict the conclusion drawn from *IJsselcentrale* that there is a close link between the two tests of necessity and indispensability. The conclusion is further supported by two cases where the Court and the Commission have accepted that Article 90(2) applied (i.e. the two tests were met) and that Article 90(2) was relevant to an analysis under Article 85(3).

The two cases are *Metropole Television S.A.*[15] and *Jahrhundertvertrag*[16]. In *Metropole Television*, the Court held that the Commission could take into account in its assessment under Article 85(3) the fact that the undertakings concerned had been entrusted with a task of general public interest and might need a derogation from the full application of the competition rules in order to carry out their task[17].

Metropole Television was an appeal by various television channels against the Commission Decision in *EBU* granting an exemption to the EBU agreement and, in particular, to the exclusive right held by the television channel members of EBU to transmit sport events.

---

[13] *IJsselcentrale* (n. 8 above), paragraph 40 of the Decision.
[14] *IJsselcentrale* (n. 8 above), paragraph 46 of the Decision.
[15] See *Metropole*, n. 10 above.
[16] See *Jahrhundertvertrag* (see n. 11 above).
[17] *Metropole Television* (n. 10 above), paragraph 118.

The Court held that the EBU members could be regarded as undertakings entrusted with a task of general economic interest. But the Commission should then have examined whether they accordingly needed exclusive rights of transmission in order to carry out their specific tasks. The necessity test should be conducted on the basis of a proper economic analysis and should not be based on broad statements only. The Court stressed that the Commission failed to carry out the necessity test properly:

> By using in this case as a criterion for granting exemption from the rules of Article 85(1) of the Treaty simply fulfilment of a particular public mission defined essentially by reference to the mission of operating services of general economic interest referred to in Article 90(2) of the Treaty, the Commission based its reasoning on a misinterpretation of Article 85(3) of the Treaty.

The Court held that the Commission should base itself 'upon a minimum amount of actual economic data, which might have consisted of figures for the investments made [by] the members of the EBU . . . and on specific calculations establishing a ratio between those investments and the income associated with the broadcasting of sports events'.

The *Metropole* case emphasizes the need to carry out a proper economic analysis. It is very much in the line of the *Corbeau* and *Almelo* cases where the Court set out the necessity test and based it on the concept of 'economic equilibrium'.

The Court decision in *Metropole* was preceded by another case, *Jahrhundertvertrag*, where the Court applied Article 90(2) to its assessment of Article 85(3). This case involved two agreements between first, coal suppliers and electricity suppliers and second, coal suppliers and industrial companies generating their own electricity, in Germany. The agreements involved long-term minimum purchases of coal even though it was more expensive than coal produced in other Member States and than other sources of primary energy. The agreements were encouraged by the public authorities but not required by them. The Commission, therefore, only examined the agreement pursuant to Articles 90(2) and 85(3).

The Commission held that the electricity companies had a task of general economic interest[18]. It found that domestic primary energy sources may need a degree of protection from competition. But this should be limited to the minimum necessary to safeguard basic energy supplies. The Commission laid down what this limit should be. It found that the purchasing agreements in question fell within the limit and, therefore, met the necessity test of Article 90(2). This analysis was then used and referred to by the Commission in its assessment of the compatibility of the agreements with Article 85(3)[19].

---

[18] *Jahrhundertvertrag* (n. 11 above), paragraph 28.
[19] *Jahrhundertvertrag* (n. 11 above), paragraph 33.

Under Article 85(3), the Commission, however, required one amendment to ensure that the indispensability test was met[20].

This case shows how an analysis under Article 90(2) may be combined with an analysis of the indispensability condition under Article 85(3).

All the cases mentioned involved agreements entered into by undertakings on their own initiative. If the agreement was taken pursuant to a law, the Commission would be likely to examine the compatibility of the law with the competition rules, pursuant to Articles 90(1) and 85 before examining the compatibility of the agreement. This was the approach adopted by the Commission in *Ijsselcentrale*[21] for part of the agreement. The Court confirmed in *Rendo*[22] that the Commission was entitled to adopt this approach. The reason for this is that by examining the validity of the agreement first, the Commission would be anticipating the question of the validity of the underlying law on which it is based.

---

[20] *Jahrhundertvertrag* (n. 11 above), paragraph 34.
[21] *IJsselcentrale* (n. 8 above), paragraph 50.
[22] *Rendo*, case T-16/91, [1992] ECR II-2417 (Rendo I), paragraph 105.

Chapter 5

STATE-RELATED UNDERTAKINGS AND THE RULES ON THE FREE MOVEMENT OF GOODS: ARTICLES 30 TO 37 AND ARTICLE 90

Articles 30 to 34

The purpose of Articles 30 to 34 is to ensure the free movement of goods between Member States, i.e the import, transit and export of goods:

- Article 30 prohibits quantitative restrictions on imports and all measures having equivalent effect;
- Article 34 prohibits quantitative restrictions on exports and all measures having equivalent effect.

Article 36

This Article provides a derogation to these rules:

> the provisions of Articles 30 to 34 shall not preclude prohibitions or restrictions on imports, exports or goods in transit justified on grounds of public morality, public policy or public security; the protection of health and life of humans, animals or plants; the protection of national treasures possessing artistic, historic or archaeological value; or the protection of industrial and commercial property. Such prohibitions or restrictions shall not, however, constitute a means of arbitrary discrimination or a disguised restriction on trade between Member States.

Article 37

This Article requires Member States 'to adjust any State monopolies of a commercial character' 'so as to ensure that no discrimination regarding the conditions under which goods are procured and marketed exists between nationals of Member States'.

This chapter considers the question of the validity of State monopolies over goods (e.g. electricity, telecoms equipment, television equipment) with the rules on free movement of goods. It examines first the general rules on free movement of goods contained in Articles 30 to 36 and second the scope of the rules on monopolies of a commercial character set out in Article 37.

General rules on free movement of goods: Articles 30 to 36

The Treaty contains two sets of rules, those prohibiting restrictions on imports and exports (Articles 30 and 34) and those permitting restrictions in exceptional cases (Article 36).

It is necessary to review briefly how these articles have been interpreted by the Court in its traditional case-law before examining their application to monopolies.

The traditional case-law of the Court

The three leading cases

These cases are *Dassonville*[1], *Cassis de Dijon*[2] and *Keck*[3].

Dassonville

The key part of the judgment provides:

> All trading rules enacted by Member States which are capable of hindering, directly or indirectly, actually or potentially, intra-community trade are to be considered as measures having an equivalent effect to quantitative restrictions.

It follows from this broad statement that restrictions on trade within the Community may include merely indirect or potential restrictions. The Court applied this reasoning to the case at issue which involved the requirement by Belgium of a certificate of designation of origin in respect of trade in Belgium of Scotch whisky bought in France. The whisky in question had first been imported into France from the UK before being traded in Belgium. The certificate of origin was difficult to obtain because the Belgian traders were not original importers and could therefore not obtain a certificate of

---

[1] *Dassonville: Commission* v. *Belgium*, case 2/78 [1979] ECR 1761.

[2] *Cassis de Dijon: REWE-Zentral* v. *Bundesmonopolverwaltung für Branntwein*, case 120/78 [1979] ECR 649.

[3] *Keck: Keck and Mithouard* C-267 and 268/91 [1993] ECR 6097.

origin made in their names by the UK suppliers, as direct Belgian importers buying directly from the UK could.

The Court found that the requirement constituted a measure having an equivalent effect to a quantitative restriction. It then examined whether it could be exempted by Article 36. It held that the two justifications put forward by the relevant party did not apply: (i) the need to protect Belgian industrial or commercial property did not constitute a valid justification because the product was of UK origin. The restriction on trade could in this case only protect the UK not Belgian industrial or commercial property; (ii) the need to protect health could also not constitute a valid justification because in practice exactly the same whisky was allowed in Belgium with no designation of origin, i.e. as 'whisky' rather than 'Scotch whisky'.

The *Dassonville* test was further applied, in another leading case, namely *Cassis de Dijon* of 1979.

Cassis de Dijon

This case involved a German law which prohibited imports into Germany of liqueur which did not contain a minimum content of alcohol. It thus prohibited the entry into Germany of the French liqueur, cassis de Dijon, which fell below the minimum 25° alcohol contents.

The law applied to domestic and imported products alike. The Court nevertheless found that it was a measure equivalent to a quantitative restriction since it hindered trade between Member States. It held that such measures infringe Article 30 unless they can be justified:

> as being necessary in order to satisfy *mandatory requirements* relating in particular to the effectiveness of fiscal supervision, the provision of public health, the fairness of commercial transactions and the defence of the consumer. (*emphasis added*)

In practice, the Court held that a restriction on trade could be justified on the ground that it is necessary to satisfy mandatory requirements provided it meets a double test: first, the restriction serves a general interest, namely an interest which the Community recognizes as taking precedence over the requirement of free movement of goods (general interest test); second, the restriction does not go beyond what is necessary to satisfy this interest (proportionality test).

In *Cassis de Dijon*, the Court held that the national rule on minimum alcohol contents did not meet the double test:

- the first alleged general interest, the protection of consumer health, was a legitimate aim which could, in principle, form the basis for a mandatory requirement; but the Court found that, in practice, the rule was not needed to meet that aim. This is because consumers could obtain an extremely wide range of weaker or moderately alcoholic products on the market; furthermore consumers frequently take high alcohol beverages in a diluted form;

- the second alleged general interest, the protection against unfair commercial practices, was also a legitimate aim which could, in principle, form the basis for a mandatory requirement; the German government argued that the rule was necessary in order to prevent beverages with a low alcohol content, and usually sold at a cheaper price, from having a competitive advantage over beverages with a higher alcohol content. The Court, however, ruled that the requirement was disproportionate to the objective sought. In order to avoid a standardization of products and of their designation, it would be sufficient to require that suitable information relating to their alcohol content and origin should be clearly displayed on the packaging of the product.

The Court held that in so far as the drinks had been lawfully produced and marketed in another Member State there was no valid reason to restrict their imports into Germany.

Cassis de Dijon paved the way for the *mutual recognition* principle whereby, in the absence of Directives harmonizing Member States' laws, products which have been lawfully produced and marketed in a Member State must normally be allowed unimpeded throughout the Community, unless a restriction is necessary to meet a mandatory requirement. The principle was the basis for much of the 1992 White Paper prepared by Lord Cockfield for the creation of a single market.

Dassonville and *Cassis de Dijon* were increasingly used by traders to challenge national commercial rules regulating internal trade and not just imports. Because of this attempt to widen the scope of Article 30 beyond its original aim – the free movement of products in and out of Member States – the Court recently restated a key principle and amended the ambit of *Dassonville*, in the *Keck* case.

Keck

Reference should be made to the two key parts of the judgment contained in paragraphs 15 and 16. Paragraph 15 refers to *Cassis de Dijon* and states that:

> Measures of equivalent effect prohibited by Article 30 include obstacles to the free movement of goods where they are the consequences of applying rules that lay down *requirements to be met by such goods* (such as requirements as to *designation, form, size, weight, composition, presentation, labelling, packaging*) to goods from other Member States where they are lawfully manufactured and marketed, even if those rules apply without distinction to all products unless their application can be justified by a public interest objective taking precedence over the free movement of goods. (*emphasis added*)

Paragraph 16 states that:

> However, contrary to what has previously been decided, the application to products from other Member States of national provisions restricting or prohibiting *certain*

> *selling arrangements* is not such as to hinder directly or indirectly, actually or potentially, trade between Member States within the meaning of the *Dassonville* judgment (. . .) provided that those provisions apply to all affected traders operating within the national territory and provided that they affect in the same manner, in law and in fact, the marketing of domestic products and of those from other Member States. (*emphasis added*)

These two paragraphs have given rise to much comment and criticism[4]. It is not the purpose of this chapter to review these but to examine the implication of the case for monopolies. The Court draws a distinction between *rules relating to requirements to be met by goods* and gives a list of examples (designation, form, presentation, labelling, etc.) and requirements relating to *selling arrangements*. The former would normally constitute a measure having an equivalent effect to a quantitative restriction, whilst the latter would not fall within Article 30 at all. The difficulty lies in the fact that the Court did not define the second category, i.e. 'selling arrangements' (*'modalité de vente'* in the French judgment) and so one must rely on court cases as they are gradually decided. The type of rules which have been held to constitute selling arrangements and which, therefore, fall outside Article 30, include for example, prohibition of resale at a loss (*Keck*); prohibition of sale on Sunday (*Sunday trading cases*[5]); restrictions on pharmacists advertising outside their pharmacies (*Hünermund*[6]); restrictions on the retail sale of tobacco in Italy (*Banchero*[7]). Further judgments will, however, be necessary to clarify the distinction, particularly in borderline cases.

The main lesson from *Keck* is that certain rules relating to selling arrangements do not fall within the scope of Article 30 at all. A rule prohibiting resale at a loss does not constitute a measure having an equivalent effect to a quantitative restriction and it is therefore not necessary, in respect of such rules, to carry out the proportionality test. The test only needs to be carried out in respect of the first type of rules, i.e. those governing the requirements to be met by goods.

---

[4] See in particular René Joliet: *'La libre circulation des marchandises: l'arrêt Keck et Mithouard et les nouvelles orientations de la jurisprudence'*, in *'L'exposé présenté lors de la visite des magistrats de Cours Suprêmes à la Cour de Justice'* on 6 June 1994; D. Waelbroeck, *'L'arrêt Keck et Mithouard: les conséquences pratiques'*, in *Journal des Tribunaux, droit européen*, 16 November 1994, p. 160; Luc Gyselen, *'Mesures étatiques restrictives de concurrence: des enseignements à tirer de l'arrêt Keck et Mithouard'*, *Journal des Tribunaux, droit européen*, 16 November 1994, p. 170; *'Bilan de l'article 30 après 20 ans de jurisprudence: de Dassonville à Keck et Mithouard'* by Richard Wainwright and Virginia Melgar in *Revue du Marché Commun et de l'Union Européenne*, September-October 1994; Article by A. Mattera in *Revue du Marché Unique Européen*, 1/1994.

[5] See in particular *Torfaen* v. *B & Q* case C 145/88 [1989] ECR 3851 and *Stoke on Trent* v. *B & Q* [1992] ECR 6635.

[6] *Hünermund* v. *Landesapothekerkammer Baden-Wüttenberg* C-292/92 [1993] ECR 6787.

[7] *Banchero* case C-387/93, 14 December 1995 [1995] ECR I-4663; this involved an Italian law which reserved the *retail sale* of tobacco to authorized outlets. The Court held that the law was not contrary to Articles 30 to 37 and 90 even if the authorization was given by a national undertaking which enjoyed a *de facto* monopoly over the *wholesale* of tobacco.

The reason for this distinction is that the rules governing certain selling arrangements, in principle, have nothing to do with the regulation of trade within the Community but only seek to regulate trade within a Member State; they apply without discrimination to imported and domestic products and do not *affect* the former more than the latter. They are made for socio-economic reasons.

Relevance of Keck to monopolies

The question which arises here is do arrangements whereby a Member State grants an undertaking the exclusive (or special) right to sell certain products constitute a 'selling arrangement' within the meaning of the *Keck* judgment? If the answer to this question is positive it would seem to follow that the grant of a monopoly over the import, export and trade in goods in one Member State falls outside Article 30 altogether and does not constitute an infringement of Treaty rules. It would follow that one of the instruments at the disposal of the Commission or Member States to force open product markets subject to national monopolies (for example, the telecommunications market; the electricity market) is thus destroyed. It has been argued that this is one of the unfortunate consequences of the *Keck* judgment[8].

There are two principal views on the application of the *Keck* judgment to state-related enterprises.

The first view is that whilst the creation of a monopoly appears to fall within the general meaning of 'selling arrangements' and to affect both domestic and foreign traders alike, it is doubtful that it would meet the other criteria contained in paragraph 16 of the *Keck* judgment:

(1) the selling arrangement must 'apply to all affected traders operating within the national territory'; it may be argued that this criterion would not apply to a monopoly situation where, by definition, there is only one national operator and potential traders, not only national, but also traders from other Member States (the latter being the principal object of this criterion), are prevented from operating on the national territory;

(2) the selling arrangement must 'affect in the same manner, in law and in fact, the marketing of domestic products and those from other Member States'. Judge Joliet[9] explained that the purpose of this criterion is to ensure that the national rule does not have the effect of giving an advantage to a domestic industry or product. Now, one of the obvious consequences of a national monopoly is to give an

[8] See article by A. Mattera, *'De l'arrêt Dassonville à l'arrêt Keck: L'obscure clarté d'une jurisprudence riche en principes novateurs et en contradictions'* in *Revue du Marché Unique Européen* 1/1994.

[9] *'La libre circulation des marchandises: l'arrêt Keck et Mithouard et les nouvelles orientations de la jurisprudence'* by René Joliet in *'L'exposé présenté lors de la visite des magistrats de Cours Suprêmes à la Cour de Justice'* on 6 June 1994'.

advantage to a domestic industry or product[10]. This criterion would, therefore, not be met either and statutory monopolies would not constitute 'selling arrangements'.

The second view is that domestic products, other than those of the monopoly enterprise, and products of other Member States are affected in the same manner, i.e. all independent traders, both national and foreign, outside the national trader benefiting from the monopoly are equally affected. They cannot market their products in the territory of the Member State where a monopoly has been granted. Under this second interpretation, statutory monopolies would constitute 'selling arrangements' and would fall outside the prohibition in Article 30.

The second view appears to be too restrictive and it is suggested that the first view should prevail.

It may be further argued in support of the first view that paragraph 16 only covers national provisions restricting or prohibiting *certain* selling arrangements and that it was not intended to apply to the grant of a monopoly to an undertaking; the word 'certain' leaves open the possibility of excluding from the *Keck* exception the grant of certain monopoly rights.

The comments in the next section are based on the first interpretation whereby the grant to a national undertaking of a monopoly over the sale of products does *not* constitute a *selling arrangement* and falls to be considered under Article 30.

The recent case-law on monopolies over goods

The meaning of goods

Article 9 of the Treaty provides that 'the Community shall be based upon a customs union which shall cover all trade in goods'; it further states that the rules set out in Articles 30 to 34 relating to the elimination of quantitative restrictions between Member States 'shall apply to products originating in Member States and to products coming from third countries which are in free circulation in Member States'. The word 'goods', however, is not defined in the Treaty.

In many cases the concept of goods and products creates no difficulty. It includes both raw materials and manufactured products. Examples of products examined in the context of Article 90 have involved products as varied as tobacco (*INNO* v. *ATAB*); oil (*Campus Oil*); coffins (*Bodson*) and telephone equipment (*France* v. *Commission*).

[10] Advocate General Cosmas in the *Electricity and Gas* cases, cases C-157/94, C-158/94, C-159/94 and C-160/94 (not yet reported) expresses the opinion that the *Keck* judgment cannot apply to national laws the object of which is specifically to regulate trade between Member States. It follows that State measures which create exclusive import and export rights must be examined solely in the light of the *Dassonville* judgment to check whether they constitute measures of equivalent effect to a quantitative restriction (see paragraph 61 of the opinion).

Products may be linked to services. This is often the case where an undertaking is granted a monopoly over certain goods and services related to the goods. In the field of terminal equipment, the monopoly over telephone handsets and other telecommunications equipment generally also involved a monopoly over the installation and maintenance of the equipment[11]. The reverse may be true: a monopoly over services may involve an ancillary monopoly over products. In *Bodson* a monopoly over funeral services also involved a monopoly over funeral items such as coffins and marble work. A monopoly over films and sound recording material may be linked to a monopoly over broadcasting services (*Sacchi, ERT*). Article 30 will apply to the 'product element' of the service.

In some cases it may be difficult to ascertain whether goods or a service are involved. The question was raised in *Costa* v. *Enel*[12] in relation to electricity but the Court stated that electricity was a product. This was confirmed in *Rendo* and *Almelo*.

The Terminal Equipment revolution

The *Terminal Equipment* case[13] constitutes a milestone in the case-law on Article 90. To understand it, it is useful to recall that up to this case, state monopolies over the supply of goods or services were presumed legal. This was indeed the view expressed by the Advocate General in *Terminal Equipment*. The Court, however, departed from the view of its Advocate General casting severe doubts on the *per se* legality of monopolies.

The Advocate General's view: monopolies over goods are presumed legal

Advocate General Tesauro explained that there is a slight contradiction between Article 30 on the one hand and Article 90 on the other, but that this contradiction was accepted by the draftsmen of the Treaty and was resolved as follows: whilst Article 30 generally prohibits restrictions on the free movement of goods nevertheless the Treaty generally accepts such restrictions if they are the result of the grant of a monopoly by the State to an undertaking.

Advocate General Tesauro's view corresponds to the *first interpretation* of Article 90 explained in the Introduction and in chapter 2 above. According to this interpretation, Article 90(1) presupposes the existence of exclusive rights, including monopolies, and therefore, the existence of complete barriers to trade. The *prima facie* legality of monopolies is further borne out by Article 222 which provides that Member States are free to decide on the system of property ownership; they may thus decide that a sector of activity,

---

[11] See the *Terminal Equipment* Directive, 88/301/EEC, OJ 1988 L 131/73, and the judgment in the *Terminal Equipment* case (*French Republic* v. *Commission* C-202/88, 19 March 1991, [1991] ECR 1223).

[12] Case 6/64 [1964] ECR 614.

[13] *Terminal Equipment* case, see n. 11 above.

for example, electricity supply is to be owned by the State and not by private operators. If there is only a single owner, it follows that it enjoys a monopoly over the sector in question and no one else, including foreign traders, may sell products falling within the monopoly. Finally Article 37 also presupposes the existence of monopolies and their compatibility with the Treaty. The Article requires Member States to '*adjust* any state monopolies of a commercial character'; it does not require Member States to *abolish* their monopolies.

Pursuant to the first method of construction of monopolies examined above, Article 90(2) would not invalidate monopolies. It could be argued that even where the Treaty requires Member States to *adjust* their monopolies of a commercial character (Article 37), this obligation does not apply if the monopoly has, in addition, been entrusted with a task of general economic interest and the adjustment would prevent it from carrying out this task. Generally, under the first method of construction, Article 90(2) is not applied to the existence of a monopoly but only to its abusive exercise, for instance in dealings with third parties such as clients or suppliers.

If a monopoly over the production, import and marketing of certain goods is presumed legal, it must logically follow that restrictions over the purchase and marketing of goods from other Member States which fall within the scope of the monopoly will also be presumed legal. Advocate General Tesauro expressed a similar view in *Terminal Equipment* (at paragraph 24):

> . . . the fact that a monopoly or a system of exclusive or special rights may not fit in perfectly with a system based on free competition must be obvious . . .
> Yet . . . it is clear that the Treaty . . . intended to 'tolerate' monopolies and systems of exclusive and special rights by regarding them as legal 'in themselves' . . .

The Court's departing view: monopolies over goods are restrictions contrary to Article 30

Contrary to the Advocate General's opinion, the Court upheld the Commission's view that Member States' monopolies over terminal equipment were contrary to the Treaty. The consequences of the judgment are that, for terminal equipment, the prohibition contained in Article 30 prevails over the *prima facie* legality of monopolies.

The Court did not, however, give any clear explanation. On the contrary it merely stated the obvious, i.e. that the monopoly constituted a restriction on the free movement of goods. It did not say why the restriction on trade, which is inherent in any monopoly over products, was contrary to Treaty rules notwithstanding the presumption of legality of monopolies which may be inferred from Articles 90(1), 37 and 222.

The Court stated at paragraphs 34 to 36:

> . . . it should be noted first that the existence of exclusive importing and marketing rights deprives traders of the opportunity of having their products purchased by consumers.

> It should be pointed out, secondly, that the terminals sector is characterized by the diversity and technical nature of the products concerned and by the ensuing constraints. In those circumstances there is no certainty that the holder of the monopoly can offer the entire range of models available on the market, inform customers about the state and operation of all the terminals and guarantee their quality.
>
> Accordingly, exclusive importation and marketing rights in the telecommunications terminal sector are capable of restricting intra-community trade.

The Court simply stated that exclusive rights over terminal equipment had a restrictive effect on traders who could not market all their own products; and on consumers who were not guaranteed to have access to the full range of goods available on the market. These restrictions, however, flow from the very existence of a monopoly over terminal equipment; they should not, therefore, automatically be held to constitute a measure equivalent to a quantitative restriction contrary to Article 30.

The Court went on to say that the restrictions were not *justified* by the need to ensure user safety, i.e. the safety of network operators, the safety of the network itself and inter-operability, since proper safeguards had been provided in the Commission Directive[14].

But the application of the justification test goes no further than in relation to a 'traditional' Article 30 case, one not involving a monopoly. This is surprising since the existence of a statutory monopoly presupposes the existence of barriers to trade which are, in principle, valid in themselves and do not need to be justified pursuant to Article 36 or the mandatory requirements doctrine.

Terminal Equipment casts a serious doubt over the per se legality of monopolies

It seems to follow from this judgment that the existence of a monopoly over the import and marketing of goods is no longer presumed legal, in itself, but on the contrary that it will be regarded as illegal unless it can be justified under Article 36 or by the mandatory requirement doctrine.

This judgment is surprising because a monopoly is normally the result of a choice based on economic, political or social reasons. By upholding the key part of the Commission Directive – the abolition of exclusive rights over terminal equipment – the Court accepted that the Commission could substitute itself for and prevail over decisions by government in socio-economic or political areas[15].

The precise reason why a monopoly may in itself be contrary to EC law is far from clear from a straightforward reading of the Treaty. It is a pity that the Court did not provide a full analysis.

---

[14] Directive 88/301/EEC on competition in the markets in telecommunications terminal equipment, OJ 1988 L131/73.

[15] In his opinion in the *Höfner* case (C-41/90 [1991] ECR I-1979, delivered in January 1991 i.e. before the judgment of the Court in *Terminal Equipment* (March 1991) (n. 11 above), Advocate General Jacob's view was that Articles 90 and 222 did not impose a general limitation on Member States' power to place certain sectors of the economy under public ownership (see paragraph 43 of his opinion).

Lessons from Terminal Equipment

Two factors seem to have played an important part in the *Terminal Equipment* judgment:

- the fact that consumers could *not* have *access to the full range of products* available on the market and might, therefore, be denied access to the most technologically advanced products;
- and the fact that other *economic operators with a wider range of equipment were present* elsewhere, in particular in the US and Japan. Efficient potential competitors in Europe were denied access to the market, which could have the effect of holding back the EC in the field of telecommunications equipment.

The evolution of technology in other parts of the world was no doubt a determining factor. The aim of the EC Treaty is to ensure the economic development of the Community (Article 2), the abolition of barriers between Member States within the internal market (Article 3(c)), and the introduction or maintenance of competition (Article 3(g)) to the benefit of consumers.

The existence of monopolies in the field of terminal equipment had the effect of maintaining barriers between Member States, thus preventing competition between them and slowing the pace of development to the ultimate detriment of consumers who would be offered less choice than in other countries.

The interest of the majority of consumers seems to have been the determining factor. In other monopoly cases[16], the Court has allowed the maintenance of monopolies or part monopolies where it could be proved that this was in the interest of the majority of consumers. But even in those cases the maintenance of a monopoly is not definite. It is subject to further changes, including technological changes, which might make it necessary to limit further the scope of the monopoly.

Article 30 type restrictions

There have been several cases dealing with Articles 30 and 90. Some cases[17], although clearly involving state-related undertakings or State measures, hardly mention Article 90. Yet they remain important because they explain when Article 30-type restrictions may or may not be permitted.

---

[16] Voice telephony (see *Telecom Services*, C 271/90, C 281/90 and C-289/90, [1992] ECR I-5833); in the field of electricity (see *Almelo*, C-393/92, [1994] ECR 1477) the Court gave no definite answer to the question of whether exclusive purchasing rights were compatible with the Treaty. It held that they would be legal if they were necessary in the general economic interest.

[17] See, for example, *Campus Oil* (C-72/83, [1984] ECR 2727) where the Court did not examine Article 90(1) and rejected the possibility that Article 90(2) might apply; in *Waste Oils* (C-172/82, [1983] ECR 555), the Court briefly examined Article 90(1); it did not apply Article 90(2) which was regarded as not having direct effect.

Relevant types of restrictions include absolute import or export bans resulting from the grant of import and export monopolies; compulsory pricing and certain forms of type approval.

Import and export restrictions

Absolute bans on imports or exports resulting from the grant by public authorities of exclusive rights to certain undertakings will normally be contrary to Article 30 and will not be exempted by Article 36 or the doctrine of mandatory requirements based on the *Cassis de Dijon* principle. An example of this is *Waste Oils*[18]. This case related to an obligation imposed by a French law to dispose of waste oils with approved disposal firms on the national territory. It resulted in a total prohibition on the export of waste oils and gave a clear advantage to approved French disposal firms. An EC Directive on the disposal of waste oils required Member States to ensure the safe collection and disposal of waste oils. It allowed Member States to provide for a system of disposal through one or more disposal centres *where they could not otherwise achieve the aim of the Directive to ensure safety*. This suggests that other methods of disposing of waste oils should be set up by the Member States wherever possible; the grant of exclusive rights to one or more undertakings should be the exception. In addition the Directive did not give Member States the right to restrict exports to approved disposal centres established outside the national territory. The total barrier on export introduced in the French law was therefore contrary to Article 34. It was not exempted by any of the provisions of Article 36.

In the field of imports *Crespelle*[18] is a case which bears some resemblance to *Waste Oils*, but only on the face of it. There are, in fact, certain fundamental differences between the two cases which resulted in a finding that the restriction on imports was lawful. *Crespelle* concerned a French law obliging users of semen for agricultural purposes to obtain their supplies from approved centres and obliging importers to deliver the semen for storage in an approved centre. As in *Waste Oils* the law made the use of approved centres compulsory; *Waste Oils* related to the disposal of certain goods, whilst *Crespelle* related to the purchase of certain goods. Both laws purported to implement an EC Directive. But one major difference was that the national law at issue in *Crespelle* did not ban imports completely: on the contrary, it expressly provided that imports of semen were possible as long as the semen was stored in the approved centres in France. Another major difference is that the Court found that the relevant EC Directives in *Crespelle* did not cover all relevant health issues and that Member States retained a degree of freedom to ensure that health was fully protected.

The Court found that the French law constituted a restriction on imports but was based on a legitimate requirement, namely the protection of animal

[18] *Waste Oils*, see n. 17 above; *Crespelle* [1994] ECR 5077.

health, which applied equally to domestic operators[19] and did not exceed what what was necessary to meet that requirement on a proportionality test. It followed that the law infringed neither the Directive nor Articles 30 to 36 of the Treaty.

Whilst a national law which imposes an absolute ban on imports or exports is likely in many cases to be held to infringe Articles 30 to 36, it appears that a restriction on import or export – short of such an absolute barrier – may be found to be lawful.

Cases on absolute bans include, besides *Waste Oils* (absolute ban on export), *Terminal Equipment* which involved an exclusive right on the import and marketing of products resulting in an absolute ban on import and marketing by independent traders. It may be questioned, however, whether the Court might not allow absolute bans on imports or exports in certain limited fields of activity even though they result in a partitioning of the Common Market. One such field could be electricity. At the time of writing, there are four cases pending before the Court relating to gas and electricity. The Commission started proceedings against the Netherlands, Italy, France and Spain for granting exclusive import and/or export rights to national undertakings for the supply of electricity resulting in a complete ban on imports and/or exports. The case against France also concerned a ban on import and export of gas. The judgment of the Court in these cases should clarify[20] the law in this area and confirm whether *absolute* bans on imports and exports are indeed prohibited as contrary to Article 30 where they may significantly affect trade between Member States.

Cases on restrictions on imports and exports, not resulting in an absolute ban, include, besides *Crespelle* (restriction on imports of semen), *Campus Oil*. This case related to an Irish law obliging oil companies in Ireland to take 35 per cent of their requirements from the national refinery. This meant that imports were restricted up to a corresponding amount. The Court held that the obligation was clearly a restriction on imports but that in the absence of EC Directives dealing with all possible problems of shortage, Member States could rely on Article 36, provided they satisfied the proportionality test (see p. 118 below).

Compulsory Pricing

INNO v. *ATAB*[21] involved compulsory pricing at the retail level. It concerned a Belgian law providing that tax on cigarettes must be calculated on the basis

---

[19] The Court, however, left it to the national judge to examine whether in practice there was discrimination against foreign imports (*Crespelle*, n. 18 above, ground 39).

[20] The question remains unanswered at the time of writing, particularly in the light of *Almelo* (see n. 16 above) where the Court left it to the national judge to say whether an exclusive purchasing requirement resulting in a restriction on imports might be compatible with the competition rules on the basis of Article 90(2); see Stop Press section.

[21] *INNO* v. *ATAB*, see C-13/77, [1977] ECR 2115.

of the price stated on the tax label. The law further required that the price on the label must be the compulsory selling price to the customer. Belgium thus ensured that manufacturers and importers would not voluntarily undervalue the product for the purpose of paying a lower tax whilst later applying a higher price to consumers. It also ensured that the product could not be sold at a lower price which could undermine sales by small retailers to the advantage of large outlets such as chains of supermarkets.

The Court found that the law in question applied equally to domestic manufacturers and importers. It held at paragraph 52 of the judgment that a law which imposed maximum prices on both domestic and foreign products could still be contrary to Article 30 if it had the effect, because of the level at which the price was fixed, of making sales of imported products more difficult than sales of domestic products.

The Court, however, found at paragraph 53 that under the Belgian law in question prices were freely chosen by the manufacturer or the importer and only subsequently fixed at that level by the national measure: 'such measure generally has exclusively internal effects'. The Court left it to the national judge to say whether on the facts the law at issue was in itself likely to hinder trade between Member States. The Advocate General's opinion was that the law could not be regarded in itself as infringing Article 30.

It follows that imposed maximum retail price, in circumstances where the price has first been freely chosen by the manufacturer or importer, is unlikely to infringe Article 30. On the other hand, such a price could be contrary to Article 30 and affect trade between Member States if it is fixed at a level such as to make imports more difficult than the sale of domestic products.

Type-approval

The Court dealt with the issue of type-approval in a number of telecoms cases.

Many Member States have granted public undertakings the exclusive power to lay down specifications for telephone equipment intended for connection to the public network and to grant type-approval.

Applying the *Cassis de Dijon* approach, the Court held in *RTT* v. *GB* that, pending harmonization of Member State legislation, disparities between legislation relating to specifications must be accepted if the legislation (i) applies without distinction to national and imported products, (ii) can be justified as being necessary to satisfy imperative requirements of Community law, and (iii) is proportionate to the object to be achieved.

The Court accepts that, in the absence of Community rules for the establishment of public telecommunications networks, Member States may reserve to undertakings of their choice the right to lay down specifications relating to the type and quality of telephone equipment for connection to the public network and to ensure that the specifications are met. The exercise of this right creates restrictions on trade within the meaning of Article

30 because manufacturers must take foreign specifications into account when selling abroad, which is likely to involve additional delay and expense. Nevertheless, the restrictions are necessary to satisfy mandatory requirements relating to the protection of users and of the network.

The Court, however, held in *RTT* v. *GB* that restrictions on trade resulting from type-approval will only be compatible with Article 30 if decisions refusing to grant type-approval can be challenged before the courts. Otherwise, there would be a risk that the procedure could be used to deny access to imported telephone equipment as the authority exclusively authorized to grant type-approval could behave in an arbitrary and discriminatory manner in favour of national goods.

This approach was confirmed in *Lagauche*, *Decoster* and *Taillandier*[22]. The law relating to type-approval of telecommunications terminal equipment was further extended by Directive 88/301 and clarified by the Court in these three cases as well as in *Tranchant*[23]. The Court made it clear that the authority granting type-approval for terminal equipment should be independent and should not also be involved in the sale of such equipment (*Lagauche* and *Decoster*). The Court also held that laboratory tests requested for the grant of type-approval of terminal equipment should be carried out by an independent entity which should not be linked to a public undertaking involved in the sale of competing equipment (*Tranchant*).

In summary, national laws setting up type-approval systems may involve measures having an equivalent effect to quantitative restrictions contrary to Article 30. They may nevertheless be lawful if they fall within one of the exceptions set out in Article 36 or fulfil a mandatory requirement, provided that (i) they meet the proportionality test and (ii) they include a right to appeal against decisions refusing type-approval[24]. National law may need to fulfil other conditions in order to comply with the competition rules as interpreted in *RTT* v. *GB* and with any specific Directive regulating the sector concerned such as the Terminal Equipment Directive.

Conditions for exemption: justification and proportionality

Justification

Once it is clear that a measure constitutes a direct or indirect restriction on imports or exports, either because it imposes more stringent requirements on foreign goods than on domestic goods, or because it affects imports or exports more than domestic products, it is necessary to examine whether the measure can be justified under Article 36 in the case of direct restrictions or, alternatively, by a mandatory requirement in the case of indirect restrictions and whether it meets the proportionality test.

---

[22] *Lagauche*, joint cases C-46/90 and C-93/91 [1993] ECR I-5267; *Decoster*, case C-69/91 [1993] ECR I-5335; *Taillandier*, case C-92/91 [1993] ECR I-5383.

[23] *Thierry Tranchant* v. *Telephone Store SARL*, case C-91/94, 9 November 1995, [1995] ECR I-3911.

[24] *RTT* v. *GB*, C-18/88, [1991] ECR I-5941; *Lagauche*, (n. 22 above); *Decoster*, (n. 22 above); *Taillandier*, (n. 22 above).

Permitted justifications have included, in particular: the safety of users (consumers, operators and vendors of a product) and the safety of the network to which the goods are connected; protection of animal health; protection of the environment; efficient operation of a product; the achievement of sufficient profit in order to meet the aim of a Directive; public security; and socio-economic reasons. These justifications can only be used where there is no harmonizing Directive dealing with the specific issue.

Safety of users; safety of the network

These have been held to constitute legitimate requirements. They have been relied on, in particular, in the field of telecommunications equipment[25]. In *Terminal Equipment*, the Court held that an absolute restriction on imports could not be justified on the basis of the safety of users and protection of the network because these essential requirements were in fact already covered by the Terminal Equipment Directive. Member States had clearly accepted that these essential requirements were adequately covered in the Directive. In *RTT* v. *GB*, on the other hand, the Court held that a restriction on imports, in the form of an obligation to obtain type-approval subject to an appeal procedure, could be justified on those grounds. The reason was that the existing Directive did not entirely harmonize Member States' specifications and expressly permitted the use of an approval procedure (Articles 5 and 6 of the Directive). The Court held that:

> In the absence of Community rules on the establishment of public telecommunications networks, and in view of the *technical diversity* of the networks in the Member States, the Member States retain, on the one hand, the power to lay down technical specifications which telephone equipment must meet to be capable of being connected to the public network and, on the other, the power to examine whether the said equipment is fit to be connected to the network in order to satisfy the *imperative requirements* regarding *the protection of users* as consumers of services and *the protection of the public network* and its proper functioning. (*emphasis added*)

The existence of technical diversity between Member States was a significant factor: in order to ensure the safety of users or the proper operation of the network it was *necessary* to arrange for some *control* of the equipment in use within each Member State.

Animal health

The protection of animal health is expressly recognized as a legitimate objective by Article 36 of the Treaty. It was relied upon in *Crespelle*. In defence of its restriction on imports, the French government had argued that the restriction was necessary (i) on health grounds and (ii) to improve bovine stock genetically. The Court rejected the second reason on the ground that the EC

[25] *Terminal Equipment* case (see n. 11 above); *RTT* v. *GB* (n. 24 above); *Lagauche* (n. 22 above); *Taillandier* (n. 22 above).

had issued a Directive dealing with the problem. It held that the Directive had fully harmonized zootechnical and pedigree requirements. The Court, on the other hand, accepted the first reason based on health grounds because the Directive had not fully covered this problem. Member States accordingly retained some power to deal with it. The Court further held that the measures at issue did not in themselves create discrimination in favour of domestic products and met the proportionality test.

Protection of the environment

This can be a legitimate requirement. The protection of the environment is an issue which is taken seriously at Member State and EC levels. The need to protect the environment has been relied on as a defence to restrictions on the free movement of goods. It was accepted in *Almelo*[26] (a case based on Articles 86 and 90); it was rejected in *Waste Oils* because the environment was clearly not better protected by the restriction at issue, namely an export ban preventing the disposal of waste oils in approved centres in other Member States; in other words, France failed to prove that the protection of the environment was an essential requirement which justified its export ban.

Efficient operation of a product

The efficient operation of telecommunications equipment and of the network to which it may be connected has been held to constitute a mandatory requirement in *Terminal Equipment* and the *RTT* v. *GB* case. Only in *RTT* v. *GB*, however, did the Court accept that the requirement could constitute a valid justification to the restriction in question. In *Terminal Equipment*, Member States could not rely on this ground to justify restrictions on trade because the efficient operation of the telecommunications network was fully taken into account by the Directive. Maintenance of the monopoly over telephone equipment was therefore not necessary in order to ensure the operation of the network. In *BT*, (see chapter 2 n. 6) Italy claimed that private operators acting as message forwarders using the British Telecommunications' network were abusing the network and could harm it. The Court, however, found no evidence of possible harmful effects to the network as a result of the introduction of new technology by the message forwarders.

The Court will not rely on mere assertions; it will seek clear evidence substantiating the allegations made. It is for the party relying on the essential requirement to prove its case before the Court by demonstrating that the requirement is objectively necessary.

Meeting the aim of a Directive

This may constitute a legitimate justification. In *Waste Oils*, the French government argued that in order to fulfil the objective of the Directive, namely

[26] *Almelo* (n. 16 above): protection of the environment was one of the issues expressly mentioned by the Court to justify the application of Article 90(2).

the safe disposal of waste – it was essential that approved disposal centres should collect *all* the oils in the region where they were situated and should thus have an exclusive right to collect oil even if this resulted in a ban on exports. The argument put forward was of an economic nature. The Court said that the relevant EC Directive on waste oils actually dealt with the economic problem by providing for the payment of an indemnity to the disposal centres under the 'polluter pays' principle, which should ensure their profitability.

This case shows that purely economic arguments to justify a restriction can probably only be used in support of another non-economic factor[27], such as the protection of the environment.

Public security

This is one of the justifications expressly permitted by Article 36. It was relied on by the Irish Government in *Campus Oil*, to justify an obligation on oil companies to buy 35 per cent of their requirements from the Irish national refinery. The Irish government argued that it needed to have its own resources in refined oil in order to be able to meet severe shortages in times of crisis at world level. The Advocate General accepted that the argument fell within the concept of public security or possibly public policy but he doubted whether the proposed restrictions on imports were necessary to meet that objective and whether their real purpose was to meet an economic objective rather than to safeguard public security. The Court, however, took the view that the restriction was introduced essentially for non-economic reasons and was objectively necessary in terms of public security.

The Court reiterated (in ground 35) the principle that Article 36 refers to matters of *a non-economic nature*:

> A Member State cannot be allowed to avoid the effect of measures provided for in the Treaty by pleading economic difficulties caused by the elimination of barriers to intra-community trade.

The Advocate General gave examples of economic considerations: protection of employment, improvements in the balance of payments or financial returns, keeping in operation a domestic industry[28].

The Court also clearly stated that Article 36 cannot be relied on if Community rules, for example, in the form of directives, provide means to ensure the protection of the interests mentioned in the article. It recalled that certain directives were in force in the oil sector. There were also measures taken at world level in the context of the International Energy Agency. Nevertheless, it held that these measures did not provide absolute protection and that there could still be a real danger in the event of crisis. It followed

---

[27] See also *Campus Oil*, C-72/83, [1984] ECR 2727.

[28] *Campus Oil* (n. 27 above) see page 2764 of the Advocate General's opinion.

that Member States were not precluded from relying on Article 36 despite the existence of certain measures at EC level.

The Court thus found that the maintenance of adequate supplies in petroleum products fell within the concept of public security. Petroleum products were exceptionally important as a source of energy and were of fundamental importance for a country's existence: they were essential to the maintenance of a country's institutions, its essential services and the survival of its inhabitants. An interruption of supplies could constitute a significant danger seriously affecting the public security. The Court concluded that 'the aim of ensuring a minimum supply of petroleum products at all times is to be regarded as *transcending purely economic considerations* and thus as capable of constituting an objective covered by the concept of public security' (emphasis added) (ground 35).

The fact that a measure might also serve economic objectives did not, in such circumstances, exclude the application of Article 36 because the essential objective was clearly of a non-economic nature[29]. The restriction envisaged by Ireland could therefore be allowed provided it met the proportionality test (see p. 118 below).

Socio-economic reasons

The protection of a national socio-economic interest may also constitute a valid ground for justification under the subsidiarity principle but it will be severely scrutinized to ensure that it does not undermine superior Community interests.

In *INNO* v. *ATAB*, the Court recognized that Member States could impose certain restrictions on imports where this was necessary for a socio-economic reason, in this case, the maintenance of retail trade. The restriction also met fiscal objectives namely the collection of tax revenue.

It is not clear how far socio-economic reasons will be permitted. This is likely to vary depending on the economic sector in question. In *Höfner* (a case decided under Articles 90 and 86) the Court effectively rejected Germany's claim that it was entitled to maintain the placement of employees under State control for socio-economic reasons. The Court has also held that telecommunications equipment may not be within an absolute state monopoly[30]. On the other hand, it accepted that semen for animal reproduction could temporarily be subject to exclusive rights but for health reasons only[31]. When health issues are harmonized at EC level it seems that Member States will no longer be able to maintain such exclusive rights on semen even if they pursue non-profit making objectives and merely seek to increase the quality

[29] In contrast, Article 90(2) allows defences based on economic arguments; see the second test set out in *Corbeau* (C-320/91, [1993] ECR I-2533): the application of the competition rules must not be such as to jeopardize the financial equilibrium of the holder of the monopoly, and hence the performance of its task of general economic interest (see chapter 2 above).

[30] *Terminal Equipment* (see n. 11 above).

[31] *Crespelle* (see n. 18 above).

of their product by centralising and controlling production. A doubt, however, remains in relation to electricity[32]; the Court may be prepared to accept that exclusive rights can be maintained as long as there is no clear evidence that a universal service can adequately be provided under conditions of full competition (see further details on the concept of universal service in chapter 8).

Proportionality

Where a requirement appears to fall within Article 36 or the mandatory requirement exception, the Court will check whether the proportionality test is met. This was clearly stated by the Court in *Campus Oil* (in ground 44):

> Member States may have recourse to Article 36 to justify a measure having equivalent effect to a quantitative restriction on imports only if no other measure, less restrictive from the point of view of the free movement of goods, is capable of achieving the same objective.

The Court thus held (in ground 51) that a:

> Member State may rely on grounds of public security for the purpose of requiring importers to cover a certain proportion of their needs by purchases from a refinery situated in its territory at prices fixed by the competent minister on the basis of costs incurred in the operation of that refinery . . . The quantities of petroleum products covered by such a system *must not exceed the minimum supply requirement* without which the public security of the state concerned would be affected or the level of production necessary to keep the refinery's production capacity available in the event of a crisis and to enable it to continue to refine at all times the crude oil for the supply of which the State concerned has entered into long-term contracts. (*emphasis added*)

The Court left it to the national judge to apply these parameters to ensure that the proportionality test was met.

Link between Article 30 infringements and Article 86 infringements

In *Port of Genoa*[33] the Court held that a law which infringes Article 86 because it facilitates the abuse of a dominant position is also likely to infringe Article 30 in so far as it makes imports from other Member States more expensive.

The telecommunications cases offer examples of the combined application of Articles 30 and 86. They also show that the consequences of an infringement of Article 86 are likely to be more far-reaching than the consequences of an infringement of Article 30. Thus in *RTT* v. *GB* the Court held

---

[32] *Almelo* (see n. 16 above).

[33] *Port of Genoa* (C-179/90, [1991] ECR I-5923), ground 21; see also *INNO* v. *ATAB* (see n. 21 above, ground 35.

that the combination within one entity of exclusive rights (i) to set telecommunications standards, (ii) to ensure compliance with those standards, (iii) to grant type-approval, and (iv) to operate a telecommunications network or (v) to sell telecommunications equipment, constitutes an abuse of a dominant position contrary to Article 86. Supervisory and regulatory functions must be separated from operational functions in order to avoid an infringement of Article 86. This separation of activities is called *unbundling*.

RTT v. *GB* also raised issues of compatibility with Article 30 because the establishment by a national undertaking of technical standards applicable throughout a Member State made imports of telecommunications equipment from other Member States more difficult. Importers must first adapt their standards to the national specifications of the country of importation. However the restriction on imports might be justified and therefore upheld in certain cases: the Court decided that pending harmonization of Member States' legislation, the restriction may be necessary and therefore allowed pursuant to the mandatory requirements doctrine (see p. 101 above) provided that it is proportionate to the objective to be achieved. The Court nevertheless stressed that Article 30 would be infringed if the law granting an undertaking exclusive supervisory and regulatory functions does not also provide for the possibility to appeal against a refusal to grant type-approval.

Unlike Article 86, the effect of Article 30 is not to require the unbundling of activities but only the possibility to appeal against negative decisions.

In many cases therefore it will be in the interest of claimants to make a claim under both Article 86 and Article 30 in order to obtain maximum results: Article 30 may lead to the introduction of an appeal procedure against refusals to grant type-approval; Article 86 may lead to a reduction in the scope of the monopoly and to greater access for potential competitors.

Rules governing State monopolies of a commercial character: Article 37

Article 37 requires Member States to adjust any State monopolies of a commercial character so as to ensure that there is no discrimination in the way domestic goods and goods from other Member States are procured and marketed.

The questions arise as to what – if anything – Article 37 adds to Articles 30 to 36 and how does Article 37 tie in with Article 90(1) and (2)?

Before examining these questions in detail it is relevant to recall the precise wording of Article 37, define the type of monopolies covered and the purpose and scope of the obligation imposed by the Article. This section will focus on Article 37(1) and refer to Article 37(2) to (6) only where necessary.

Article 37(1) provides as follows:

Member States shall progressively adjust any State monopolies of a commercial character so as to ensure that when the transitional period has ended no discrimination regarding the conditions under which goods are procured and marketed exists between nationals of Member States.

The provisions of this Article shall apply to any body through which a Member State, in law or in fact, either directly or indirectly supervises, determines or appreciably influences imports or exports between Member States. These provisions shall likewise apply to monopolies delegated by the State to others.

Monopolies covered by Article 37

Monopolies may be defined, in the economic sense, as situations where an individual or a group of individuals entirely controls the offer of certain products. As far as Article 37 is concerned, only state monopolies are taken into account, that is, monopolies set up by the State in conformity with national law. The exclusive rights granted to these relevant monopolies are based on national law.

Commercial monopolies

Commercial monopolies refer to undertakings involved in trade in goods[34]. They do not cover services which fall within Article 59. They cover imports, exports and the marketing of products.

In *Costa* v. *Enel* the Court held that Article 37 applies to monopolies which on the one hand have as their object 'transactions [on] a commercial product liable to be subject to competition and trade between Member States' and which, on the other hand, play 'an effective role in trade between Member States'.

It is not necessary that the product in question should actually be traded between Member States. It is sufficient that it can be so traded.

Examples of commercial monopolies include the following (some of the monopolies listed have now been abolished):

- tobacco and tobacco products such as cigarette lighters
- matches
- alcohol
- potash
- newsprint
- powder and explosives
- petroleum products
- bananas

[34] See *Almelo* (n. 16 above) paragraph 27; *Sacchi* (C-155/73, [1974] ECR 409) paragraph 10; *Lagauche* (n. 22 above) paragraph 33.

- salt
- sulphur
- quinine
- natural gas for export

Form of commercial monopolies

Article 37(1), paragraph 2, provides that the obligation to adjust monopolies of a commercial character applies to any body through which the State exercises an influence over EC imports or exports[35].

State monopolies may be exercised by a ministry, a division of a ministry, a government body, or an autonomous public undertaking. Article 37 covers situations where the monopoly is exercised by an undertaking, a group of undertakings or by territorial units of the State such as local authorities[36].

Article 37, however, does not apply to a situation where – as in *Bodson* – the State entrusts local authorities with the provision of a service including the sale of products and the local authorities, in turn, entrust one or any number of independent private undertakings with the task. The reason for this is that any effect on trade that may follow does *not* stem from the behaviour of the State or the local authorities but from the private undertakings themselves[37].

Purpose and scope of Article 37

Purpose of Article 37

Article 37 was included to complete Articles 30 to 35. These Articles deal with *normative measures*, i.e. the legal rules imposed by Member States to regulate trade by undertakings. Thus Article 30 prohibits *laws* by which Member States impose quotas and charges on imports of foreign goods. The problem was that the Article might not apply to *actions* (as opposed to laws) *taken by state agencies* having an exclusive right over imports or exports. These agencies were likely to control imports or exports by deciding what products to import and in what quantities, and what price to sell them on the domestic market and, because they acted on their own initiative and not pursuant to a law, their decisions might have fallen outside the scope of Articles 30 to 35.

Article 37 was thus introduced to prevent Member States discriminating in favour of domestic products through their national import and export agencies. Given the considerable extension of Article 30, following the *Terminal Equipment* case, it might now be argued that Articles 30 to 35 provide a complete set of rules and that the mere creation of a State agency together with the grant of exclusive rights over imports and exports is covered by these Articles.

---

35 See also *Almelo* (see n. 16 above), paragraph 29; *Bodson* (C-30/87, [1988] ECR 2479), paragraph 13.

36 *Bodson* (n. 35 above), paragraph 13.

37 *Bodson* (n. 35 above), paragraph 14.

Because of the political interests at stake resulting from the economic importance of commercial monopolies to certain Member States, in particular, France and Italy, Article 37 provided for the *gradual adjustment* of these monopolies.

Article 37(6) gave the Commission the power to issue recommendations to encourage the Member States to carry out the necessary adjustments.

Scope of Article 37

Article 37 includes an obligation on Member States to *adjust* their state monopolies of a commercial character in order to avoid all *discrimination*.

Adjustment of State monopolies

The obligation to 'adjust' does not go as far as to 'abolish'. It seems, on the contrary, that Article 37 allows, in principle, the maintenance of monopolies provided that they do not result in discrimination in the trading of products between Member States. This is confirmed by the case-law of the Court which has repeatedly held that Article 37 does not require commercial monopolies to be abolished, but merely adjusted[38].

In certain cases, however, the application of Article 37 has resulted in an obligation to abolish a monopoly. Examples include *Manghera* and *the Greek oil monopoly* case[39] (see p. 123 below), where the Court considered that there was no alternative available in order to meet the obligation imposed by Article 37 to bring all discrimination between Member States to an end.

Removal of discrimination

Article 37 requires Member States to 'adjust' their monopolies of a commercial character so as to bring to an end all discrimination. Relevant discrimination is that which results in a restriction on imports or marketing of foreign goods. Discrimination can take various forms:

- straightforward limitations of imports resulting from the existence of a State monopoly holding a discretionary power over imports. Limitations are likely to be particularly significant where the State monopoly covers not just import but also production. The national undertaking benefiting from such a monopoly is likely to limit imports to the strict minimum so that they do not compete with its own products (*Manghera*);
- obligation to purchase minimum quantities from the national undertaking (*Greek oil*[40]);
- application of higher charges on imports than on domestic goods (*Miritz*[41]);

---

[38] *Hansen* [1979] ECR 935; *Manghera* [1976] ECR 91; *Commission* v. *Italy* [1983] ECR 1955.

[39] *Greek Oil*: case 347/88, 13 December 1990, [1990] ECR 4747.

[40] See n. 39 above.

[41] *Miritz*: case 91/75, 17 February 1976, [1976] ECR 217.

- application of different rules on marketing, pricing etc., for imported products.

Discrimination may either result directly from action taken by the monopoly itself or from a State measure related to the monopoly (as in *Miritz*).

Discrimination arising from exclusive import rights

Article 37 does not require Member States to abolish their state monopolies of a commercial character but simply to adjust them. Yet in the leading case, *Manghera*, the Court held that in order to meet the obligation in Article 37 to remove all discrimination, it was necessary to abolish the exclusive right over import as it was liable to give rise to discrimination[42].

The Court held at paragraph 13 of *Manghera*:

> Article 37(1) of the EEC Treaty must be interpreted in the sense that as from 31 December 1969, every State monopoly of a commercial character must be adjusted so as to *eliminate the exclusive right to import* from other Member States. (*emphasis added*).

The principle set out in *Manghera* is clear on the face of it. But, how far does this principle extend? *Manghera* involved a State monopoly over tobacco products where the monopoly undertaking also had an exclusive right over production – a situation which was all the more likely to result in discrimination.

It is uncertain whether the principle could be extended to *all* commercial monopolies including monopolies set up for reasons of general economic interest such as monopolies over telecommunications equipment or electricity. It seemed that the principle was limited to exclusive rights over imports and would not, for example, extend to other rights such as the marketing, connection, bringing into service and maintenance of telephone equipment. This was certainly the view of the Advocate General in *Terminal Equipment* (at paragraphs 18, 19 and 22): 'The judgment (in *Manghera*) is limited to the exclusive right of import and it is not certain that the statements which it contains would similarly apply to other exclusive or special rights'.

Discrimination arising from exclusive marketing rights

The question of whether Article 37 can, in certain cases, be used not only to require the abolition of exclusive rights over imports but also of exclusive marketing rights is controversial. The Advocate General in *Terminal Equipment* clearly thought that it could not. The Court avoided the question in *Terminal Equipment* by basing its entire judgment – including its confirmation

---

42 Sweden and Finland were recently reminded by the European Commission to abolish their State monopolies over the import, export, production and sale of wholesale alcohol; Austria was reminded by the Commission to abolish its monopolies over alcohol and salt, as well as to adjust its monopoly over manufactured tobacco (see European Commission XXVth Report on Competition Policy 1995, paragraphs 125 to 129).

that exclusive import rights were contrary to the Treaty – on Article 30. The Court did not mention Article 37. It is unclear whether by doing so the Court implicitly recognized that Article 37 could not be used to deal with exclusive rights other than those relating to imports[43].

The question remains unanswered. In principle, there seems to be no reason, following the reasoning in *Manghera*, why exclusive marketing rights should not also be prohibited by Article 37 if they inevitably result in the discrimination which the Article seeks to prevent. However, given the considerable extension given to Article 30 in *Terminal Equipment* it is not certain, in any event, that the question remains meaningful; it may be argued that Article 30 can be used to assess the validity of *any* exclusive rights granted by the State over products which have a restrictive effect on imports and exports of Community products.

Comparison between the scope of Articles 30 to 36 and Article 37

Articles 30 to 34 and Article 37

As mentioned above, Article 30 relates to normative measures issued by the State whilst Article 37 relates to the activity of state agencies or other bodies benefiting from an exclusive right over trade in products.

Following *Terminal Equipment*, it is clear that Article 30 can now be used to assess the compatibility with EC rules of State measures creating trade monopolies governing the import or export of products. It follows that it may be argued that Article 37 situations fall within the scope of Article 30 in the sense that actions or decisions of monopolies of a commercial character may in fact be prohibited by Article 30 pursuant to a ruling prohibiting the existence of the monopoly in all or in part. Furthermore, Article 37 is more limited than Article 30 in the sense that it only deals with State *monopolies* and does not cover other measures such as the grant of *special rights* as opposed to exclusive rights; nor does it cover restrictive rules imposed by the State on certain state-related undertakings such as the obligation to notify programmes of supply and to obtain authorization in the event of changes[44].

The question, therefore, arises as to whether Article 37 has become redundant. The answer is 'no', for practical reasons, principally due to the abundant case-law on the Article, such as *Manghera*, which offers useful precedents.

In practice, whilst Article 37 and its supporting case-law may be helpful in seeking to argue that an exclusive right over trade in products is contrary to Treaty rules, it will be necessary in most cases also to use Article 30. The

---

[43] This is the view expressed by Wachsmann and Berrod in *Revue Trimestrielle de Droit Européen* Jan-March 1994 p.41, but see the *Franzén* case in the Stop Press section.

[44] See the obligation imposed by Greece on purchaser of petroleum products in the *Greek Oil* (n. 39 above) case.

Terminal Equipment Directive[45] refers to both Article 30 and Article 37; the *Greek Oil* case bases its finding of infringement on both articles.

Advocate General Cosmas examined the relationship between Article 37 and Article 30 in his recent opinion[46] in the cases concerning *Exclusive Rights to Import and Export Electricity and Gas* in various Member States[47]. He considers that certain State measures may fall under both Article 37 and 30. His recommendation is to examine whether a State monopoly falls within Article 37; if the answer is negative, the compatibility of the monopoly with EC rules should be examined solely under Articles 30 to 34; if the answer is positive, the analysis should consist in examining first whether the monopoly is discriminatory and therefore contrary to Article 37; second whether the monopoly amounts to a measure equivalent to a quantitative restriction on imports or exports, contrary to Articles 30 to 34.

Advocate General Cosmas's proposed a two-step approach is helpful in relation to cases which raise issues of compatibility with both Article 37 and Articles 30 to 35. One will need to check whether the Court follows his approach. Once the two steps analysis has been carried out, he recommends examining whether a monopoly which is contrary to both Article 37 and Articles 30 to 34 may nevertheless be exempted by Article 36. This further recommendation which establishes a link between Article 37 and Article 36 is more controversial. It is examined in the following paragraphs.

Article 36 and Article 37

The question arises as to whether Article 36 can be used in relation to Article 37.

Advocate General Cosmas considers that Article 36 may apply to monopolies covered by Article 37 which also infringe Articles 30 to 34 provided they satisfy the conditions of Articles 36[48].

The question may be asked, however, whether Article 37 imposes an obligation on Member States to which there can be no exception and, in particular, no exception based on Article 36. This is because the text of Article 36 only refers to Articles 30 to 34 and would therefore seem to provide an exemption only to situations covered by these articles.

In the *Greek Oil* case which involved infringements of both Articles 30 and 37 the Court examined whether – as claimed by Greece – Article 36 could apply. But it did so only in connection with the alleged Article 30 infringement.

The answer to the question of the application of Article 36 to State monopolies falling within Article 37 remains uncertain. One possible view would be to say that State monopolies of a *purely commercial* character *must* be

[45] See recitals 3 and 4 of Directive 88/301/EEC (OJ 1988 L 131/73).

[46] Cases C-157 to 160/94 (not yet reported); AG Cosmas Opinion, paragraph 24.

[47] The Court judgment is awaited at the time of writing, see the Stop Press section.

[48] Cases C-157 to 160/94, Advocate General opinion, paragraph 26; see also Leigh Hancher, *EC Electricity Law* (Wiley Chancery, 1992), p. 41, paragraph 2.75.

adjusted as required by Article 37 so as to ensure that there is no discrimination regarding the purchase and sale of goods from other Member States and that the Article 36 derogation cannot apply. However, where the monopoly does not have a pure commercial character but has also been established for so-called *'non-economic reasons'* - in accordance with the Sacchi test[49] – the Article 36 derogation might apply in certain cases. A distinction might thus be made between a monopoly over the import of tobacco and a monopoly over the import of electricity. The monopoly over tobacco being a 'pure' commercial monopoly whose principal task is to generate revenue for the State could not benefit from the Article 36 derogation; whilst the monopoly over electricity which meets other interests of a more general nature and satisfies vital needs of the population of a non-economic nature might, in certain cases, benefit from the Article 36 derogation.

In the cases concerning *Exclusive Rights to Import and Export Electricity and Gas*, Advocate General Cosmas examined whether the Article 36 applied and found on the facts that the conditions for the application of the derogation were not met. This was because the Member States concerned had not proved that the exclusive rights were absolutely necessary to ensure minimum supplies and therefore public security. According to the Advocate General, proof had not been made out that abolition of the exclusive rights would jeopardize Member States' minimum supplies[50].

Link between Article 37 and Article 90

Application of Article 90(1) to commercial monopolies

There is no doubt that commercial monopolies under Article 37 constitute public undertakings or undertakings benefiting from exclusive rights under Article 90(1) (see chapter 1, above).

Application of Article 90(2) to commercial monopolies

It is not entirely clear, however, whether Article 90(2) can apply to State monopolies of a commercial character and, if it can, how it applies.[51]

To the extent that Article 37 imposes a 'clear'[52] obligation to adjust commercial monopolies which may, in certain cases, result in the abolition of exclusive rights over imports, it might be argued that there is no scope for the application of the derogation in Article 90(2)[53]. This, however,would be an extreme and unrealistic approach.

---

[49] *Sacchi*, case 155/73 [1974] ECR 409, ground 14.

[50] See Advocate General Cosmas's opinion (n. 46 above), paragraphs 82 to 85; see the Stop Press section.

[51] See Opinion of Advocate General Rozes in *Commission* v. *Italy*, case 78/82, [1983] ECR 155, p. 1985.

[52] 'L'obscure clarté de l'article 37 du Traité de Communauté Economique Européenne', C-A Colliard, Dalloz, 1964, Chronique, p. 266.

[53] See Advocate General Rozes and the references made in her opinion in *Commission* v. *Italy* (n. 51 above), at page 1986.

Article 90(2) expressly mentions revenue producing monopolies. These have been defined as monopolies the purpose of which is maximization of the revenue of the State; they include monopolies over tobacco or alcohol. These monopolies clearly fall within the definition of commercial monopolies; they constitute a type of commercial monopoly.

Pursuant to Article 90(2), the rules of the Treaty (this must include Article 37) apply to commercial monopolies in so far as they do not obstruct the performance of their particular tasks, for instance the task of raising revenue for the State. The derogation from the application of all Treaty rules permitted by Article 90(2) must be the exception. Advocate General Rozes adopts this approach in her opinion in *Commission* v. *Italy*. She considers that the Article 90(2) derogation will only apply to commercial monopolies to exempt them from the obligation arising from Article 37 in exceptional circumstances, namely if Article 37 makes impossible the specific task entrusted to the monopoly. If the task is only made more difficult, as opposed to impossible, the derogation provided in Article 90(2) will not apply. Advocate General Rozes expressed the view that the test was not met in *Commission* v. *Italy*.

Article 90(2) was not mentioned in the *Greek oil* case. Greece only used Article 36 in its defence arguing that the restrictions over imports resulting from an import monopoly and a compulsory purchasing quota were justified by reasons of public security. The same argument could probably also have been based on Article 90(2).

The analysis of the case-law suggests that it must be possible in certain circumstances to apply the derogation contained in Article 90(2) to Article 37 cases[54] in the case of commercial monopolies. It may thus apply to revenue-producing monopolies, whether or not they are also entrusted with a task of general economic interest, provided they meet the necessity test provided in the Article; but the test is more likely to be met in cases where the revenue-producing monopolies also have a task of general economic interest.

Article 90(2) might thus apply to some exclusive rights relating to imports or exports of electricity and gas. In his opinion on the cases currently pending before the Court, Advocate General Cosmas considers that the exclusive rights held by certain Member States over the *import* of electricity and gas are covered by the derogation contained in Article 90(2). This is because it seems to him that Member States might not otherwise be able to carry out their task of general economic interest which includes supplying electricity and gas to all customers throughout their territory at the best possible price. On the other hand the Advocate General stated that he was not convinced by the Member States' arguments that an exclusive right over *exports* was also necessary in order to achieve that aim. The Advocate General's analysis is helpful and his conclusions are interesting. It is to be regretted, however, that he did not fully state the reasons why Member States' arguments in

---

[54] See opinion of AG Rozes in *Commission* v. *Italy*, n. 51 above; see also Commentaire Megret *le Droit de la Communauté Economique Européenne* volume 1 (1970) p. 122.

defence of their exclusive rights over export did not meet the test of Article 90(2).

The Court judgment on these electricity and gas cases is likely to bring further key replies to some of the questions raised by the application of Article 90(2) to State monopolies[55]. On the basis of Advocate General Cosmas's opinion one may already conclude that detailed economic analyses of the impact of exclusive rights are likely to be increasingly important in the future to prove or disprove that exclusive rights over import and export of products are necessary to carry out a task of general economic interest. The same conclusion will also apply to exclusive rights over the supply of services which are examined in the next chapter.

---

[55] See the Stop Press section.

Chapter 6

STATE-RELATED UNDERTAKINGS AND THE RULES ON THE FREEDOM TO PROVIDE SERVICES: ARTICLES 59 AND 60 AND ARTICLE 90

Article 59

. . . Restrictions on the freedom to provide services within the Community shall be progressively abolished during the transitional period in respect of nationals of Member States who are established in a State of the Community other than that of the person for whom the services are intended.

. . .

Article 60

Services shall be considered to be 'services' within the meaning of this Treaty where they are normally provided for remuneration, in so far as they are not governed by the provisions relating to freedom of movement for goods, capital and persons.

'Services' shall in particular include:

(a) activities of an industrial character;

(b) activities of a commercial character;

(c) activities of craftsmen;

(d) activities of the professions.

Without prejudice to the provisions of the Chapter relating to the right of establishment, the person providing a service may, in order to do so, temporarily pursue his activity in the State where the service is provided, under the same conditions as are imposed by that State on its own nationals.

Article 56(1)

The provisions of this Chapter and measures taken in pursuance thereof shall not prejudice the applicability of provisions laid down by law, regulation or administrative action providing for special treatment for foreign nationals on grounds of public policy, public security or public health.

This chapter examines the compatibility of State monopolies with the rules on freedom to provide services. It analyses first the general application of Article 59 and the possible application of the derogation contained in Article 56 as well as other permitted restrictions on the freedom to provide services pursuant to the case-law of the court.

It then deals with the application of these rules to state-related monopolies.

General application of Article 59

Article 59 lays down one of the fundamental principles of the Treaty, namely the freedom to provide services.

Services

Services are defined in Article 60 as a residual category consisting of activities not governed by the rules on the free movement of goods, capital and persons. They include activities of an industrial or commercial character; activities of craftsmen and those related to the professions.

Services rendered for remuneration

Article 59 only applies to services which are normally rendered for remuneration, such as insurance services. In contrast, purely socio-economic activities, such as services falling within a national social security system, are not covered (see, for example *Poucet*[1], paragraph 19); or activities exercised by a public authority acting in the exercise of its powers (see, for example, *Eurocontrol*[2], paragraph 28). Another exception, interpreted strictly, is that contained in Article 55, namely activities connected even occasionally with the exercise of official authority.

The scope of 'services rendered for remuneration' may be difficult to determine. What is an activity normally provided for remuneration as opposed to an activity meeting a social need and provided free?

The *Höfner* case (see chapter 1 n. 8) illustrates the problem even though it was not decided on the basis of Article 59. Germany had removed employment agency services from the field of commercial business activities. The

---

[1] *Poucet et Pistre*, joined cases C-159/91 and C-160/91, 17 February 1993, [1993] ECR I-063.
[2] *Eurocontrol*, case C-364/92, 19 January 1994, [1994] ECR I-0043.

services were provided free by an organization emanating from the State. The German government considered that the services were not an activity for remuneration falling within the definition of services in Article 60. But the Advocate General[3] considered that the activity was 'normally' provided for remuneration in other Member States and even in Germany, to some extent, since the State organization did not always enforce its monopoly. He concluded that the activity was therefore a service within Article 60. The Court did not rule directly on the issue because it found that, on the facts, the case was entirely confined to a single Member State and, therefore, did not fall to be considered under Article 59; but it implicitly agreed with the Advocate General that the services in question fell within the definition in Article 60 since it held that they were subject to the competition rules and were not exempt from the application of the provisions of the Treaty.

Cross-border element

Services covered by Article 59 must involve a cross-border element. All the elements of the case in *Höfner* related to one Member State only: the defendant service providers and the users of the service were situated in Germany, as was the State organization which had an exclusive right to provide the services in question. The Court accordingly ruled that the case did not fall to be considered under Article 59[4].

The Treaty expressly provides for only one type of cross-border situation in Article 59, namely where the service provider is situated in a State other than that of the recipient of the service. An example of this is the German *co-insurance* case[5] where the users of the service, i.e. the policy holders, were situated in Germany whilst the insurance companies providing the service were situated in other Member States. But there may be many other cross-border situations and these are well summarized in the *Alpine Investments* case[6] by the Advocate General (at paragraph 28 of his opinion). The various possibilities are:

(1) the provider of services moves to another Member State in order to provide services;
(2) the recipient of services moves to another Member State in order to receive services;

---

[3] *Höfner*, C-41/90, [1991] ECR I-1979, Advocate General Jacobs, paragraph 20.

[4] See also *Debauve*, case 52/79 [1980] ECR 833, paragraph 9. But against this, it may be argued that *ERT*, *Mediawet*, and *Telecom Services* examined below did not clearly involve a cross-border element (see footnote 34 below).

[5] *Commission* v. *Germany*, case 205/84, [1986] ECR 3755; see also the French, Danish and Irish co-insurance cases: *Commission* v. *France*, case 220/83 [1986] ECR 3663; *Commission* v. *Denmark*, case 252/83 [1986] ECR 3713; *Commission* v. *Ireland*, case 206/84 [1986] ECR 3817.

[6] *Alpine Investments BV and Minister van Financien* case C-384/93, 10 May 1995, [1995] ECR I-1141.

(3) both the provider and the recipient are established in the same Member State but the provider moves to another Member State in order to offer its services;

(4) neither the provider nor the recipient moves physically and services are provided by post or telecommunications, such as telephone, fax or electronic mail.

In the *Tourist Guide* cases[7], the Court held that there was a cross-border element because the tourist guides were seeking to provide services in a Member State other than their Member States of establishment. In *Alpine Investments*[8] there was a cross-border element because the service provider and the recipient were situated in different Member States; the former was established in the Netherlands and the latter in other EC Member States.

The provision of cross-border services may include the temporary pursuit of activities in the Member State where the service is provided (Article 60, paragraph 2). In that case the 'host' Member State (the State where the service is provided) may not impose all its domestic laws on the service provider as if it was permanently established in its territory (see *Van Wesemael*[9] and *Webb*[10]). *Van Wesemael* related to a provision of Belgian law requiring entertainment agencies, i.e. agencies engaged in the placement of entertainers, to hold a licence in Belgium. The Court held that the requirement constituted an unnecessary restriction if the entertainment agency held a similar licence in its Member State of establishment and was subject to similar conditions and controls, in particular on its activities exercised abroad.

Direct effect and direct applicability

The rules on the freedom to provide services have direct effect and became directly applicable at the end of the transitional period (see *Van Binsbergen*[11] and *Van Wesemael*). Although Article 63 of the Treaty provided for the preparation of a general programme to be followed up by Directives which would implement the abolition of all restrictions on the freedom to provide services, the Court held that, in the absence of implementing Directives, Article 59 could be relied upon by individuals and companies to resist restrictions on the fundamental freedom given to them by the Treaty.

The absence of Directives, or the existence of only draft Directives to regulate a specific service has often, in practice, constituted an impediment to the unrestricted provision of services. This is because, in the absence of

---

[7] *Commission* v. *Italy*, case C-180/89, [1991] ECR-709, see also C-198/89 *Commission* v. *Greece* and C-154/89 *Commission* v. *France*.

[8] See n. 6 above.

[9] *Van Wesemael*: joined cases 110/78 and 111/78, [1979] ECR 35.

[10] *Webb*: case 27/80, [1981] ECR 3305.

[11] *Van Binsbergen*: case 33/74, [1974] ECR 1299.

harmonising Directives, Member States have often been able to justify restrictions on the fundamental freedom provided by Article 59.

Justification of restrictions on the freedom to provide services

There are two types of justifications:

(1) justifications based on Article 56;
(2) justifications based on the general public interest or the 'public good'.

Article 56 justification

Restrictions on the freedom to provide services including the application of discriminatory measures may be permitted if they are justified on the ground of public health, public security or public policy. These justifications have been narrowly interpreted by the Court. The public policy defence, in particular, may only be used in extreme cases where one of the fundamental interests of society is in jeopardy (*Bouchereau*[12]).

Public interest justification

The great majority of cases where the Court has permitted a restriction on the freedom to provide services relate to non-discriminatory measures and are based on the public interest justification. The *co-insurance* cases of 1986 (see n. 5 above) summarize the four conditions which the restriction must satisfy for the justification to apply:

(1) it must be based on the need to protect the *public interest* (also referred to as the *public good*);
(2) it must be *non-discriminatory*;
(3) it must *not duplicate* a requirement already imposed in the Member State of establishment;
(4) it must be *proportionate* to the aim to be achieved.

First condition: the concept of public interest and public good

As one of the fundamental principles of the Treaty, the freedom to provide services may only be limited by overriding reasons relating to the public interest. What are overriding reasons relating to the public interest or the public good?

A first example is the *protection of consumers*. In the *co-insurance* cases the Court held that insurance was a sensitive sector from the point of view of

[12] *R* v. *Bouchereau*, case 30/77, [1977] ECR 1999.

consumers. It involved risks against which they needed to be protected by adequate legislation. This was particularly the case for what have been called 'mass risks'; these are risks relating to individuals (the 'mass' of the population) as opposed to 'large risks' relating to large industrial or business risks. Mass risks relate to consumers, acting outside their professional knowledge, who are generally more vulnerable because less knowledgeable of potential risks. In the absence of legislation at EC level to harmonize the rules under which insurance services are provided, the Court held that Member States were entitled to maintain their own legislation and impose it on foreign insurers wishing to offer services in their territory, provided it met the three other conditions set out above. The obligation to observe the legislation of a number of Member States was likely to constitute a significant restriction on the freedom to provide services. Such a restriction could, however, be permitted because it satisfied superior or overriding objectives relating to the public interest.

In *Mediawet*[13], the Court summarized the types of objectives constituting overriding reasons relating to the public interest and which could, therefore, justify a restriction on the freedom to provide services:

- the protection of professional standards and the protection of users of professional services; for example, by means of professional rules imposing obligations relating to ethical conduct, professional qualifications, organization of the professions, supervision and liability[14];
- the protection of intellectual property[15];
- the protection of workers[16];
- consumer protection[17];
- the conservation of the national and historic heritage[18];
- the dissemination of knowledge on the artistic and cultural heritage of a country[19];

To this list must be added certain other objectives:

- in *Mediawet* the Court held that a cultural policy aimed at preserving pluralism, by ensuring that various opinions whether social, cultural, religious and philosophical can be expressed, may also constitute a valid reason[20];

---

[13] *Mediawet*: *Commission* v. *The Netherlands*, case C-353/89, 25 July 1991, [1991] ECR I-4088, ground 18.
[14] *Van Wesemael* (n. 9 above); *Van Binsbergen* (n. 11 above).
[15] *Coditel*: case 62/79, [1980] ECR 881.
[16] *Webb* (n. 10 above); *Seco* v. *Evi*: joined cases 62 and 63/81 [1982] ECR 223; *Rush Portugesa*, case C-113/89, [1990] ECR I-1417.
[17] See the co-insurance cases n. 5 (above) and also *Commission* v. *Italy* C-180/89 [1991] ECR I-709, *Commission* v. *Greece* C-198/89 [1991] ECR I-727.
[18] *Commission* v. *Italy* C-180/89 [1991] ECR I-709.
[19] *Commission* v. *France*, case C-154/89 [1991] ECR I-659.
[20] *Mediawet*, n. 13 above, ground 30.

- the protection of the coherence of a tax system[21]. The Court held in *Bachmann* that a Member State could restrict the provision of cross-border services (in this case insurance services) by providing that premiums would only be deductible if paid to an insurance company established in the national territory. Member States could do so if this was necessary to ensure that they received the tax payable by insurance companies established in their territories. *Bachmann* was a controversial case because it allowed a significant restriction on the freedom to provide services based on tax grounds;
- another controversial objective is the protection of the good reputation of the financial sector[22]. This could justify the prohibition of cold calling as a marketing method to provide certain financial services in other Member States.

All these objectives have been held to constitute valid reasons permitting certain restrictions on the freedom to provide services, although not all restrictions. In many cases the Court has held that whilst the objective put forward by a Member State was, on the face of it, in the public interest, it did not justify the specific restriction that the Member State sought to impose because the other conditions set out in the co-insurance cases were not met.

The other conditions

The three other conditions are as follows:

- the restriction must be *non-discriminatory*. A restriction which seeks to give an advantage to national service providers would be discriminatory and prohibited under Article 59. An example of this can be found in *Van Binsbergen*: this involved a Dutch law whereby the Netherlands imposed a restriction on legal advisers from other Member States which it did not impose on its own nationals;
- the restriction must *not duplicate* a requirement already imposed by the Member State of establishment. *Van Wesemael* offers a good example of duplication: Belgium imposed a licence on entertainment agencies irrespective of the licence that service providers from other Member States might already hold and which might already provide for the guarantees afforded by the Belgian licence for situations arising in Belgium;
- the restriction must be *proportionate* to the aim to be achieved. This is a key condition and one of the principal reasons why Member States often fail to prove that restrictions are compatible with Article 59. Only if they can convince the Court that there can be no less restrictive means of achieving their objective, will the restrictions be compatible with the

---

[21] *Bachmann*: case C-204/90, 28 January 1992, [1992] ECR I-249.
[22] The *Alpine Investments* case, n. 6 above.

Treaty[23]. In the *co-insurance* cases the Court held that the obligation on insurers to be authorized in the Member State of provision of services was valid; on the other hand, an obligation to be established locally was disproportionate and the very negation of the right guaranteed by the Treaty. An obligation on tourist guides accompanying tourists in other Member States to be licensed in the Member State visited was regarded as wholly disproportionate to the aim sought, namely the protection of the national and historic heritage[24]. An obligation on patent agents to be licensed in the Member State of provision of services irrespective of the limited scope of their activity was held to be disproportionate to the aim sought, which was the protection of the patent holder[25].

The four conditions examined above, i.e. fulfilment of a public interest need, non-discrimination, non-duplication of requirements and proportionality[26] also play an important role in the case-law on the application of Articles 59 and 90 to state-related monopolies. Of particular importance are the conditions relating to the objective of public interest and the requirement of proportionality. In relation to the first condition, Article 90(2) modifies the application of Article 59 in one significant respect by allowing Member States to claim that a restriction is necessary for *economic reasons*. Such reasons are normally prohibited in cases decided pursuant to Articles 59 and 56[27].

The application of Article 59 to monopolies

Type of monopoly services affected by Article 59

State-related monopolies in the field of services are liable to be affected by Article 59 if they fall within the definition of services in Article 60. The following services have been examined by the Court under Articles 59 and 90:

- telecommunications services[28]
- television broadcasting[29] and television advertising[30]

---

[23] At ground 19 of *Mediawet* (n. 13 above), the Court held: 'The application of national provisions to providers of services established in other Member States must be such as to guarantee the achievement of the intended aim and must not go beyond that which is necessary in order to achieve that objective. In other words, it must not be possible to obtain the same result by less restrictive rules'.

[24] *Tourist Guide* cases: see n. 18 and n. 19 above.

[25] *Säger* v. *Dennemeyer*: case C-76/90, [1991] ECR I-4221.

[26] See the *co-insurance* cases, n. 5 above.

[27] See ground 15 of *Mediawet* (n. 13 above) which refers to *Bond Van Adverteerders* 352/85 [1988] ECR 2085 (paragraph 34).

[28] *The Telecom Services* case, joined cases C-271/90, C-281/90 and C-289/90, [1992] ECR I-5833.

[29] *ERT*, C-260/89, [1991] ECR I-2925.

[30] *Mediawet* (n. 13 above).

- postal services[31]
- maritime services[32]
- national insurance bureaux.[33]

A number of cases on services, involving state-related enterprises, have been decided pursuant to Article 86 rather than Article 59. Thus *Höfner*, which concerned the placement of employees, was not based on Article 59 because it lacked a cross-frontier element. *Corbeau*, which concerned postal services, was decided pursuant to Article 90 mainly and Article 86 subsidiarily. Article 59 was not mentioned, presumably because the competition rules were sufficient to deal with the case; in addition all the elements of the case related to one country only[34] as in *Höfner*. Port services in *Port of Genoa* were examined only pursuant to Article 86, even though the case involved a cross-frontier element, again presumably because Article 86 was sufficient to deal with the issues raised.

Such services as the placement of employees, port services involving loading and unloading of goods, and postal services are, however, all liable to be examined under Articles 59 and 90 if they are provided by state-related enterprises.

Many other services carried out by state-related enterprises could be examined under Articles 59 and 90, for example, transport services (air transport, rail or maritime transport). It may be argued that certain services relating to the provision of energy could also be covered by Article 59 even though electricity and gas have themselves been held to constitute goods. Whilst the import, export and distribution of gas and electricity have been held to relate to the supply of goods falling within Article 30, one might argue that the *transit* of such goods, short of distribution, constitutes a service within Article 59. In this case, there is no sale and purchase of goods but merely the provision of a service relating to the goods.

It thus seems possible to argue that not all cases relating to gas and electricity fall within Article 30 and/or Article 86 only; they may also fall within Article 59, if the essential part of the activity concerned constitutes a service rather than the sale of a product. Similarly, in the telecoms field, whilst the Court has held that the sale of terminal equipment falls within the rules on free movement of goods, connection and installation of the equipment are clearly services. They were only examined under Article 30 in the *Terminal Equipment* case because they were *ancillary* to the sale of the product; had they constituted the core activity at issue, it is likely that they would have been examined under Article 59.

---

[31] *Dutch messenger services* case: *The Netherlands* v. *Commission*, joined cases C-48/90 and C-66/90 [1992] ECR I-565.

[32] *Corsica Ferries*, C-18/93, [1994] ECR I-1783.

[33] *Van Ameyde*, C-90/76, [1977] ECR 1091.

[34] But note that in *ERT*, *Mediawet* and *Telecom Services* examined below, the Court examined the application of Article 59 even though the cases did not clearly involve a foreign element.

Restrictions caught by Article 59

Article 59 prohibits restrictions on the provision of services by foreigners established in other Member States. In other words, it prohibits discrimination on grounds of nationality.

As far as state-related monopolies are concerned, one may distinguish between two different forms of discrimination. The second clearly calls in question the existence of the monopoly:

- first, discrimination on grounds of nationality resulting from the *behaviour* of the monopoly enterprise *towards third parties*;
- second, discrimination arising from the *existence of the monopoly* itself.

Discrimination resulting from the behaviour of the monopoly enterprise

Decisions taken by state-related enterprises in favour of national service providers or of locally established foreign service providers may be contrary to Articles 59 and 90 if they are taken pursuant to a law and not covered by one of the permitted justifications.

Three principal cases show how the Court has applied the concept of discriminatory behaviour by service companies.

Van Ameyde[35]. This case involved the services of loss adjusters in relation to car accidents, in Italy. Loss adjusters evaluate claims for insurance purposes. Italy changed its law on car insurance in 1963, making insurance compulsory. The operation of its national car insurance bureau, which had been set up for the application of the Green Card system, changed in accordance with the new law so that foreign insurers could no longer appoint a loss adjuster of their choice in Italy to evaluate claims involving foreign cars. Instead, the Italian bureau would appoint a local insurer who alone had the power to settle claims and left it to him to select a loss adjuster either on his own initiative or following the suggestion of the foreign insurer involved. The law had the effect of reducing the activity of foreign loss adjusters established in Italy.

The Court held that:

> the effects of reserving to the national bureau of a Member State or to its member insurance companies or insurance companies established there, the final decision as to the payment of damages to victims of accidents caused in the territory of that State by vehicles normally based in another Member State are not discriminatory within the meaning of Articles 52 and 59 of the Treaty.

The Court held that the law was not discriminatory *if* the exclusion of other categories of undertakings was not based on the ground of nationality.

---

[35] See n. 33 above.

The obligation to use only the bureau or insurers established locally was imposed with a view to protecting the victims of car accidents. The Court thus found that the rule made by the Italian bureau pursuant to the new Italian law on car insurance was not discriminatory even though it imposed the obligation to use the services of insurers established in Italy. *Van Ameyde* was decided in 1977. Today, following *Mediawet* and *Telecoms Services*, the Court might adopt a different view and find that the obligation to use the services of a local enterprise constituted a discrimination against companies situated abroad, which is only compatible with Treaty rules if it is covered by one of the permitted justifications.

One may draw a distinction between discrimination resulting from a *legal obligation* to behave in a certain way (for example, the obligation to use a service provider locally established as in *Van Ameyde*) and discrimination resulting from the *behaviour* freely adopted by a state-related enterprise and subsequently authorized by a law or a regulation. An example of this second type of discrimination is *Corsica Ferries*[36].

In *Corsica Ferries*, the Court held that the setting of different tariffs for identical piloting services (these consist in guiding maritime vessels in and out of ports) depending on whether or not the vessels carry the national flag, is contrary to Article 1(1) of Council Regulation (EEC) 4055/86 of 22 December 1986 which applies the principle of freedom to provide services contained in Article 59 to maritime transport.

The Court found that the measure taken by the Italian public authorities[37] following a decision by a state-related undertaking discriminated on grounds of nationality and was not justified by Article 56 or on public interest grounds.

Another case on discriminatory behaviour is *ERT*[38] which involved the *possibility* for the monopoly television broadcaster in Greece to give preference to national services. The Court held that the monopoly could lead to a discriminatory behaviour in the form of a preference given to national programmes over foreign programmes contrary to Article 59.

The Court stated that the combination of functions within the monopoly TV broadcaster could lead to discrimination. The functions consisted in (i) the right of ERT to broadcast its *own* programmes and (ii) the right to receive and retransmit *programmes from other Member States*. The Court said (at ground 22):

> The concentration of the monopolies to broadcast and retransmit in the hands of a single undertaking gives that undertaking the possibility both to broadcast its own programmes and to restrict the retransmission of programmes from other Member States.

---

[36] *Corsica Ferries*, see n. 32 above.

[37] The tariffs were set by the Corporation of Pilots of the Port of Genoa and approved by the Minister for Merchant Shipping and brought into force in each port by a Decree of the appropriate maritime authority.

[38] *ERT*, see n. 29 above.

The Court finally held (at ground 23) that whether discrimination to the detriment of programmes from other Member States *actually* took place was a matter of fact, which only the national court had jurisdiction to determine. It appears from the ruling of the Court that it is only if the combination of functions *actually* gives rise to discrimination that the 'double' monopoly will be contrary to Article 59.

The lessons from *ERT* are not entirely clear in relation to Article 59. The question is whether a finding by the national judge that discrimination actually took place should result in the abolition of the monopoly or part of the monopoly; or whether it should only result in an obligation on the Member State to eliminate the discriminatory behaviour by setting up appropriate safeguards but without being required to abolish the monopoly or part of it.

The most likely interpretation is that if discrimination is proven[39], the law granting *ERT* a double monopoly is contrary to Article 59 and the infringement must be cured by putting an end to at least one of the two exclusive rights, either that to broadcast national programmes or that to retransmit foreign programmes.

ERT is not only a case of discriminatory behaviour by a State monopoly resulting from actions decided by the monopoly on its own initiative; it is also a case where discrimination may be the inevitable result of the creation of a monopoly, a situation which is examined in more detail in the following paragraphs.

Discrimination arising from the existence of the monopoly itself

In the cases examined above, the Court condemned discriminatory behaviour in the form of discriminatory prices to the detriment of non-nationals (*Corsica Ferries*); or of a preference given to national programmes to the detriment of foreign programmes (*ERT*). Such behaviour made it more difficult, although not impossible, for foreign operators to provide their services in the Member State where the discrimination took place.

In the two cases referred to below, *Mediawet* and *Telecom Services*, the Court examined the validity of the obligation on consumers to use only the services of the national monopoly undertaking. In other words, it examined not just the difficulty but the impossibility for foreign operators to provide their services in the country of the monopoly undertaking.

Mediawet makes it clear for the first time that the creation of a monopoly undertaking and the legal obligation imposed on customers to use exclusively or principally the services of that undertaking constitutes a form of discrimination which operates to the detriment of foreign service providers contrary to Article 59.

---

[39] It appears from *ERT* that Article 59 will only be infringed if discrimination is proven, whilst under Article 86, it is sufficient if the monopoly is liable to create a situation in which the undertaking is led to discriminate. In *RTT* v. *GB*, the Court made it even clearer that there was no need to prove that discrimination actually took place under Article 86 (see paragraphs 23 and 24 of the judgment).

Previously, it was commonly thought that because the obligation to use the services of one national undertaking affected all potential service providers – not only those established abroad but also those established nationally – there was no discrimination on the grounds of nationality. In *Mediawet*, the Court made it clear that a finding of discrimination will not only exist where *all* the undertakings of a Member State are advantaged in comparison with foreign undertakings; the Court held that 'it is *sufficient* that the preferential system set up should benefit *a* national provider of services'[40] (emphasis added).

Mediawet related to the monopoly of a Dutch undertaking in the field of radio and television broadcasting. A Dutch law, the 'Mediawet', reserved to a broadcasting undertaking all or most orders for the preparation of programmes from Dutch national broadcasting bodies and restricted the retransmission in the Netherlands of programmes from other Member States containing advertising specifically intended for the Dutch public.

The Dutch law contained two types of restrictions:

- first, an obligation on radio and TV broadcasting companies in the Netherlands to make all or at least part of their programmes through a specific Dutch company, i.e. the Nederlandse Omroeepproduktie Bedrijf (NOPB). They had to use its recording studios, set workshops, technicians etc;
- second, an obligation to observe certain rules relating to advertising.

The first obligation, namely the absolute or partial obligation to use the technical resources of the NOPB for the production of radio or television programmes was held to have 'a protective effect for the benefit of a service undertaking established in the national territory and, to that extent, disadvantages undertakings of the same kind established in other Member States' (ground 23). The obligation to use exclusively, or to a significant extent, the services of a national service provider as opposed to foreign service providers was held to fall within the prohibition in Article 59 and was not covered by one of the permitted justifications.

The second obligation relating to the advertising contained in radio and television programmes broadcast in other Member States and transmitted into the Netherlands imposed two sets of requirements:

- first, *foreign broadcasters must be structured* in a specific way: they must entrust advertising to a legal person independent of the suppliers of programmes; use all their advertising revenue for the production of *programmes*; and they may *not* permit third parties to make a *profit*. This obligation on foreign broadcasters matched a similar obligation on national broadcasters who were not allowed to make profits from commercial advertising as this was reserved to a third party, the *Stichting Etherreclame (the STER)* (the television advertising foundation);

[40] *Mediawet* (n. 13 above): paragraph 25.

- second, it imposed conditions relating to the *form of the advertising* on both national and foreign broadcasting: advertising must be clearly recognizable as such and separated from other parts of the programme; it must not exceed 5 per cent of air time; and must not be broadcast on Sundays.

The Netherlands considered that in order to maintain pluralism in the media and keep a balance between entertainment and cultural programmes, as well as to avoid the overwhelming influence of advertising, it should reserve a predominant role for its national programme maker, NOPB. This body would be under a statutory obligation to observe the balance between programmes and to ensure that advertising was carried out by a non-profit making body. The law thus obliged Dutch broadcasters to use NOPB to a large extent even though this might have been to the detriment of other programmers including foreign programmers; and it imposed conditions on advertising (relating to the structure of the advertising body and the form of advertisements) on broadcasters.

The Court considered that *all* these rules had the effect of limiting competition in the Netherlands from foreign service providers and safeguarding the unique status and privileges given (i) to NOPB, the principal national programmer in the Netherlands, and (ii) the STER, the television advertising foundation, a non-profit making body. The Court held that the obligations imposed by the Dutch law fell within Article 59 and were not covered by one of the permitted justifications.

Mediawet makes it clear for the first time that for services, as for goods, it is no longer possible for a Member State to decide by itself which economic activities it will withdraw from free competition. Even if maintenance of pluralism was a legitimate objective which might be achieved by maintaining TV and radio broadcasting as non-commercial activities, the Court held that there were other ways of achieving this objective. It followed that the Netherlands could not maintain this sector of economic activity outside *competition from other Member States* and outside the rules on freedom to provide services.

Although this is not expressly stated by the Court, it appears that Mediawet constitutes, at least, a partial reversal of the judgment in *Sacchi*. Paragraphs 35 and 36 of *Mediawet* provide:

> . . . in order to establish whether a Member State may exclude the provision of certain services from free competition, it is a matter of determining whether the restrictions on freedom to provide services thereby created can be justified on the grounds relating to the general interest set out above.
>
> As was stated above, in the present case, the privileged system afforded to the NOPB by the Dutch authorities is *not justified on any ground relating to the general interest.* (*emphasis added*)

In *Sacchi*, the Court said that a Member State could withdraw certain economic activities from the field of competition '*for considerations of public interest, of a non-economic nature*'[41]. On the face of it, the reasoning of the Court is similar. But the key difference is that in *Sacchi*, a Member State could decide on the basis of its own interest alone that the creation of a monopoly was the best way to achieve its aim. Following *Mediawet* it can no longer take such a decision if it will be contrary to the Community interest. The Court will control the compatibility of a Member State's decision with Community interest and, in particular, whether the Member State could achieve the same objective with *less restrictive means.*

The lessons from *Mediawet* are that a Member State cannot:

(1) shelter a national undertaking from competition from abroad unless there are overwhelming reasons compatible with Community interest;

(2) seek to impose its rules on other Member States if this is to protect the revenue of one of its undertakings.

It follows that Member States may not, in many cases, ring-fence their national monopoly undertakings for considerations of a non-economic nature relating to their national public interest. Even though Article 59 does not stop them creating a monopolistic structure, this structure must generally permit competition from foreign companies. In practice, such foreign competition may ultimately force the monopoly undertaking to disappear. This could be the case if, over the years, national consumers choose to obtain their services from abroad because they are more adapted to their needs. But this may be time-consuming.

In *Telecom Services*[42], the Commission and the Court went a step further. In its Directive on Telecom Services[43], the Commission not only required Member States to accept competition from abroad through the application of the rules on freedom to provide services, it also required Member States to abolish the monopoly of their national telecommunications undertakings over defined value-added services. The Commission justified its Directive on the basis of Articles 59 and 86 and the Court upheld the Commission's proposal[44]. The Commission considered that there was no reason why Member States should maintain such a monopoly which was contrary to Article 59 in that it restricted competition from abroad and was also contrary to Article 86 in that it restricted internal competition and could affect trade between Member States. The objective of general economic interest could be achieved by less restrictive means. The case illustrates the fact that the frontier is narrow between (i) the obligation for a national monopoly undertaking to withstand competition from abroad (in accordance with Article 59);

---

[41] *Sacchi*, C-155/73, [1974] ECR 409.

[42] *Telecom Services*, joined cases C-271/90, C-281/90 and C-289/90, [1992] ECR I-583.

[43] *Commission Directive of 28 June 1990 on competition in the markets for telecommunications services*, OJ 1990 L192/10.

[44] Subject only to the provision of the Directive on Telecom Services (n. 43 above) relating to special rights as opposed to exclusive rights.

and (ii) the obligation to meet competition from national companies as well (in accordance with Article 86).

Following *Terminal Equipment*, the question was still open as to whether the Court would take the same approach for services as it did for goods and go as far as upholding the abolition of a monopoly over services. In theory, there seemed to be no reason why it should not take the same approach, which is, indeed, what the Advocate General advised in *Telecom Services*. After *Terminal Equipment* all the key issues relating to the power of the Commission to issue Directives under Article 90(3) in order to reduce or abolish monopolies had been considered and resolved. Yet it was not absolutely certain that the Court would be prepared to adopt the same reasoning for services. Belgium had raised the issue of the complexity of telecommunications services and stressed that they entailed more complicated issues than terminal equipment. In addition the history of Community case-law showed that progress had always been slower for services than for goods. Therefore the possibility remained that the Court would take a more cautious approach for monopolies over services than for monopolies over goods.

The Court, however, upheld the essential part of the Commission's approach and, in particular, its finding that monopolies over value-added telecommunications services restricted the ability of foreign service providers to provide their services in another territory contrary to Article 59. This had the effect of limiting the variety of services offered to consumers. There were apparently no reasons why the full variety of value-added telecommunications services available in each Member State should not be offered throughout the Community. Essential requirements such as consumer protection could normally be secured by less restrictive means.

But the application of the principle of freedom to provide services could not by itself create a truly integrated Community market. The reason for this is that unless customers are situated near the border it may not always be possible for the foreign service provider to provide the necessary after sales service. In many cases it will need to be established in the country where it provides services. A purchaser may be reluctant to buy a telephone or a television from a foreign vendor if the vendor is not established locally and able to repair the product in the event of damage.

In order to enable customers in a Member State to have access to the variety of services provided abroad, it was necessary to abolish the exclusive right created in favour of a national undertaking and enable foreign service providers to provide their services from abroad or by setting up a company in the territory of the monopoly undertaking. Article 86 was therefore an essential tool in addition to Article 59 to secure full competition throughout the Community.

Telecom Services confirms, together with *Mediawet*, that Member States can no longer decide entirely for themselves to keep an area of activity sheltered from free competition, not only from abroad (on the basis of Article 59 as in *Mediawet*) but also from within (on the basis of Article 86).

What are the possible justifications of restrictions on the freedom to provide services?

Various justifications have been put forward but only some of them have been regarded as valid by the Court. Moreover, whilst certain types of justifications could in theory constitute valid grounds for maintaining restrictions on the freedom to provide services, they have nevertheless been rejected by the Court. The following types of justification have been examined by the court.

- socio-cultural reasons;
- socio-political reasons;
- environmental reasons;
- pursuit of Community objective;
- lack of capacity;
- need for a smooth transition towards competition;
- the provision of a universal service;
- integrity of the network; security and safety of network operations; interoperability and data protection.

Socio-cultural reasons

The case which best illustrates this type of reasons is *Mediawet*. The Dutch government had argued that the reservation to its national body, namely the Bedrijf, of all or most of the preparation of TV and radio programmes in the Netherlands was done for cultural reasons, namely to establish a pluralist and non-commercial radio and television broadcasting system. This was an objective of general interest which, in its view, provided a proper justification for taking a sector of activity – the preparation of TV and radio programmes – wholly or partially outside the field of competition. The Bedrijf, being a national body, could be made to observe the diversity of cultures in the Netherlands and reflect the various social, cultural, religious and philosophical components making up the Dutch society. 'By making technical resources available to the various national broadcasting bodies, the NOPB enables the pluralist and non-commercial nature of the Dutch audio-visual system to be maintained.' The government also mentioned that the NOPB performs cultural tasks such as managing a sound library, keeping film archives and managing an orchestra. Finally, in addition to securing cultural objectives, the system helped avoid the collapse of the NOPB and mass redundancies.

The Court held that the cultural objective put forward did not constitute a valid justification. Further, it did not even consider the social point, namely the avoidance of mass redundancies. As far as the cultural point was concerned, it indicated that this could constitute a reason of general economic interest which could justify certain restrictions on the freedom to provide services. It did not, however, justify reserving the preparation of TV and radio programmes to one national body. The introduction of competition from foreign service providers would not, according to the Court, prevent the attainment of the Dutch objective, namely pluralism. Thus, the fact that this objective might be reached under a monopoly system was not sufficient to justify maintaining such a system when competition might equally permit the attainment of the same objective. It follows that, in such circumstances, competition must prevail over a monopoly system.

The fact that other systems, based on competition, have been shown to exist in other Member States is certainly a factor which must have played a part in the conclusion reached by the Court. This is true both in the case of socio-cultural reasons and socio-political reasons (see below).

Socio-political reasons

An example of this type of justification is *Höfner*. Even though the case was decided on the basis of Article 86 of the Treaty, it deals with the provision of a service and the arguments put forward can apply to an Article 59 situation.

In *Höfner*[45], representatives of the German government explained that they did not want employment procurement services to fall within the field of competition: they wished to reserve this activity to the State to ensure that the public interest was best served. Thus in granting a monopoly to a German body entrusted with the task of employment procurement, the German government did not pursue a purely economic objective but a social objective. The government representative explained that in the past, private employment procurement agencies had given priority to the placement of highly qualified specialists and had sometimes encouraged people to breach their contract in order to obtain their commission as intermediaries and generally made decisions which were in their favour: for example, they took decisions which resulted in higher profit to themselves rather than decisions which would solve unemployment problems, in particular, among less specialized and skilled workers. The German government considered that impartial and free of charge services offered by the Bundesanstalt was the fairest system. The German government also explained that it was important for the government to deal with all employment activities, not just low-skilled employees but also high-skilled executives, so as to have the widest possible view and knowledge of the situation on the market and thus be in the best position to find solutions to unemployment problems. The Court, however, did not accept that this socio-political/economic reason constituted a valid justification of the monopoly.

The Court stressed that the monopoly service provider had effectively allowed many competitors to operate. It did not find the arguments of the

---

[45] *Höfner*, see n. 3 above.

German government convincing, in particular, it did not accept that the existence of competition would prevent the German government from carrying out its service. There was a clear demand for competing services which could be met by other service providers. The existence of potential and actual competition combined with (or perhaps resulting from) the manifest inability of the German body to meet the demand on the market was sufficient to convince the Court that competition should prevail over the wish of the German government.

It follows from this case that a Member State may not withdraw an activity from the field of competition even if its principal motive is to fulfil a socio-political aim (in this case, the search for a solution to unemployment problems). Although the Treaty, as interpreted in *Sacchi*, seems to allow such objectives and the ensuing withdrawal from competition of an entire field of activity, it is now clear that the Commission will check whether the objective could not be met by other means which would better fit the aim of the Common Market and that the Court will apply similar reasoning.

Environmental reasons

Environmental issues play an increasingly important role and have been recognized by the Commission and the Court as constituting a legitimate concern and, possibly, even an overriding reason relating to the public interest. In *Almelo*, the Court said that in determining whether a restriction on competition was necessary in order to enable a state-related undertaking to perform its tasks of general interest, it was necessary to take into consideration *inter alia* the environmental legislation[46] which applied to it. If a state-related undertaking is subject to the strict observance of environmental rules, it may be able to rely on Article 90(2) to resist the competition of foreign undertakings which are not subject to such rules; for example, if a state-related undertaking is obliged to pay heavy taxes on account of pollution, it may be able to rely on Article 90(2) to argue that it should not be forced to face the competition of foreign undertakings which are not subject to such taxes.

The environmental justification must be applied with caution. It is necessary to show that it is essential for the performance of a task of general interest. In *Almelo*, the exclusive right in question related to the distribution of electricity by regional distributors. It prohibited the import and distribution of electricity through the public network by independent distributors. In other words, the exclusive right had the effect of preventing third party access to the network. Pending harmonization at EC level, the environmental reason might be relied on to restrict third party access to the network if it would otherwise be impossible for the undertaking benefiting from the exclusive right to perform its task of general interest.

[46] *Almelo* judgment, C-393/92, [1994] ECR 1477, paragraph 49; AG opinion, paragraph 165.

The environmental reason cannot be relied on in the abstract. Like other overriding public interest reasons, it must be closely connected to the facts of the case and may not be relied on to justify any type of exclusive right. This can be seen from the *Corsica Ferries* case which is examined in the following paragraph.

Pursuit of Community objective

The pursuit of a legitimate Community objective will not necessarily constitute a valid justification if it appears that the objective is not in fact the principal reason for the restriction on competition and/or the objective could be attained by other means which would not restrict competition.

In *Corsica Ferries*, the Court gave a warning against misuse of so-called Community objectives:

> the Corporation and the Italian government are *wrong* in attempting to justify the different tariffs on grounds of *navigational safety*, national transport policy or *protection of the environment*. Even if these objectives were capable of justifying intervention by the public authority in the transport sector, a discriminatory tariff system such as that at issue before the national court does not appear necessary for attaining these objectives.[47] (*emphasis added*)

During the Oral Hearing the Italian government had emphasized that the application of different tariffs was in line with Community principles: the preferential tariff applied to vessels carrying on cabotage thus making this transport mode competitive compared with road transport which is environmentally unfriendly. The measure was, therefore, consistent with environmental policy.

The Advocate General and the Court emphasized the differences in tariff between vessels carrying on cabotage (which mostly bore an Italian flag) and other vessels (mainly foreign vessels). These were not necessary to achieve the stated aim. The Italian authorities could have achieved the same result, namely the encouragement of maritime cabotage over road transport on grounds of navigational safety and protection of the environment, by extending to foreign vessels the tariff applied to Italian vessels carrying on cabotage.

All theoretically legitimate reasons of overriding public interest must therefore be closely adapted to the facts of the case. This is further confirmed by the further type of justification examined below.

Lack of spare capacity

This ground may be put forward by the owner and/or operator of a network, for example, a television broadcasting network; a telecommunications network; a gas or electricity network; an airline or railway network. It may also be

---

[47] *Corsica Ferries* (see n. 32 above), paragraph 36.

used by the operator of an essential facility such as an airport or a port, in relation to slots.

It was unsuccessfully applied in the television broadcasting field by the Greek authorities in the *ERT* case. They had submitted that the reason for the grant of a monopoly to ERT over both broadcasting of national programmes and retransmission of foreign programmes was to avoid disturbances due to the restricted number of channels available. The Court, however, held that this did not constitute a valid justification for the purpose of Article 56 because the undertaking in question, ERT, only used a limited number of the available channels[48].

The lack of spare capacity did not constitute a proper justification, on the facts of the case, because there were still many channels available and therefore no lack of capacity.

Need for a smooth transition towards competition

This is a difficult defence to apply because it runs counter to the general case-law of the Court. The defence is that a Member State needs more time to introduce competition in a particular service sector and that pending the introduction of full competition, it must be entitled to maintain its monopoly. The Court has, however, repeatedly said that Article 59 has direct effect and becomes *immediately* applicable following the end of the transitional period. The leading case on this point is *Van Wesemael*[49].

In *Mediawet*, the Dutch government argued that it needed to ensure a smooth transition between the previous broadcasting system, in which the technical resources required for the productions of programmes belonged to the public sector, and the introduction of a system of competition. One of the main reasons was the need to avoid mass redundancies. As mentioned above, however, the Court is not normally sensitive to this argument.

If it can be shown that competition is possible throughout the Community in a specific field of activity, the Court will not normally tolerate that certain Member States should be given more time and allowed to derogate from the full application of Article 59. Yet, in practice, when the Commission itself introduces a new Directive, it usually provides for a transitional period which may vary between Member States. Such transitions are introduced even in areas covered by Article 59 and which, in theory, should have been fully liberalized since the end of the transitional period. The question of the possible reliance on this defence by a Member State seems therefore worth examining on a sector-by-sector basis.

The defence is unlikely to be available in sectors which are already open to competition in most Member States and where there is no need to introduce a harmonizing Directive as in the TV and radio broadcasting sector. The

---

[48] *ERT* (see n. 29 above), paragraph 25.

[49] *Van Wesemael*: Joined cases 110/78 and 118/78, [1979] ECR 35.

defence is also unlikely to be effective if used on its own; conversely it might possibly be used with success if it is linked to the need to provide a universal service (see below) in areas which have not been the object of a harmonizing Directive. This could be the case, in particular, of the energy and railway transport sectors.

The need to provide a universal service

The concept of 'universal service' has not yet been mentioned by the Court as such. It has continued to use the words 'service of general economic interest'. However, when defining the postal service or the distribution of electricity, the Court mentioned criteria which are specific to certain types of services of general economic interest referred to by the European Commission as 'universal services'. The relevant criteria have been laid down in *Corbeau* and *Almelo* and consist in the provision of an essential service or the supply of essential goods, in the following manner:

(1) throughout a territory;
(2) to all customers/users on that territory;
(3) at uniform tariffs;
(4) on similar quality conditions;
(5) on terms which may not vary except in accordance with objective criteria applicable to all customers/users.

The term 'universal' is used to reflect the fact that all relevant services are offered in the territory. Some of the services which are considered as absolutely essential and which therefore must be offered to all consumers/users are further referred to as 'basic services'. The concept of 'basic services' requires the services to be offered at prices affordable, in principle, to all.

'Universal services' normally relate to what may be regarded as constituting essential or vital goods or services and include the principal utilities, namely water, energy, transport, post and telecommunications. They are one type of services of general economic interest.

The Court considers that certain restrictions of competition or even the exclusion of competition contrary to Article 59, *must* be permitted where this is necessary to enable the performance of a universal service.

In *Corbeau*[50], the Court accepted the exclusion of competition where this was necessary to ensure that consumers could have access to a basic postal service consisting in the collection and delivery of mail on equal quality conditions, at similar and affordable prices throughout the territory.

The derogation from the application of Treaty rules, including Articles 86 and 59, on the basis of Article 90(2) did not, however, extend to the provision of other services, also called *specific* services, if competition for these

---

50 *Corbeau* (see Introduction n. 6).

services did not jeopardize the provision of the service of general economic interest. In other words, the provision, for example, of express courier services – which could be distinguished from the basic service – would be subject to Articles 86 and 59 and therefore to competition from within and outside the territory because such competition did not jeopardize the performance of the basic service. Competition may have to be limited to certain geographic zones and/or certain minimum prices in order to ensure that it does not jeopardize the universal service.

Where a third party wishes to provide a service which cannot be regarded as specific and which, therefore, falls within the universal service, it remains possible to try to use the universal service defence to restrict or exclude competition from the third party. Thus in *Almelo*, the Court, following to a large extent its Advocate General, held that an exclusive right over the distribution of electricity was compatible with Treaty rules if the exclusivity was necessary for the performance of the service of general economic interest. The Court left it to the national judge to decide if the exclusive right was indeed necessary, taking into account the economic conditions and the legislative framework within which the undertaking had to operate.

Pending harmonization of the rules on the supply of electricity at Community level (to which the Advocate General refers but not the Court), Member States are free – to a large extent – to impose environmental rules, tax rules, conditions relating to the quality and security of the service, the obligation to make long-term investments etc. The Court seems to recognize that Member States remain free to organize their system of supply of electricity so that such rules and conditions are fully met: this may entail the grant of exclusive rights to an undertaking and the corresponding exclusion of competition, at least at the level of distribution if there is no other way of ensuring that the national rules and conditions are met.

This situation may evolve and it is not certain that such exclusion of competition will be allowed in the future if the progress of technology, the development of the infrastructure network and of Community legislation make it possible for competition to take place at the level of the basic universal service. For the moment, the Court does not seem to have sufficient data to enable it to decide that competition must be allowed.

Security and safety of network operations, integrity of the network, inter-operability and data protection

These various justifications apply principally to networks such as gas and electricity, telecoms and transport. They have been expressly defined in the telecoms services Directive and mentioned as possible grounds for derogating from the application of Article 59[51]. It must always be remembered that these derogations are subject to changes as the technology evolves. What

[51] Recital 9 of the Telecoms Services Directive 90/388/EEC, OJ 1990 L192/10.

constitutes a justification on the ground of safety at a certain point in time may no longer be so a few years later if technological progress has been such that there is no longer a need to restrict competition by limiting the number of operators in order to ensure the security of the network.

The *security* of the network has been defined as the availability of the public network in the event of emergency. The concept of security is linked to the concept of permanence of the network in all circumstances.

Technical *integrity* of the network means the operation and interconnection of public networks throughout the Community on the basis of common technical specifications.

Inter-operability of services means complying with such technical specifications so as to increase the provision of services and the choice available to users.

Data protection means measures taken to warrant the confidentiality of communications and the protection of personal data.

Chapter 7

STATE-RELATED UNDERTAKINGS AND OTHER TREATY ARTICLES

This chapter deals with the application of Article 90 in the light of the following other Treaty rules: Article 222 on the permitted systems of property ownership, Articles 92 and 93 on state aids, Article 52 on freedom of establishment and the rules contained in the EC Merger Regulation.

Article 222

This Article provides:

> This treaty shall in no way prejudice the rules in Member States governing the system of property ownership.

Article 222 has been taken to mean that the Treaty is neutral as far as property ownership is concerned. In other words, the Treaty would leave Member States free to decide whether an industrial company, a piece of land or any moveable or immoveable property, should be in public or private ownership.

The European Commission continues to insist that its move towards liberalization and the introduction of more competition, principally in the utilities sector such as transport, telecommunications, postal services and energy, must not be seen as a requirement for Member States to privatize these sectors of activity. The rules of competition apply equally to public and private enterprises and the Commission has no policy requiring Member States to carry out privatization.

Yet, in practice, there is often an inconsistency between competition and public ownership: where the State is the one and only owner of an entire industry, there can be, by definition, no competition in that industry. For example, in the field of telecommunications, until the first Directives based on Article 90(3), the State was often the only owner of the telecommunications industry and used to control all purchases of foreign telecommunications equipment, thus restricting the free movement of goods contrary to Article 30[1]; it would also be the only provider of telecommunications services in the territory covered by the monopoly thus restricting and even excluding the possibility for foreign operators to provide their services contrary to Article 59[2]; and it would prevent all competition within the territory by potential domestic telecoms operators or suppliers of equipment in a way that could affect trade between Member States contrary to Article 86[3].

By abolishing the monopoly over telecommunications equipment and services – an act which was upheld by the Court – the Commission encouraged competition in this sector both from foreign and domestic operators. This meant that there could no longer be only one national operator, the State, and that sole ownership would be shared with others, these being, by definition, not state-owned but private parties.

It follows that the effect of the telecommunications Directives has been to render illegal the system of state monopoly or sole ownership by the State of the entire telecommunications industry, notwithstanding the terms of Article 222.

The consequence is that Article 222 must now be re-interpreted. It can no longer be taken to mean, together with Article 90(1), that State monopolies are, in principle, legal. What it means is that a Member State may own, wholly or partly, certain enterprises but in many cases the State will no longer be entitled to be the sole owner of entire industries such as telecommunications, postal services, air transport etc. If the State remains the owner of an infrastructure, for example, the railway infrastructure, it cannot reserve to itself the entire railway industry but must allow foreign service providers on its tracks who will not only use the tracks but decide by themselves how they will commercialize their services on the territory of the infrastructure's owner.

In the field of energy, which remains the subject of intense debate, it is uncertain whether Member States will be allowed to remain the sole owners of the infrastructure as well as providers of all or most services based on this infrastructure, as is currently the case in several Member States. To the extent that a monopoly over both infrastructure and operations is not essential for the provision of a good quality universal service to users, bearing in mind the need to make long-term investments, it is likely that sole public ownership of the infrastructure and/or sole provision of services will have to make way for the introduction of third party competitors, both private and public, and including State enterprises from other Member States.

---

[1] This gave rise to the Terminal Equipment Directive (OJ 1988 L 131/73).
[2] This gave rise to the Telecom Services Directive (OJ 1990 L 192/10).
[3] See the Terminal Equipment and Telecom Services Directives (see n. 1 and 2 above).

State aid rules

Questions arise as to the relationship between Article 90(2) and Articles 92 and 93 on State aid: how do the State aid rules apply to state-related undertakings? Does the Article 90(2) derogation apply to exempt State aids which would otherwise be illegal from the prohibition under Article 92?

Article 92(1) provides that State aids which distort or threaten to distort competition between undertakings and affect trade between Member States are incompatible with the common market.

Article 92(2) and (3) provide exceptions to the prohibition in Article 90(1) in respect of certain types of State aid which *are* or *may be* considered compatible with the Common Market, for example, 'aid having a social character, granted to individual consumers, provided that such aid is granted without discrimination related to the origin of the products concerned' (Article 92(2)(a)).

Article 93 lays down the procedure applicable to State aids and gives the Commission exclusive competence to declare an aid compatible or incompatible with the Common Market. The Commission has a duty to examine the compatibility of both new aids and existing aids. Member States have a duty to notify new aids to the Commission. National courts have concurrent jurisdiction in respect of the concept of an aid, which means that they may determine whether a financial advantage constitutes a State aid within Article 92(1); they may also declare that a new aid that has not been notified, contrary to Article 93(3), is illegal and if required to do so, they must order the repayment of the aid unless, for exceptional reasons, repayment is inappropriate[4].

The key issue is whether, irrespective of Articles 92 and 93, a Member State may continue to make available an existing aid (one which existed prior to the Member State's entry into the Common Market) or introduce a new aid, in favour of a state-related undertaking to which it has granted a task of general economic interest. The question was examined in *Banco de Credito Industrial SA*, now *Banco Exterior de Espana SA Ayuntamiento de Valencia*[5].

The city of Valencia claimed a 'municipal establishment tax' from the bank, Banco de Credito Industrial. The tax was payable by any undertaking established in the City and exercising industrial, commercial or professional activities. The bank refused to pay on the ground that it was a public credit institution and was, as such, expressly exempted from the payment of the tax by a law of 1971. The *rationale* behind the exemption granted by the Spanish law to all public credit institutions was that these institutions are subject to strict rules on the types of loans which they can make. They must give priority to the objectives of the Economic and Social Development Plans and, in

[4] *Syndicat Français de l'Express International (SFEI) and others* v. *La Poste and others*; case C-39/94; 11 July 1996, [1996] ECR I-3547.

[5] Case C-387/92, 15 March 1994, [1994] ECR I-877.

particular, select loans on the basis of the profitability of projects having regard to the economic and social nature of the sectors concerned[6]. Thus, according to the Spanish government, public credit institutions covered by the Spanish law fulfil a function which meets the general interest.

The Spanish government claimed that the procedure laid down in Article 93 did not apply to State aid exempted by Article 90(2); in its view, this procedure is fundamentally different from the procedure under Article 90(3) which should apply to the case. The Treaty could have provided – if this had been the intention of its draftsmen – that state-related undertakings may fall within the ambit of Article 92(3); only in this case would the procedure in Article 93 apply to them.

The Court did not accept this argument and largely followed the reasoning of the Advocate General on the issue of the relationship between Articles 90(2) and Articles 92 and 93. The Advocate General first examined whether Article 90(2) had direct effect so that it could be relied upon directly by the Member States or the state-related undertakings concerned. He recalled the ambiguous and even conflicting case-law on the subject and concluded that Article 90(2) is directly applicable but must be examined in the context of other relevant Treaty articles, for example, Article 90(1) and Article 86[7].

According to the Advocate General, Article 90(2) must therefore be examined within the context of Articles 92 and 93, in this particular case.

He explained that the issue of the applicability or non-applicability of Article 93 to state-related undertakings is difficult and is not made any clearer by the statement of the Court in the *Steinike and Weinlig* case[8]:

> save for the reservation in Article 90(2) of the Treaty, Article 92 covers all private and public undertakings . . .

The Advocate General in *Banco Credito* said that the statement 'does not enable an unambiguous reply to the question concerning the relationship between those provisions'. The statement can be interpreted in two ways:

(1) Article 92 covers all private and public undertakings *except* undertakings falling under Article 90(2), in other words except state-related undertakings which have been entrusted with a task of general economic interest; or

(2) Article 92 covers all private and public undertakings *including* undertakings falling under Article 90(2) but in the latter case, the derogation from the application of Treaty rules may apply to legitimize State aids which would otherwise be contrary to Article 92(1).

---

[6] Article 2 of the Spanish law as quoted in AG Lenz's opinion at paragraph 13.

[7] AG opinion, paragraph 43.

[8] *Steinike and Weinlig* v. *Germany*, case 78/76, [1977] ECR 595, quoted by AG at paragraph 60, note 41.

If the first interpretation prevails then, in keeping with the Spanish government's approach, the State aid rules would not apply to state-related undertakings. It would still be necessary to ensure that the undertaking meets the criteria set out in Article 90(2)[9] and, in particular, that the State aid was indispensable to enable the state-related undertaking to carry out its task of general economic interest. Provided a Member State was satisfied that the Article 90(2) requirements were met, it would not need to apply the Article 93(3) procedure.

If the second interpretation prevails, Article 90(2) undertakings are covered by the State aid rules. It follows that any new aid must be notified by the Member States to the Commission under Article 93(3). Existing aids do not have to be notified but remain subject to supervision by the Commission. Both in the case of new aid and of existing aid, it is for the Commission to ascertain whether, having regard to Article 90(2), an aid which is not exempted by Article 92(2) and (3) may nevertheless be exempted on the ground that the aid meets the 'necessity test' of Article 90(2).

The Court, following to a large extent the Advocate General, chose the second interpretation. It held (at paragraphs 17 and 18):

> That power of the Commission (i.e. its power to supervise aids) also covers State aid granted to the undertakings referred to in Article 90(2), in particular, those which the Member States have made responsible for the management of services of general economic interest.
>
> It follows that the distinction which Article 93 of the Treaty draws between existing aid and new aid is equally applicable to State aid granted to the undertakings covered by Article 90(2).

The consequence is that the State aid rules apply in their entirety to undertakings covered by Article 90(2) and the notification procedure under Article 93(3) must be fully observed. The fact that the undertakings may benefit from a derogation from the Treaty rules only becomes relevant in a second stage (and not *a priori*) when the Commission assesses the compatibility of the aid with the Common Market. The assessment cannot be made *a priori* by the Member State and the undertaking concerned in order to avoid the application of Article 93(3). The judgment in *Banco Credito* makes this absolutely clear; it was further confirmed in *Fédération Française des Sociétés d'Assurances and others* v. *Commission*[10].

The Court in *Banco Credito* finally held that the matter referred to it related to an existing aid which, therefore, predated the accession of Spain to the Common Market. The consequence of this is that the aid should be regarded as valid and may be implemented as long as the Commission has not found it to be incompatible with the Common Market. It also followed that the Court did not have to examine the question whether the aid could be justified by

[9] See chapter 2.

[10] *Fédération Française des Sociétés d'Assurances* v. *Commission*, case T-106/95, 25 February 1997, [1997] ECR II-229, grounds 167 and 169.

Article 90(2) of the Treaty in accordance with the 'general economic interest' arguments put forward by the Spanish government.

The question of whether an aid could be justified pursuant to Article 90(2) was fully examined in *Fédération Française des Sociétés d'Assurances* v *Commission.* The answer given was positive on the facts of the case. The *Fédération* and other French insurance companies had alleged that a tax rebate given by the French government to the French Post Office gave the latter an advantage which it could use to compete unfairly in the field of insurance. The Commission disagreed with the claimant and the Court of First Instance held that although the tax rebate constituted State aid pursuant to Article 92 it could be exempted by Article 90(2) because the level of the aid did not exceed what was *necessary* for the French Post Office to carry its task of general economic interest. The amount of the rebate only aimed to compensate for the additional costs incurred by the French Post Office in providing its task of public service[11] and did not enable it to cross-subsidize activities falling in the competitive sector such as insurance activities[12].

It follows from this case that a State aid granted to a state-related enterprise may be legal pursuant to Article 90(2) if it goes no further than is necessary to enable the enterprise to carry out its task of general economic interest and does not enable it to cross-subsidize activities in the competitive sector.

Parallel to the link between Article 92 and Article 90(2) which is demonstrated by the *Fédération Française des Sociétés d'Assurances,* there is a link between Article 93 and Article 90(3). The latter two articles give the Commission a power of supervision over the relationship between Member States and state-related enterprises. They enable the Commission to ensure compliance with the State aid rules (Article 93) and the rules of the Treaty in general (Article 90(3)).

The link between the supervisory power of the Commission under both Articles was stated in the Transparency Directives and upheld by the Court in *France, Italy, UK, and Ireland* v. *Commission*[13].

In the latter case, the Court referred to the recitals of the Transparency Directive and stated that the purpose of the Directive issued under Article 90(3), was to 'promote the effective application to public undertakings of the provisions contained in Articles 92 and 93 of the Treaty'[14]. It agreed that the relationships between public undertakings and the State were often complex and therefore difficult to supervise and that there was 'an undeniable need for the Commission to seek additional information on those relations'[15]. The Directive would increase the transparency by forcing the Member States to maintain certain data relating to public funds granted to state-related undertakings and to disclose the information upon demand from the Commission.

---

[11] See n. 10 above, ground 178 of the judgment.
[12] See n. 10 above, grounds 188 and 189.
[13] *France, Italy, UK* v. *Commission,* cases 188 to 190/80; [1982] ECR 2545.
[14] See n. 13 above, paragraph 2.
[15] See n. 13 above, paragraph 18.

In addition to the general State aid rules contained in Articles 92 and 93 of the Treaty, specific rules apply to specific industry sectors which may concern state-related undertakings. It is, therefore, necessary in practice to check whether an aid granted to an undertaking covered by Article 90(2) falls under any specific regulations such as those applying to the coal industry; the textile industry; transport by rail, road, or waterway; or maritime transport[16].

Article 52

This article requires the abolition of restrictions on the freedom of nationals of a Member State to establish in the territory of another Member State. It states that:

> . . . Freedom of establishment shall include the right to take up and pursue activities as self-employed persons and to set up and manage undertakings, in particular companies or firms within the meaning of the second paragraph of Article 58, under the conditions laid down for its own nationals by the law of the country where such establishment is effected, subject to the provisions of the Chapter relating to capital.

This article provides for the right of an individual to settle in another Member State and carry out a business activity there as a self-employed person or by setting up a company, a subsidiary, an agency, a branch or other legal entity.

There has been one case dealing with Article 52 and Article 90. This related to the insurance in Greece of public property[17]. A Greek law required that all public property in Greece including all the assets of Greek public undertakings should be exclusively insured by public sector insurance companies. In addition, the law required state-owned banks to recommend in writing to their customers to take out insurance with an insurance company owned and controlled by the public banking sector. Given the importance of the state-owned bank in Greece (representing 80 per cent of the credit market at the time of the law) the recommendation was likely to have significant impact on the insurance market.

The Commission held that the Greek law infringed Articles 52 and 53 and 90(1) of the Treaty: it constituted a measure which favoured Greek public insurance companies, in particular, those controlled by state-owned banks, to the detriment of private insurance companies including foreign insurers established in Greece. The Commission stated that the measure taken by Greece 'makes it impossible for insurance companies from other Member

[16] See the list of sectoral aids in EC competition law, volume II A, rules applicable to State aids, European Commission 1994. In contrast, no specific rules apply to the postal sector: see *Syndicat Français de l'Express International [SFEI] and La Poste*, case C-39/94, 11 July 1996; [1996] ECR I-3547 and *SFEI and others* v. *Commission*, case T-77/95, 15 January 1997, [1997] ECR II-1.

[17] Commission Decision of 24 April 1985, OJ 1985 L152/25.

States to set up in business in Greece as public property insurers'. The restriction on the freedom of establishment for foreign insurers was significant because of the importance of the public market in Greece which represented about 25 per cent of annual premium income.

The Greek law did not establish a monopoly in favour of one national insurer, rather it directed a large part of insurance business towards state-controlled insurers. The Commission's decision, issued on the basis of Article 90(3) of the Treaty, aimed at restoring the balance between state-owned insurance companies and other insurers, in particular foreign insurers established or seeking to establish in Greece.

Another type of measure might raise a question of compatibility with both Articles 90 and 52, namely the creation of a monopoly in favour of one undertaking. Unlike the *Greek Insurance* case, the measure would not favour a number of state-controlled undertakings but only one state-related undertaking.

There is little doubt that the existence of a monopoly undertaking covering the entire territory of a Member State constitutes a restriction on the freedom of establishment since it precludes a foreign national from establishing a business falling within the field of activity of the monopoly in that State.

The question arises as to whether such a restriction is contrary to Article 52. There appears to have been *no* case to date holding that such a monopoly was contrary to Article 52. The reason may lie in the precise wording of Article 52, paragraph 2. This provides that the right of establishment *shall include* the right to take up activities and manage undertakings *under the conditions laid down for its own nationals* by the law of the country where such establishment is effected. The emphasized words 'under the conditions laid down for its own nationals' express the idea that there should be no discrimination against a national from another Member State. It might be argued that, where a monopoly exists, the prohibition on setting up a competing business applies equally to the nationals of the Member State where the monopoly is established and to the nationals of foreign Member States. On that basis, rules prohibiting establishment would not be contrary to Article 52.

A different interpretation may be adopted: the right of establishment *includes* the right to set up and manage an undertaking 'under the conditions laid down for its own nationals'; where most nationals are restricted from establishing because of the presence of a monopoly undertaking, it would still be possible for a company from another Member State to argue that it may establish locally, under the conditions applying to the monopoly undertaking, on the basis of the general principle of freedom of establishment.

A third interpretation may be found in an early draft of the proposed Directive on postal services dated 16 December 1994, where the Commission took the view that the granting of special or exclusive rights to one or more national postal organizations to carry out postal services constituted a restriction contrary to Article 52. The Commission's view was based on the concept of discrimination:

> the fact that one Member State, in the selection of one organization favours a national undertaking without taking into account the merits of possible applicants from other Member States to provide the relevant public service tasks, constitutes a violation of its obligation under Article 52. Even when based on the criterion of public ownership and not nationality, such a measure actually constitutes a form of discrimination which results in discrimination connected with the nationality of the operator.

The Commission considered that Member States could decide that only one undertaking should carry out certain activities on their territory, but in making this decision – they should consider the possibility of a foreign, and not just a national, undertaking carrying out the activities.

It is not certain, however, that the notion of discrimination as mentioned in the third interpretation can be applied to the creation of a legal monopoly over postal services. In its revised draft dated 26 July 1995, the Commission no longer referred to Article 52 or to the possible infringement of this Article by Member States' postal monopolies.

It might be argued that the second interpretation based on the general freedom of establishment could form the basis of a legal challenge against monopolies. Given the lack of precedents, it is however very uncertain whether such an interpretation could prevail. Any challenge to a monopoly on the basis of Article 52 should also make use of other legal arguments based, for example, on Article 59 or Article 86.

Article 90(2) and the Merger Regulation

Article 90(2) and state-related undertakings are expressly mentioned in the Merger Regulation and in two of its accompanying notices. These are:

- recital 12 of the Merger Regulation[18];
- paragraphs 55 and 56 of the Commission Notice on the notion of 'undertakings concerned'[19];
- paragraph 8 of the Commission Notice on the notion of 'concentration'[20].

The Merger Regulation applies to state-related undertakings but questions arise (i) as to the calculation of the turnover of these undertakings for the purpose of determining whether the Regulation's thresholds are met; (ii) whether there is a concentration within the meaning of the Regulation when the merger is between two state-owned enterprises of the same Member State.

[18] Council Regulation 4064/89, EEC, OJ 1990 L257/14.
[19] Commission Notice 94/C 385/03, OJ 1994 C385/12.
[20] Commission Notice 94/C 385/02, OJ 1994 C385/5.

The application of the Merger Regulation to State-related enterprises

There have been several Commission Decisions relating to mergers involving state-owned enterprises pursuant to the Merger Regulation, in particular, *Péchiney/Usinor* of 24 June 1991; *CEA Industrie/France Télécom/SGS Thomson* of 22 February 1993; *Air France/Sabena* of 5 October 1992; and *Kali und Salz/MKD/Treuhand* of 14 December 1993.

Recital 12 of the Merger Regulation provides:

> Whereas the arrangements to be introduced for the control of concentrations should, without prejudice to Article 90 (2) of the Treaty, respect the principle of non-discrimination between the public and the private sectors; . . .

The precise meaning of 'without prejudice to Article 90(2)' in Recital 12 is ambiguous. On a literal reading, it means that there should be no discrimination between public and private enterprises in the application of the Regulation, provided this is compatible with Article 90(2). In other words, it seems to leave open the possibility that a different treatment of public and private enterprises may arise if this is justified by Article 90(2). Such an interpretation, however, would be contrary to the principle consistently repeated by the Commission and based on Article 222 of the Treaty, namely that the Treaty is neutral in terms of public and private ownership. Furthermore, the principle of non-discrimination between public and private enterprises was implicitly recognized in the *Transparency Directive* case[21].

The correct interpretation of Recital 12 must be that the Merger Regulation applies equally to public and private enterprises, thus respecting the principle of non-discrimination between the two forms of property ownership, but that there may be cases where the Merger Regulation would not apply, *irrespective of whether the merger involves public or private undertakings*, where this would prevent the performance of a task of general economic interest.

There may, for instance, be circumstances involving public policy or public security where Article 90(2) could be relied on successfully to exempt a merger which would otherwise create or strengthen a dominant position and significantly impede effective competition in the Common Market contrary to Article 2(3) of the Merger Regulation. An example of this would be in the field of energy where resources are scarce. A Member State could decide to take control of the entire supply of energy throughout a territory through the acquisition and merger of all relevant entities. Such an operation would create or strengthen a dominant position in a substantial part of the

[21] *France, Italy and UK* v. *Commission*, cases 188 to 190/80, 6 July 1982, [1982] ECR 2545 (paragraphs 20 and 21).

Common Market. It might, nevertheless, depending on the circumstances, be justified as being necessary for the carrying out of a task of general economic interest in conformity with Article 90(2).

The defect in the wording of Recital 12 is that it seems to provide a derogation from the principle of non-discrimination, based on Article 90(2). In fact, the Article 90(2) derogation does not apply to *the principle of non-discrimination* between public and private undertakings but to *the application of the Merger Regulation* to undertakings (public and private) which carry on a task of general economic interest.

Except in the rare cases where the Article 90(2) derogation is applicable, the Merger Regulation will apply to concentrations involving state-related enterprises if the relevant turnover thresholds are reached.

Calculation of turnover

This may be difficult to determine in the case of a state-owned undertaking where it is difficult to disentangle the results of the undertaking from other activities of the State.

Recital 12 of the Merger Regulation and paragraph 55 of the Commission Notice on the notion of undertakings concerned provide that:

> . . . in the public sector, calculation of the turnover of an undertaking concerned in a concentration needs [. . .] to take account of undertakings making up an economic unit with an independent power of decision, irrespective of the way in which their capital is held or of the rules of administrative supervision applicable to them.

A difficulty is likely to arise in the case of a merger between state-owned enterprises. The question is whether the merger constitutes a concentration within the regulation.

Concentration or internal restructuring within a group

For the Merger Regulation to apply, there must be a concentration within the meaning of Article 3. A concentration occurs when there is a change in the control of an undertaking. It follows that internal restructuring within the same group without a change of control is not a relevant concentration.

When two state-owned entities, i.e. two entities ultimately controlled by the State merge, the question may arise as to whether the merger involves a change of control and accordingly constitutes a concentration.

Guidelines may be found in paragraph 8 of the Commission Notice on the notion of concentration. The Notice provides that whether the operation is to be considered as an internal restructuring depends on whether both undertakings were formerly part of the same economic unit with an independent power of decision:

> Where the undertakings were formerly part of different economic units having an independent power of decision the operation will be deemed to constitute a concentration and not an internal restructuring.

The Notice refers to two examples, namely *Péchiney* and *CEA Industrie/France Télécom/SGS Thomson.* It adds that 'such independent power of decision does not normally exist, however, where the undertakings are within the same holding company'.

Chapter 8

KEY ARTICLE 90 CONCEPTS

This chapter deals with key concepts associated with Article 90. Most have already been mentioned in the earlier chapters of this book but they are examined here in detail so as to clarify their scope of application and the way in which they inter-relate.

Certain concepts have a wider application than their application to state-related undertakings, for example, access to an essential facility, transparency, proportionality and non-discrimination. Others apply almost exclusively to state-related undertakings; for example, unbundling, universal service, basic service, reserved and unreserved service, services of general economic interest, public service, liberalization and deregulation.

Access to an essential facility

The concept of access

A right of access is the right of a third party undertaking to use the service, the product or the network of another undertaking which is often an undertaking benefiting from a monopoly, an exclusive or special right.

The concept of 'access' is often linked to that of an essential facility.

The concept of essential facility

This is something which is indispensable for the third party to provide its own services or products to customers. Essential facility is defined in Community case-law by the Commission in *Stena Sealink*[1] as:

> a facility or infrastructure, without access to which competitors cannot provide services to their customers.

---

[1] *Sea Containers* v. *Stena Sealink* (interim measures) Commission Decision of 21 December 1993; OJ 1994 L15/8 (paragraph 66).

The case in question provides a good illustration of the concept. Stena Sealink Ports is a private company which owns and operates the port of Holyhead in Wales. Stena Sealink Line, a related company, operates ferry services between the United Kingdom, Ireland and France. Sea Containers Ferries Limited, a competitor of Stena Sealink Line, operates passenger, car and freight ferry services. It complained to the Commission that the two Stena Sealink companies were denying it reasonable access to the port of Holyhead, thus restricting it from providing competing ferry services between Holyhead in Wales, Dublin and Dun Laoghaire in Ireland.

The Commission considered that the port of Holyhead constituted an essential facility and that there was a *prima facie* case of abuse by Stena Sealink in that it was restricting access to the facility by its competitor, Sea Containers Ferries Limited. The Commission, however, rejected the request for interim measures made by Sea Contaners Ferries because following positive action[2] by the Stena Sealink group, risks of serious and irreparable damage had disappeared.

On the same day as its interim decision in Stena Sealink, the Commission took a final decision relating to the port of Rødby in Denmark[3] which was also regarded as an essential facility. DSB, a Danish public undertaking and a department of the Danish Transport Ministry was the owner of the port of Rødby. It also had the exclusive right to organize rail traffic in Denmark and the non-exclusive right to operate ferry services between Denmark and neighbouring countries, in particular, between Rødby and Puttgarden in Germany. The Rødby–Puttgarden route is operated jointly by DSB and the German railway, DB. A third party, Stena Rederi AB, of Swedish origin, wished to operate a ferry service on that route and requested either the right to build a new terminal near the port of Rødby or to have access to the existing terminal at the port. The Danish authorities rejected Stena Rederi's request. The Commission held that this was done to prevent the activity of a competitor and thus protect the current joint dominant position of DSB and DB on the route.

It held that the port of Rødby constituted an *essential facility* for the provision of *maritime transport services*, because access to the port was essential for shipowners to provide competing ferry services on the Rødby–Puttgarden route. The Commission issued a Decision based on Article 90(3) of the Treaty ordering the Danish government to put an end to the infringement by giving access to Stena Rederi.

---

[2] Stena Sealink had eventually offered suitable slots to Sea Containers (paragraph 79 of the Decision).

[3] Access to the facilities of the port of Rødby (Denmark): Commission Decision of 21 December 1993; OJ 1994 L55/52.

Origin of the concept of access

The concept was first developed in American case-law. There are also many cases in EC law which rely on the concept of access without referring to it by name and which are at the roots of the concept. The two principal cases are *Commercial Solvents*[4] and *Telemarketing*[5].

Commercial Solvents was a company dominant in the field of raw materials which were used to make a medical drug against tuberculosis. It used to supply Zoja, a company which depended on its raw material to produce the drug which it would then sell to its own customers. Commercial Solvents decided to enter into the field of drug manufacturing itself, and in order to keep the market for itself, refused to continue supplying its previous client, Zoja, active in that field. The Court held that the refusal by Commercial Solvents constituted an abuse of a dominant position.

Applying the *Stena Sealink*[6] definition, the raw material may be regarded as constituting an essential product without access to which competitors (in this case drug manufacturers) could not provide their products to their customers.

Telemarketing concerned the broadcasting sector. The company known as Compagnie Luxembourgeoise, controlled a television channel and dominated the market in television advertising aimed at viewers in French-speaking Belgium. It refused to grant CBEM, a competitor in telemarketing services, access to its broadcasting channel unless CBEM refrained from referring its customers to its own telemarketing services, but instead referred them to those of Compagnie Luxembourgeoise. The Court said that Compagnie Luxembourgeoise was thus trying to reserve to itself an ancillary activity, telemarketing, on a neighbouring but separate market from that of television advertising with the possibility of eliminating all competition on that market.

This case is also one of the principal cases mentioned to illustrate the concept of access to an essential facility. Compagnie Luxembourgeoise controlled an essential facility, namely a broadcasting channel including television advertising, which was indispensable to the provision of telemarketing services to French-speaking viewers. It refused access to CBEM, a competitor active in the field of telemarketing, thus preventing competition in this sector.

The implications of the concept of access in competition law terms

The concept of access applies to undertakings which occupy a dominant position in the market as a result of rights – often exclusive – which they have been granted.

[4] *Commercial Solvents* v. *Commission*, cases 6 and 7/73, [1974] ECR 223.

[5] *Telemarketing*, case C-311/84, [1985] ECR 3261.

[6] See n. 1 above.

Refusal by these undertakings to grant access to companies which could compete with them in the provision of services or in the supply of goods could constitute an abuse of a dominant position contrary to Article 86. If the State is party to the refusal, either because it took the actual decision to refuse access or it encouraged or supported the decision taken by the undertaking, the refusal may constitute an infringement of Article 90(1) together with Article 86.

In *Stena Sealink*[7], the Commission refers to several judgments of the Court in addition to *Commercial Solvents* and *Telemarketing* and to some of its own decisions to support the concept of access. The following sections examine both the principal case-law, sector by sector, and the relevant EC legislation, which apply the concept.

Transport sector

Port

The concept of access was expressly applied for the first time in Commission Decisions relating to ports. Mention was made above of *Stena Sealink* and *Rødby*. The other relevant decision is *B & I* v. *Sealink*[8].

Airport

The question of access to airports has also been raised in relation to the liberalization of air transport. Airports may be regarded as constituting an essential facility without access to which competing airlines could not provide their services. They play a role similar to ports in the maritime sector.

Access to airports means, first of all, the right to wait (or park) at, take off from, and land at an airport. Because of the limited capacity of the air traffic system the right of access depends on the availability of slots, in particular, take-off slots and landing slots[9].

Access to airports also involves the right to use the underlying infrastructure to provide ground handling services such as the loading and unloading of luggage and passengers and access to catering facilities. In many airports, foreign airlines may be obliged to use the existing ground handling facilities provided by the airport authority or national airline rather than their own equipment. The airport authorities or the national company appointed by them often have a monopoly over such activities and may not be prepared to accept any competition. The Ground Handling Directive[10] provides that Community airports, to the extent that they have sufficient capacity, should gradually give access to third party airlines, in particular for the purpose of

[7] *Stena Sealink* (n. 1 above). See footnote to paragraph 66 of the Decision.
[8] [1992] 5 CMLR 255.
[9] See Council Regulation 95/93 EEC, OJ 1993 L14/1; and Commission Regulation 1617/93, OJ 1993 L155/18.
[10] Directive 96/67/EC on access to the ground handling market at Community airports, OJ 1996 L272/36.

providing ground handling services for themselves and for other airlines. The Directive nevertheless recognizes that conditions of access need to be be regulated so that security, in particular, is not endangered.

A further issue relating to access to airports relates to the supply of fuel. Access to fuel is obviously essential in order that aircraft may fly. The issue arose in the context of the new fuelling system at Malpensa airport[11]. A number of oil companies had set up a joint venture with the company operating Milan airport to build a new fuelling facility at the airport. There was only room for one such facility. The joint venture included preferential terms for the oil companies party to the joint venture over third party oil companies which had to pay higher tariffs. All oil companies needed to use the new fuelling system in order to provide services to their clients, i.e. the airlines. Third party oil companies were at a clear disadvantage. The Commission held that access to the facility should be granted on similar non-discriminatory terms to all its users whether or not they were members of the joint venture. If profits were made by the joint venture these should be distributed as profits and not as discounts on fuel price to the members alone.

An important feature of the concept of access to an essential facility is that access should be granted on equal terms to all the parties which qualify for access and which request the right to use the facility. The owner and/or operator of the facility may not benefit from preferential terms over third parties.

Airline Booking System

Airline booking is done by computer. It has given rise to what is known as computerized reservation systems or CRS. Computerized reservation systems display all the airline flights offered on a route, schedules, prices and the number of seats available. They are generally set up by a number of airlines but they also display information relating to third party airlines. There was obviously a danger that the airline companies which had set up their own system would seek preferential treatment by, for example, disclosing their flights in preference to those of their competitors.

The Commission has required in various decisions[12] and in a block exemption Regulation[13] that access to a CRS should be open to all airlines seeking it on a non-discriminatory basis. Such equal access means that all airlines seeking it may have their data entered into a CRS on a non-discriminatory basis. CRS have thus been regarded as constituting essential booking facilities without access to which airlines cannot effectively compete with each other.

Airline's ancillary services

Where an airline is dominant on a route, the obligation to grant access to a third party airline may sometimes mean something more than giving it the

---

[11] European Commission Press Release: IP/93/684; 30 July 1993.

[12] See, for example, London European/Sabena, OJ 1988 L317/47; *Amadeus/Sabre* IP (91) 784.

[13] Regulation 83/91, OJ 1991 L10/9.

right to operate on the route by granting it the necessary departure, arrival and flight slots. It may mean granting it additional services such as interlining which is the right for air travellers who have been issued a ticket, valid on one airline, to switch to another airline if they wish to change flights.

In *British Midland* v. *Aer Lingus*[14], the Commission held that Aer Lingus, the dominant airline on the London–Dublin route, could not withdraw interlining facilities from its competitor British Midland, even if this meant that it might lose some traffic to its competitor. The reason for this was that it hindered competition in the market. Without the interlining facility, it would be difficult for the company to continue to compete efficiently in the market. The Commission, therefore, held that the obligation to interline should be maintained for two years until British Midland's position as a competitor could be secured.

Railway infrastructure

In most European Member States, the railway infrastructure – which constitutes a network of rail tracks, stations and signalling systems – is in the hands of one national owner, the State or a state-owned undertaking. Railway infrastructure may also be in the joint ownership of the railway operator. The owner usually has special and/or exclusive rights to provide railway transport services on the tracks.

Access in this context is the right for a third party railway operator to provide freight or passenger services on the tracks.

The Railway Access Directive 91/440[15] provides for minimum rights of access to third party train operators (see chapter 13 below). It is supplemented by Directives[16] which reinforce this right by laying down conditions for the licensing of new railway undertakings as well as conditions for the charging of fees for access to the infrastructure.

As in the case of airlines, access to the track may sometimes mean more than the mere right to use the track subject to the payment of a fee. It may also mean the obligation for the owner and/or operator of the infrastructure to provide ancillary services to third party transport operators such as the provision of a locomotive and train crew. Such additional obligations were imposed by the Commission, in the course of proceedings pursuant to Article 85(3), on the railway transport operators participating in the European Night Passenger Service joint venture because, according to the Commission, access could otherwise be impossible for new operators[17]. This decision is currently subject to appeal before the Court of First Instance[18].

---

[14] *British Midland* v. *Aer Lingus*, Commission Decision 92/213/EEC OJ 1992 L96/34.

[15] Council Directive 91/440/EEC of 29 July 1991 on the development of the Community's railways, OJ 1991 L 237/25.

[16] Directive 95/18/EC on the licensing of railway undertakings, OJ 1995 L143/70; Directive 95/19/EC on the allocation of railway infrastructure capacity and the charging of infrastructure fees, OJ 1995 L143/75.

[17] Commission Decision on Night Services, 21 September 1994, OJ 1994 L259/20.

[18] Case T-374/94.

Telecommunications

Telecommunications are based on physical cable networks and on networks made up of satellites with non-physical links.

In the case of physical networks, access may mean different things. It can be the right to lease an existing line with a view to providing a telecommunications service to one's own customers. It can also mean the right to connect a piece of equipment to the network.

In many Member States, the telecommunications infrastructure has long been in the hands of one national undertaking which not only owns and runs the infrastructure but is also the only provider of telecommunications services based on this infrastructure (including voice telephony and value-added services such as fax services); and the only provider of telecommunications equipment connected to the infrastructure such as telephone handsets.

In *RTT* v. *GB*[19], the Court held that the production and sale of terminal equipment – in particular, phones intended for connection to the public network – is an activity which must be open to any enterprise, in other words it may not be limited to the operator of the public telecommunications service. The Terminal Equipment Directive had already established that State monopolies over terminal equipment were illegal and must be abolished. *RTT* v. *GB* made it clear that the connection of such equipment to the network must not be discriminatory.

The Telecommunication Services Directives (examined in chapter 10 below) set out the scope of the obligation to grant access to the network to third party operators which meet certain minimum criteria.

Postal services

Access is defined in the green paper on postal services[20]. It has two different meanings:

(1) access is often taken to mean the right of consumers to be offered a service; for example, the right of all customers in a territory to use post boxes and have their mail collected and delivered throughout the territory, at an affordable price. Access should be similar for everyone throughout the territory in the sense that the service should be provided on similar terms and irrespective of distance despite the difficulty of delivering mail in certain remote areas. This form of access is common to all major utilities and is part of the concept of 'universal service' (see chapter 6 above, and see later in this chapter);

(2) a second form of access is the right of access for operators who wish to provide a service in connection with the owner of the essential facility

[19] *RTT* v. *GB*, C-18/88, [1991] ECR I-5941.

[20] Green Paper on the development of the Single Market for postal services COM (91) 476 final, pp. 179 and 249.

or in a neighbouring market, or in an upstream or downstream market. This form of access involves, for instance, the right of third parties to consolidate items of mail from large customers such as banks, sort them and transport them to a post office near the point of delivery. Such third parties are often referred to as consolidators. Consolidators normally operate at the collection and sorting stage of postal operations, i.e. before transport and delivery. They will often enter into a contract with the postal administration for use of its services and invoice their customers direct. Consolidators and post offices may thus be competing for certain services to customers. The draft Notice[21] on the application of the competition rules to the postal sector mentions the possibility, but not the obligation, for postal administrations to provide access to their services to consolidators in the reserved sector.

Energy

The two sources of energy examined in this section are gas and electricity. They both involve physical networks: pipes in one case, cables in the other.

There is often only one network for each source of energy in each Member State. Access, in this context, is the right of a third party, for example a distributor, to be connected to the network independently and to distribute gas or electricity to customers. This is known as third party access (TPA). It also involves the right to import energy independently and to distribute it to customers via the existing network or the right to use the existing network to export energy. It may also involve the right of local producers to sell their energy through the network. These issues are the subject of the Directive on electricity (see chapter 9 below).

Another form of access in the energy sector is the right to set up a parallel network, for example the right to build additional pipelines to transport new sources of gas through a country with a view to delivering the gas in that country or in a third country. This has, historically, often been prohibited in national laws in order to protect the monopoly or exclusive national supplier. In addition, it may not always be possible or desirable for objective technical reasons to build a new pipeline in any Member State. The question could be determined, in certain cases, on the basis of Treaty Articles, in particular Articles 86 and 90. It may be argued that the right to build and/or use a parallel network may be permitted where this does not jeopardize the performance of the task of general economic interest of the owner and operator of the existing network.

TV broadcasting

Access to broadcasting may be the right for a foreign producer of TV programmes to have its programmes transmitted in another country. In *ERT*, the

[21] Draft Notice on the application of the competition rules to the postal sector, 3 January 1997, section 5(b)(vii).

Court held that the monopoly held by the Greek undertaking over the broadcasting of its own programmes as well as foreign programmes created a risk that preference would be given to national programmes. Applying the concept of access, this meant that the monopoly might render access by foreign programmers to broadcasting in Greece more difficult.

Access may also be the right for a provider of ancillary television services such as telemarketing to be granted airtime to provide its services. This was the subject of *Telemarketing* examined above.

Computer interface

The issue of access to computers was raised in the *IBM* case[22]. The Commission considered that IBM was abusing its dominant position by refusing to make available to manufacturers of software capable of being linked to IBM hardware (commonly known as IBM compatibles) the detail of its new interface sufficiently promptly so that they could adapt their products to IBM's technology. The Commission argued that by doing this IBM was trying to reduce competition in the software market in which it also had interests. The case was settled following an undertaking given by IBM to provide competing undertakings which request it, details of changes to its interface in good time.

This case shows that private companies which have not been granted a monopoly by law may nevertheless constitute an essential facility, access to which is necessary in order to stimulate downstream competition in new products. This situation is more likely to arise in the field of new technology. This raises the question of the scope of property rights which is examined in the following section.

Copyright

To what extent must a copyright owner give access to its creative work to third parties? The issue is close to that raised, in the field of patents, in the *Volvo Veng* case[23]. It was examined in detail in *Magill*[24]. The three televisions stations RTE, ITV (through ITP) and BBC each produced television listings relating to their weekly programmes and incorporated them in their television programme magazines. RTE, ITP and BBC published their own listings one week in advance but they made them available to newspapers only the day before. This meant that nobody could publish a comprehensive *weekly*

---

[22] Bull. EC 10–1984, p. 96.

[23] Case 238/87 [1988] ECR 6211. The Court held that the proprietor of a registered design for car parts who had ceased to produce spare parts for a model of which there were still many examples in circulation, would be abusing his dominant position if he refused to authorize others to manufacture spare parts based on his design.

[24] *Radio Telefis Eireann (RTE), Independent Television Publications Ltd (ITP) supported by Intellectual Property Owners Inc. (IPO)* v. *the European Commission supported by Magill TV Guide Ltd*, cases C-241/91 P and C-242/91 P, 6 April 1995.

magazine covering the programmes of the three television stations in competition with RTE, ITP and BBC individual weekly magazines.

The Court, confirming the judgment of the Court of First Instance, held that by relying on their intellectual property rights to refuse to provide advance information relating to their programmes, the three television stations prevented access by competitors to the market for weekly television guides; this was to the ultimate detriment of consumers. This refusal to grant access constituted an abuse of a dominant position.

The judgment is controversial because it calls in question the right of creators and inventors to remain sole proprietors of their creation (by the usual law of copyright) and to exploit it commercially as they wish. It shows that the concept of access has extensive application.

Lessons from the case-law and legislation

The analysis of the above cases and Directives shows that the concept of access applies to a variety of situations:

(1) it applies to physical and non-physical networks. Physical networks are based on interconnecting material such as gas pipes, telecom cables or railway tracks; non-physical networks involve, for example, airline routes;

(2) it also applies to essential facilities, notwithstanding the absence of a network, for example, to port and airport facilities, and to certain products or services which constitute an essential technological base for the supply of further products and services.

The concept is of wide application and extends beyond the field of state-related undertakings. However, its application is particularly relevant to these undertakings, notably following the deregulation of previous monopoly sectors.

In competition terms, the concept of access has the following consequence: the general principle – laid down in the *Stena Sealink* case – is that refusal by an undertaking in a dominant position which owns and/or operates an essential facility to grant access to that facility purely with a view to restricting competition constitutes an abuse of that dominant position.

The obligation to grant access to an essential facility carries with it the obligation to grant *fair* access. The consequences of this are as follows:

- the dominant undertaking and the owner or operator of the essential facility should grant third party access on non-discriminatory terms:
  - (i) tariffs should be similar to all third parties placed in a similar situation; discounts should be granted on the basis of objective criteria;
  - (ii) conditions of access, i.e. the type and quality of service rendered and of product supplied, should be similar;

(iii) the dominant undertaking should not grant preferential terms to itself or to any other public or private parties with which it owns or operates the facility. In other words, the obligation of non-discrimination applies not only to third parties but to the dominant undertaking itself and to any companies linked to it;

- in order to give proper access to a facility enabling the third party to compete efficiently in the market of the dominant undertaking, the latter may have to provide additional services. For example, the owner or operator of rail track may have to provide ancillary services to third party providers of railway transport services; an airline may have to provide interlining facilities to a competitor at least for a limited period;
- the obligation to give access to third parties on a fair basis will have repercussions on the structure of the owner or operator of the essential facility: it will often be subject to an obligation to unbundle its various activities, maintain separate accounts for each activity, and generally to operate in a transparent manner. This may include an obligation to disclose its selection criteria for the grant of access rights.

Universal service

This concept has been developed by the Commission in the field of utilities such as postal services, telecommunications services and energy services. The words 'universal service' have not yet been used, as such, by the Court but the underlying features of the concept may be found in *Corbeau* and in *Almelo.*

Universal service is defined by the Commission as an:

> evolutionary concept, developed by the Community institutions, [which] refers to a set of general interest requirements which should be satisfied by operators of telecommunications and postal services, for example, throughout the Community. The object of the resulting obligations is to make sure that everyone has access to certain essential facilities of high quality at prices they can afford[25].

The services are essential in the sense that they are necessary to the average consumer in their everyday life. Examples include the supply of energy, transport services (for example, railway services), the collection and delivery of mail, telecommunications.

The responsibility for *defining* the scope of these essential services has traditionally been assumed by the Member States because of their importance to consumers and because they are either based on the existence of only one network or so-called 'natural' monopoly[26] (as in railway transport)

---

[25] Communication from the Commission on Services of General Interest COM (96) 443 final.

[26] A monopoly is defined as natural where it constitutes the most efficient way of providing goods or services.

or involve scarce resources (as for gas and electricity). In addition, Member States usually regarded it as part of their responsibility to assume the task of *providing* the relevant services or products directly by themselves or through a state-related undertaking.

Responsibility for defining the scope of these services is now partly shared with the Commission. The latter must supervise not only the scope of universal service activities pursuant to Article 90(3) in order to ensure a proper service throughout the EU, but also the organization of the universal service because of the effect that the organization of such activities by Member States may have on competition in the Common Market. If a universal service is entrusted to one undertaking only, the Commission must ensure that this is not contrary to the rules of the Treaty on competition, free movement of goods and freedom to provide services. In addition, the Commission regards it as its duty – as the initiator of legislation – to define the scope of certain universal services in order to secure the interconnection of utilities across frontiers and the growth of international exchanges.

Universal services are defined in terms of the nature of the service or product to be provided across the EU, and in terms of minimum standards and quality.

The precise content and standard of the universal service in each relevant sector is likely to evolve as technological progress is made and consumers' needs change. The standard required for the provision of good quality universal services across Europe will thus be higher today than it was a few years ago, as a result of improved technology and higher consumers' expectations.

The role of the Commission is to define a minimum universal service so that minimum quality and standards are observed throughout the Community whilst Member States will still be free to define more extensive universal services. For example, in the field of telecommunications, France imposed on its national operator the obligation to supply and install free of charge to all consumers asking for it, a hardware known as *Minitel* enabling them to have access to a data base. This compulsory service went beyond the definition of the universal service at EC level. Member States may also, in some cases, reduce the scope of the compulsory services provided by their utilities. In the railway sector, for example, several Member States have reduced the number of railway stations and railway lines which they used to have, for economic reasons, in order to have a more profitable railway network. Such policies may be permitted as long as they do not undermine the creation of a pan-European railway service, as defined at EC level.

The concept of universal service is linked to other key concepts such as basic services, reserved and unreserved services. This is best illustrated by the postal example.

In its draft Directive on postal services[27] the Commission defines the universal postal service[28] as consisting principally of:

- the provision of a good quality postal service for all users at all points on their territory at affordable prices;
- the universal service includes the following minimum facilities both at national and cross-border level: the collection, transport, sorting and distribution of addressed postal items up to two kilograms and postal packages up to ten kilograms;
- services for registered items and insured items.

This definition sets out the scope of services that must be included in the universal service in each Member State so that adequate postal services are offered across the Community. Member States may, however, require that the universal service be more extensive and apply not just to mail items up to two kilograms but, say, to mail items up to three kilograms.

Basic, reserved and unreserved services are connected to the concept of universal service.

Reserved services mean services which are protected from competition. Monopoly activities are reserved activities. They involve those services within the universal service which must be protected from competition in order to be provided under conditions of financial equilibrium.

The reserved area is the area which may be subject to exclusive or special rights and thus taken out of competition. It can extend as far as but no further than is necessary to ensure the provision of a universal service. Reserved services are also sometimes referred to as **basic services** and are distinct from other services, including new services, which must be open to competition.

Reserved services have been defined in the proposed Directive on postal services as including[29]:

(1) the clearance, sorting, transport and delivery of items of correspondence below 350 grams and the price of which is less than five times the public tariff for an item of correspondence in the first weight step, i.e. below 350 grams.

(2) the distribution of incoming cross-border mail and direct mail.

This definition is best understood by means of Figure 8.1.

---

[27] Draft Directive on postal services, 30 January 1997.
[28] Draft Directive on postal services (n. 1 above), Chapter 2, Article 3.
[29] Draft Common Position on a Direct setting out common rules for the development of Community postal services, 30 January 1997, Article 21.

Services

Non-universal ← need *not* be available *to all users* on similar and affordable conditions throughout a Member State	**Unreserved:** open to competition	– Books, catalogues, newspapers, and periodicals above 2 kg – Postal packages above 10 kg – Express mail – Document exchange – New services
Universal: ← good quality postal service for *all* users on similar and affordable conditions throughout a Member State	**Unreserved:** open to competition	– Items of correspondence *above* 350 grams and/or *above* **tariff (a)** – Books, catalogues, newspapers and periodicals *up* to 2 kg – Postal packages up to 10 kg
	Reserved/basic: may be subject to a monopoly	– Items of correspondence up to 350 grams and up to 5 times the public tariff for an item of correspondence in first weight step referred to as **tariff (a)**

Figure 8.1

This diagram shows that:

(1) the *universal* service, which covers essential services must be provided to all customers throughout a Member State; it comprises both *reserved* services sheltered from competition and *unreserved services* open to competition. It follows that not all the services that the Commission

regards as essential to consumers may be reserved; some of these essential services, for example the conveyance of postal packages up to ten kilograms, must be carried out in competition with others;

(2) certain services may temporarily be maintained in the reserved area: all items of correspondence up to 350 grams and up to five times the public tariff for an item of correspondence in the first weight step. This includes, for instance, *direct mail*[30] and *international incoming mail*[31]. The draft Directive provides that they may be reclassified as falling in the unreserved area (whilst still forming part of the universal service) after 2002. The reason for this is that the Commission considers that it may no longer be essential for postal administrations to reserve these services to themselves in order to provide a universal postal service.

(3) *express mail* is a non-universal and unreserved service. This has the following consequences:
   (i) 'non-universal' means that it is not a service which must be offered to all users throughout the territory of the Member States at affordable prices; it may be offered primarily to companies for business purposes and to individuals who occasionally need to convey mail at a significantly increased speed; prices are significantly higher than for ordinary mail;
   (ii) 'unreserved' means that express mail is a service, the provision of which is not exclusively reserved to one operator; it is open to competition. Because the activity is in the competitive sector, the State will not normally be able to use revenues from the reserved sector to cross-subsidize it.

The fact that the Community itself defines certain services as non-reserved significantly reduces the power of decision of Member States. They may only *extend* the scope of universal services; they may *not*, in principle, extend the scope of reserved services.

The consequence of this reduction in the sovereignty of Member States may be appreciated using, for example, the case of Minitel in France. France Telecom decided that the Minitel service should be offered to everyone throughout the territory at affordable prices and that the equipment would be supplied free of charge. These features seem to qualify Minitel as a universal service. France may also have wished to include it in the reserved sector so that it could finance its development from the revenue of other more profitable reserved services. The Commission, on the other hand, might decide that such a service cannot be universal, let alone reserved, throughout the Community because not all Member States can afford to provide such an extended and 'non-essential' service throughout their territory. In such a case France would

---

[30] See the definition provided in Article 2, point 8 of the draft Directive.

[31] This is a form of cross-border mail which involves mail from another Member State or from a third country (see Article 2, point 11 of the draft Directive).

be able to treat the Minitel service as universal throughout *its* territory but it could not reserve it. It follows that, in principle, it could not cross-subsidize its development using the revenue of recognized reserved activities.

The list of services which may be reserved will be limited in Directives. It will not normally be open to Member States to add to the list except in exceptional circumstances, for example, if a Member State had no other means to provide a universal service. The burden of proving that it needs a wider reserved area would, however, lie on the Member State and would be difficult to discharge.

Pending the adoption of a Directive, the question arises as to how a new service, i.e. a service not previously provided by the State or a state-related entity, should be classified and whether it should fall in the non- reserved service.

Using the *Corbeau* approach, this should be so if the service:

- is dissociable from the service of general economic interest;
- meets the special needs of certain economic operators;
- calls for additional services not offered by the traditional postal service;
- does not, by its nature and the conditions in which it was offered, compromise the economic equilibrium of the service of general economic interest offered by the State or a state-related entity.

There are as yet no clear precedents to indicate how the *Corbeau* test would apply. It seems that it might apply in the following hypothetical situation. A private or public postal operator has set up a new postal service involving a faster delivery of bills in the pharmaceutical industry across Europe, based on new computer technology and specific delivery services.

(1) The service could be regarded as dissociable from the service of general interest because it would meet the needs of specific economic operators, namely the pharmaceutical industry. It would involve additional services not previously offered by the traditional postal service consisting of a faster service based on new technology and special delivery services.

(2) The risk of jeopardizing the financial equilibrium of the traditional postal service can probably be avoided, if the following conditions are met:

(i) the service is not merely an improved or enhanced existing service. The fact that new technology and new delivery services are used suggest that the service is more than an improved existing service;

(ii) it does not take away existing business from the traditional activities of post offices, but *adds to existing business* from which postal administrations may also benefit: in other words, it is a new product involving, for example, a new billing system and a new tailor-made delivery system. It thus brings additional revenue to the postal operators in the same way as express mail constituted a new source of postal revenue both to private and public operators;

(iii) the mail is sufficiently distinct in its format and the service sufficiently distinct in its form to avoid the risk that ordinary mail is diverted from its normal method of distribution by public operators and passed off as falling under this new service in order to be transported by private operators;

(iv) the revenues from this new service are not necessary to cross-subsidize activities currently in the reserved area.

If these conditions are met, the new service is likely to fall in the unreserved area and therefore be subject to competition on the basis of Article 90 as interpreted in *Corbeau*.

Universal service and service of general economic interest

The concept of 'general economic interest' has been described in chapter 2. It encompasses but goes beyond that of universal service. Certain services are of 'general economic interest' even though they do not constitute 'universal' services. For example, the management of a port or an important river does not constitute a universal service because it does not apply to all users throughout a territory. It is, by definition, limited in scope, at least geographically. Yet the Commission and the Court have held that they may constitute services of general economic interest (see *Rødby port*[32]; *Muller*[33] (a river constituting a service of general economic interest)).

Public service

This concept is not defined and is hardly mentioned in the Treaty. The reason for this is that it is interpreted differently in many Member States. The Community, therefore, chose to apply the concept of service of general economic interest instead (see chapter 2 above).

The concept of public service, however, is mentioned in Article 77 of the Treaty which deals with aids in the transport sector:

> Aids shall be compatible with this Treaty if they meet the needs of co-ordination of transport or if they represent reimbursement for the discharge of certain obligations inherent in the concept of a *public service*. (*emphasis added*)

The Commission tries to explain the concept in its Communication on Services of General Interest in Europe[34]. It states that it is ambiguous and that it

[32] Commission Decision OJ 1994 L 55/52.
[33] Case C-10/71, [1971] ECR 723.
[34] COM (96) 443 final, 11 September 1996.

is linked to the discharge of certain obligations[35]. The performance of a public service is close to the concepts contained in Article 90(2), namely *the operation of a service of general economic interest* and *the performance of the particular tasks assigned* to state-related undertakings. Article 77 and Article 90(2) provide that the performance of specific 'obligations' or 'tasks' may justify a derogation to the rules on state aids (Article 77) or the rules of the Treaty in general (Article 90(2)).

Member States maintain a degree of discretionary power in relation to their public services including those involving market[36] services. In the *Fédération Française des Sociétés d'Assurances*[37], the Court of First Instance held that in the absence of harmonizing Community legislation, the Commission does not have the power to determine the scope of the public service obligation entrusted to a public undertaking, nor the organization of the undertaking, as long as these do not distort competition in areas subject to competition. It follows from this judgment that Member States maintain a large discretionary power over those parts of their public service sectors which remain subject to exclusive rights and which have not been harmonized at Community level.

Proportionality

This is a well-known principle of EC law and is also found in the national law of most Member States. It is often applied in connection with Member State derogations from a fundamental right granted by the Treaty. Such derogations may be permitted to give effect to a legitimate objective (for example, an overriding national interest compatible with the Common Market) but provided they go *no further than is necessary* to achieve the objective in question.

A legitimate objective may be the fulfilment of a task of general economic interest, as provided in Article 90(2) or the protection of the interests set out in Articles 36[38] or 56[39] or the mandatory requirement[40] concept developed by the Court.

Mediawet illustrates the principle that a restriction on the freedom to provide services must be strictly proportionate to the objective sought and go no further than is necessary to achieve it. The Court held that the safeguarding of cultural pluralism could be a legitimate objective; nevertheless, the

---

[35] See also the definition of 'public service obligation' in Council Directive 2408/92/EEC on Access for Community air carriers to intra-Community air routes, OJ 1992 L240/8, Article 2.

[36] See the distinction between market and *non-market* services in Communication from the Commission on Services of General Interest in Europe (n. 34 above).

[37] *Fédération Française des Sociétés d'Assurances* v. *Commission*, case T-106/95, 27 February 1997, [1997] ECR II-229 (ground 192).

[38] See above, chapter 5, p. 99.

[39] See above, chapter 6, p. 129.

[40] See above, chapter 5, p. 113, and chapter 6, p. 133.

obligation on television or radio programmers to make all or part of their programmes through a Dutch company went *beyond* what was necessary to achieve the objective. The restriction on the freedom to provide services[41] was, therefore, disproportionate to the objective sought.

The proportionality principle is at the core of the Court's case-law on the compatibility of monopolies with Article 90, in particular, *Corbeau* and *Almelo*. It is closely linked to the *necessity test* in Article 90(2).

In *Corbeau*[42], the Court held that (at paragraph 14):

> [Article 90(2)] permits the Member States to confer on undertakings to which they entrust the operation of services of general economic interest, exclusive rights which may hinder the application of the rules of the Treaty on competition *insofar as* restrictions on competition, or even the exclusion of all competition, by other economic operators are *necessary* to ensure the performance of the particular tasks assigned to the undertakings possessed of the exclusive rights. (*emphasis added*)

The Court went on to say (at paragraph 17):

> The question . . . is therefore, the extent to which a restriction on competition or even the exclusion of all competition is *necessary* in order to allow the holder of the exclusive right to perform its task of general interest and in particular to have the benefit of economically acceptable conditions. (*emphasis added*)

The test was further applied in *Almelo*[43] where the Court largely followed the recommendations of its Advocate General. It is worth referring to the exact words of the Advocate General (at paragraph 168) to appreciate how the proportionality principle or necessity test should, in his view, apply:

> Is the existence of a monopoly *indispensable* to ensure compliance with the four essential requirements[44] mentioned above? Can the same objectives not be reached by agreements – or regulation – which would have a less restrictive effect on competition?

He went on to say (at paragraph 169) that it would be for the holder of the exclusive right to demonstrate to the national judge that 'the prohibition on imports and the exclusive purchasing obligation are the *only* means to ensure compliance with the four essential requirements above mentioned' (emphasis added).

These statements show that the test should be strictly applied. Although there remains some uncertainty as to where the burden of proof lies there is

---

[41] For an example relating to goods: see *RTT* v. *GB*, C-18/88, [1991] ECR I-5941, paragraph 33.

[42] *Corbeau*, C-320/91, [1993] ECR I-2533.

[43] *Almelo*, C-393/92, [1994] ECR 1477.

[44] The requirements mentioned by the Advocate General are: (i) uninterrupted supply of electricity to all users; (ii) equality of treatment of users; (iii) absence of distortion of competition between producers; (iv) protection of the environment.

little doubt as to the need to show that the monopoly or exclusive rights are *indispensable* to perform the task of general economic interest.

The next three concepts namely transparency, non-discrimination and unbundling are closely linked together. *Transparency* is the disclosure of relevant data to the Commission and to users or competitors; it constitutes a safeguard against distortions of competition. *Non-discrimination* is one of the positive effects that transparency should secure. The disclosure of data enables users to check that they have not been discriminated against and treated unfairly by comparisons with others. Unbundling is one of the methods of securing transparency.

Transparency

The Transparency Directive was examined by the Court in 1982[45] (see chapter 1, p. 7). The Commission took the view, and the Court agreed, that the obligation imposed on Member States to maintain certain accounting data relating to public funds and to disclose them to the Commission upon demand was necessary to ensure that competition between private and public undertakings was not distorted. It gave the Commission the means to carry out its supervisory duty and check, in particular, whether illegal State aids may have been provided to state-related undertakings.

The Court has also held that accounting transparency is essential to assess whether the tariffs charged for certain services are fair. The issue arose in *Ahmed Saeed*[46]. The Court accepted that Article 90(2) might apply to justify a restriction on competition in the establishment of tariffs by an airline. This could be the case where the airline was obliged to operate on unprofitable routes for reasons of general economic interest. But in order for the Court to assess the extent to which the competition rules may be restricted and the impact of Article 90(2) on the tariff structure it was essential that this tariff structure be transparent[47].

Transparency does not only apply in relation to accounts. It can be applied to most behaviour of monopoly undertakings. It is relevant, for example, to the way in which a port, an airport or a railway authority give access to their infrastructure. Their allocation of paths or slots should be transparent to ensure that competition is not distorted. Transparency should be observed in the use of computerized reservation systems. The Block Exemption Regulation on computerized reservation systems[48] provides that objective selection criteria should be used to determine the order of display of any data relating to competitors.

---

[45] *France, Italy, UK* v. *Commission*, cases 188 to 190/80, 6 July 1980 [1982] ECR 2545.
[46] Case 66/86, 11 April 1989 [1989] ECR 803.
[47] Note 46 above, paragraphs 55 and 56 of the judgment.
[48] Regulation 83/91 OJ 1991 L 10/9.

Non-discrimination

Non-discrimination forms part of the competition rules as well as of the rules on free movement of goods and on freedom to provide services.

It applies to the *conduct* of state-related undertakings. They have a duty not to discriminate against certain customers or suppliers. In *ERT*, the Court held that an undertaking which has a monopoly over the broadcasting of both national and foreign programmes must not discriminate in favour of the former. This line of reasoning has been constant since *Sacchi*. Such discrimination will almost never be compatible with the Treaty.

A breach of the principle of non-discrimination may also arise from the mere *creation of a monopoly* and the obligation on customers to use the services of the monopoly. Even if the obligation applies to all customers, irrespective of their nationality, it constitutes a discrimination against foreign suppliers of services contrary to Article 59 (see *Mediawet*). Such discrimination will only be compatible with the Treaty if it can be proved that it is *necessary* to carry out a task of general economic interest under Article 90(2).

Unbundling

Unbundling means separation. It is a relatively new concept which has been mostly used by the Commission in Directives and Regulations relating to utilities' accounts and activities.

The unbundling or separation of accounts by field of activities has been imposed by the Commission in order to increase the transparency of accounts and to make it easier to check whether the State or state-related enterprises cross-subsidize activities in the competitive sector with revenues from the reserved sector.

The obligation to unbundle accounts may be found, in particular, in the Directives or draft Directives relating to telecoms, postal services, energy and railway transport.

Unbundling is also used in relation to a utility's activities and entails the obligation to divide certain activities between different operators. The obligation to unbundle arises where the combination of powers within one entity would lead to an infringement of the competition rules or of the rules on the free movement of goods or the free provision of services. In *ERT* the Court held that the combination of monopoly powers to broadcast national programmes and retransmit could lead to discrimination against foreign programmes. In *RTT* v. *GB*, the Court also stated that the separation of certain powers was necessary:

> . . . the maintenance of effective competition and the guaranteeing of transparency require that the drawing up of technical specifications, the monitoring of their application and the granting of type-approval must be carried out by a body which is

> *independent* of public or private undertakings offering competing goods or services in the telecommunications sector[49]. (*emphasis added*)

In *RTT* v. *GB*, the Belgian telecommunications company, RTT was not only the manager of the infrastructure but it also set the norms for connection to the infrastructure and checked the conformity of equipment with the norms whilst, at the same time, competing with private operators in the sale of equipment. The Court held that the combination of such functions in an entity which was also a competitor on the market constituted an infringement of Article 86[50].

This analysis by the Court has had two consequences for unbundling. It has paved the way for the unbundling of:

(1) supervisory and operational activities;
(2) management of the infrastructure and the provision of services which make use of the infrastructure.

The railway Directive 91/440 requires Member States to separate the activities relating to the operation of railway services from the activities relating to the management of the infrastructure. Somewhat surprisingly, however, it does not require that they should be in the hands of totally separate legal entities[51].

The draft Directive on postal services requires Member States to designate an independent supervisory authority[52].

Liberalization/deregulation

These words are often used loosely and interchangeably to express the need to introduce more competition. They are often used in a political sense.

Liberalization means the introduction of more competition in an area which used to be primarily under the control of the State. To liberalize means to free a sector of activity from the tight control or ownership of the State and open it up to third parties. Liberalization may also mean privatization, but this is not always so: a sector of activity may be liberalized whilst maintaining the operation of a public enterprise provided the latter's activities no longer extends to the entire economic sector concerned.

Deregulation, literally speaking, means a lessening of regulations and, by extension, increased freedom resulting from the loosening of the State's grip and, hence, increased competition. This application of the word may lead to confusion. It may give the impression that the introduction of more

---

49 *RTT* v. *GB* (n. 19 above), paragraph 26.
50 See chapter 4 above.
51 Article 6.1 of Directive 91/440.
52 Draft Common Position on a Directive setting out common rules for the development of Community postal services, 30 January 1997, Article 21.

competition is linked to a lessening of regulations. In practice the reverse is true for many utilities. The introduction of more competition in sectors previously operated by monopoly undertakings often results in more regulation dealing with pricing; quality of service etc. It will also often lead to the appointment of a Regulator. This has been the case in the UK with the appointment of, for example, Regulators such as OFTEL (in the telecom sector), OFGAS (in the gas sector) and OFRAIL (in the rail sector). The Commission also proposes that Member States should appoint independent Regulators in previously state-controlled sectors.

Inter-operability

This concept refers to the link between Member State networks and the need for these networks to inter-connect. This is particularly true of all the networks based on sophisticated technologies such as the railway network and, within it, the high-speed train system. The Commission has issued an inter-operability Directive dealing specifically with the issue of the connection between railway networks. Inter-operability is also relevant to the other utility sectors such as energy; telecoms[53] and postal services.

The above concepts are further examined in Part II in respect of energy, telecoms, broadcasting, postal services and transport.

[53] See *Terminal Equipment* judgment (case C-202/88, [1991] ECR 1223), paragraph 37.

Part II

REGULATORY MEASURES COMPLEMENTING ARTICLE 90 IN THE UTILITY SECTOR

Chapter 9

COMPETITION AND EXCLUSIVE RIGHTS IN THE EUROPEAN UNION ENERGY MARKET

Introduction

In the energy sector, the Member States have traditionally pursued different and even contradictory policies, largely confined to their own territories. Energy market operators are diverse, ranging from multinationals to small and medium-sized companies, and consisting of national, public and private companies, each operator holding different rights, facilities and obligations. This diversity has led to partitioned markets with barriers to the free movement of energy products.

The concept of an internal market developed in the late 1980s with the objectives of removing the anti-competitive effects of the statutory monopolies which perpetuated market segregation, and of opening up the sector to competition, whilst safeguarding security of supply and universal service provision.

This chapter considers the three phase liberalization process undertaken by the Commission to open up the sector to competition. The process involved, as a first phase, the adoption of two Council Directives on the transit of gas and electricity and a Council Directive on price transparency. These Directives were in force by January 1992. The second phase was designed to establish common rules for the sector and to introduce an element of third party access for new market entrants. The third and final phase was intended to build up the second phase to liberalize the markets further.

In pursuance of the second phase, the Commission, in January 1992, published two proposals for Council Directives to open up the gas and electricity markets. These proposals established common principles for a licensing system, introduced the concept of unbundling of management and accounts for the different types of operations, namely production, transmission and distribution, and a limited application of mandatory network access to third party operators at reasonable rates and subject to available capacity.

Major opposition from Member States, mainly on third party access, resulted in five years of difficult negotiations. The gas proposal was put to one side and the electricity proposal was used as the working text in the negotiations. Agreement on the principles to be applied to liberalization of the electricity market was finally achieved at the end of 1996. The gas proposal is currently under revision to conform with the principles established for the electricity sector.

The most significant difference in the final text is the replacement of mandatory third party access by a new concept, a single buyer system – a model under which a single entity would purchase electricity from producers and sell to consumers, with the Member States opting for this system designating the eligible entity as single buyer in their territory. This system will coexist with a system of negotiated third party access, under which producers would sell directly to eligible customers by means of negotiated network access. Following adoption and implementation of the Directive giving effect to the new regime, electricity markets (and subsequently gas) will be opened up to new market operators over a nine-year period.

In the absence of agreement by Member States to liberalize the energy sector, the rules of the Treaty, State aid rules and Articles 30, 37, 59 and 90 have been applied by the Commission and the Court to control the anti-competitive effects of monopolies in the energy sector. The chapter also examines the enforcement of competition law under Article 90 and related Treaty articles in significant Commission Decisions and Court rulings in a number of key issues.

Setting the scene – creation of an internal energy market

In the context of creating the single European market for 1993, the Commission focused on the resources necessary for the creation of an internal energy market. Energy objectives for the Community adopted by the Council in 1986[1] referred to the need for 'greater integration, free from barriers to trade, of the internal energy market with the view to improving security of supply, reducing costs and improving economic competitiveness'.

In 1988[2] the Commission published a Working Document on the internal energy market which identified obstacles that existed in specific energy markets and suggested priorities to be addressed to achieve an internal market in energy. The Document showed the energy sector to operate within a framework of close regulation, which was often restrictive of competition, and subject to significant trade barriers.

---

[1] OJ 1986 L123/1.

[2] The Internal Energy Market (Commission Working Document), COM(88) 238 final, 2 May 1988.

The energy sector is characterized by national or regional monopolies, particularly in gas, electricity and petroleum. Analysis by the Commission contained in its Working Document on the Internal Energy Market underlined certain features common to most energy markets. These included the following:

- *Existence of exclusive rights.* In the *electricity sector* exclusive rights existed concerning the use of distribution and interconnecting transmission systems, as well as on operation of high voltage interconnection systems. In the *gas sector* the transmission and distribution companies tended to have exclusive operating concessions granted by the public authorities at national, regional or local level. In the *oil sector* exclusive rights on refining were found to exist in several Member States.
- *Imposition of obligations by the Member States.* Specific obligations were imposed by Member States. Such obligations were found to be particularly prevalent in the *oil sector.* For example, Italy and the United Kingdom imposed obligations on the landing of petrol products produced offshore at a port within the jurisdiction before transportation to other destinations. France and Spain obliged oil refiners to accept crude oil acquired by the State. France, Spain and Portugal required the use of national flag shipping for carriage of crude oil by sea; while Germany imposed obligations to use a national carrier for the inland transport of oil products.
- *Lack of price transparency.* The problem of price transparency was identified, particularly in the gas and electricity sectors. In the *gas sector* there was a particular problem concerning non-transparency with regard to off-tariff sales to industrial consumers in certain Member States. In the *electricity sector* lack of transparency was identified concerning electricity prices to large consumers, prices for electricity transfer between systems, fuel prices for electricity production, and other electricity production costs.
- *Import/export controls.* Obstacles to the free movement of energy products were caused by government controls on imports and exports. In the *gas sector* controls on imports and exports by governments and undertakings holding a monopoly or dominant position impeded the free movement of gas. In the *oil sector* several Member States imposed quantitative restrictions on the importation of Community oil products. In the *electricity sector,* infringement proceedings were taken by the Commission against *Ijsselcentrale*[3] in respect of an agreement to restrict imports of electricity which were held to be contrary to the competition rules under Article 85 of the Treaty. The Commission has also taken infringement proceedings

[3] Commission Decision OJ 1991 L28/32.

to the Court against several Member States for maintaining exclusive rights to import and export gas and electricity in breach of Articles 30 and 37 of the Treaty[4].

- *Prices and tariffs.* Lack of harmonization was found to exist on price and tariff structure.

The presence of statutory or *de facto* monopolies was identified by the Commission in its internal energy market document as giving rise to segregation of the energy market. In the *gas sector* monopolies or undertakings holding a dominant position could block the free movement of gas. In the *electricity sector* the interconnection systems are owned by monopolies and any arrangements for the transfer of electricity are carried out commercially by the electricity companies on a co-operative basis, which are, in effect, concessionary monopolies.

While the special features of the energy market in terms of its public service obligations and public security aspects are recognized, subject to two basic constraints, i.e. the requirement for security of supply and the public service obligation of universal, uninterrupted provision, the energy sector should, according to the Commission, conform to Community competition rules.

The Commission's objectives, set within these constraints, aimed to give consumers a choice of supply; to establish prices based on competition; to enable suppliers to deliver their production by means of third party networks; to limit State subsidies and make them more transparent; and to remove barriers to trade between the Member States.

Deregulation of the energy market was envisaged by the Commission to be a three phase project. Phase One involved the adoption of two Council Directives for gas and electricity transit through transmission grids (Directives 91/296/EEC[5] and 90/547/EEC[6]) and a Directive on a Community procedure to improve the transparency of gas and electricity prices charged to end users (Directive 90/377/EEC[7]). These three Directives entered into force between July 1991 and January 1992. Phase Two would involve the establishment of common rules for the internal market in electricity and gas through legislation designed to permit new entrants market access through a limited degree of third party access. Phase Three would further liberalize the energy market based on the experience gained during Phase Two.

---

[4] Case C-156/94 *Commission* v. *Ireland*; case C-157/94 *Commission* v. *Netherlands*; case C-158/94 *Commission* v. *Italy*; case C-159/94 *Commission* v. *France*; case C-160/94 *Commission* v. *Spain*.

[5] Council Directive 91/296/EEC on the transit of natural gas through grids, OJ 1991 L147/37.

[6] Council Directive 90/547/EEC on the transit of electricity through transmission grids, OJ 1990 L313/30.

[7] Council Directive 90/377/EEC concerning a Community procedure to improve the transparency of gas and electricity prices charged to industrial end users, OJ 1990 L185/16.

Regulation of the energy sector in the European Union: Phase One

Natural Gas Transit Directive

This Directive (Directive 91/296/EEC[8]) entered into force on 1 January 1992. It applies only to the entities and high pressure transmission grids listed in the Annex to the Directive. It imposes on Member States the requirement to take measures necessary to ensure the transit of gas between transmission grids in accordance with the provisions laid down in the Directive (Article 1).

The Directive is concerned with transactions involving:

- transmission by the entity/ies responsible in each Member State for high pressure natural gas grids, with the exception of distribution grids;
- where the grid of origin or final destination is situated in the Community; and
- where the transport involves the crossing of at least one intra-Community frontier (Article 2).

The Directive stipulates that gas transit contracts between grids should be negotiated between the entities responsible for those grids for the quality of service provided, and, where appropriate, with the entities responsible for the import and export of natural gas (Article 3(1)). Transit conditions must reflect EC provisions. They should (i) not discriminate between parties; (ii) not include unfair contract clauses or unjustified restrictions; (iii) not jeopardize security of supply or quality of service; and (iv) take full account of the utilization of reserve production and storage capacity and the most efficient operation of existing systems (Article 3(2)).

Member States are obliged to take the necessary measures to ensure that any entity named in the Annex to the Directive under their jurisdiction must act without delay to:

- notify the Commission and the national authorities concerned of any request for transit;
- open negotiations on the conditions of a natural gas transit and inform the Commission and national authorities concerned of the conclusion of the transit contract;
- inform the Commission and national authorities concerned of the reason for the failure of the negotiations to result in the conclusion of a contract within twelve months following communication of the request (Article 3(3)).

[8] See footnote 5, above.

The Directive provides for Commission intervention in two specific situations. First, each of the entities concerned may request that the transit conditions be subject to conciliation by a body established and chaired by the Commission upon which the entities responsible for goods in the Community are represented (Article 3(4)) (names of members have been published in the *Official Journal of the European Communities*); and second, if the parties are unable to reach agreement, and the reasons appear 'unjustified' or 'insufficient', the Commission, either acting on a complaint from the requesting body, or on its own initiative, should implement procedures provided for by Community law (Article 4).

Electricity Transit Directive

The Directive (Directive 90/547/EEC[9]) entered into force on 1 July 1991. It contains provisions identical to the Gas Transit Directive referred to above.

Transparency Directive

This Directive (Directive 90/377/EEC[10]) entered into force on 1 July 1991. It established a Community procedure designed to improve the transparency of gas and electricity prices charged to end users. The Transparency Directive imposes an obligation on Member States to ensure that undertakings which supply gas and electricity to industrial end users communicate to the Statistical Office of the European Communities (SOEC) details of the prices and terms of the sale of gas and electricity to industrial end users; the price systems in use; and the breakdown of consumers and corresponding volumes of category of consumption (Article 1).

A confidentiality clause in Article 4 prohibits disclosure by SOEC of confidential commercial data supplied to it. Confidential statistical information is accessible only to SOEC officials to be used for statistical purposes. However, publication of the data in aggregate form is permitted provided it does not enable individual commercial transactions to be identified. Annexes to the Directive set out the specific provisions for gas and electricity.

Regulation of the energy sector in the European Union: Phase Two

Electricity and Gas proposals

In January 1992 the Commission issued two proposals which went further than the Phase One proposals by establishing common rules for the internal

[9] See footnote 6, above.
[10] See footnote 7, above.

market in electricity and gas[11] through the opening up of these markets to competition by permitting market access to new entrants by means of a limited degree of third party access.

Principles inherent in the new proposals

The Commission's approach was based on principles identified in the explanatory memorandum to the proposals:

- the approach should be *gradual* to enable the electricity and gas industries to adapt to the new regime;
- in accordance with the principle of *subsidiarity* (i.e. the taking of decisions at the lowest appropriate level), a general framework should be established at Community level with detailed implementation left to the Member States;
- *excessive regulation should be avoided.* Whilst the electricity and gas sectors are characterized by monopolies and should therefore be substantially regulated, especially in the form of price controls, the Community must not regulate beyond what is strictly necessary to achieve liberalization and should leave scope for commercial negotiation between companies.

Key elements of proposals

Three elements comprising the second stage of liberalization were identified in order to implement the provisions outlined above. There should be:

(1) *A transparent, common, non-discriminatory system for the granting of licences* for electricity production and the building of electricity lines and gas pipelines.

(2) Separation of management and accounts for production, transmission and distribution operations, i.e. the concept of *unbundling*.

(3) A limited introduction of *third party access* (TPA) which would impose an obligation on transmission and distribution companies to permit access to their networks at reasonable rates subject to available capacity.

Legal base

The Commission designated as the legal base for these draft directives Articles 57(2) (on access to activities as self-employed persons), 66 (right of establishment in the services sector) and 100a of the Treaty (approximation of Member States' provisions relating to the establishment and functioning

---

[11] Proposal for a Council Directive concerning common rules for the internal market in electricity, OJ 1992 C65/4. Proposal for a Council Directive concerning common rules for the internal market in natural gas, OJ 1992 C65/14.

of the internal market). The Directives fall within the co-operation procedure of decision making, under which the European Parliament must be consulted and the Council takes the final legislative decision in adopting the legislation.

The Commission initially considered using Article 90 which would have allowed it to impose Directives or Decisions on the Member States in respect of public undertakings and undertakings which have been granted special or exclusive rights. However, while the use of Article 90 would have been expedient, it was too politically controversial.

Response to the Commission's proposals

Certain Member States and key interests in the energy industry raised fundamental objections to the introduction of mandatory third party access which they considered would be detrimental to the proper functioning of a public service in the energy sector within the framework of the market economy.

Eurogas (the European committee of the natural gas industry) argued initially that TPA would affect security of supplies; endanger environmental protection; lead to increased consumer gas prices; and generally undermine the operation of the market system, replacing it with increased regulation and a reduction in efficiency as a result of unbundling of operations in the sector.

Eurelectric (the European committee of the electricity supply industry) expressed similar concerns about the introduction of TPA[12]. It considered that 'it is without doubt the task of national governments to choose the organization of the electricity system, subject to the interests of the Community which best meet the public service obligations as they consider necessary'.

Member States such as France, Germany, Spain and the Netherlands have been reluctant to change the status quo. Only the United Kingdom and Portugal have been fully supportive of legislation in favour of full energy liberalization.

The European Parliament criticized the Commission proposals. A European Parliament report, the Desama Report[13], analysed the issues and concluded that the Commission's proposals were vague and unacceptable since they did not, *inter alia,* take a realistic view of the operation of national markets, which would be dominated for a long time to come, if not by monopolies, then by companies that lead the market in generation, distribution, and, most importantly, transmission.

Parliament's opinion, delivered in November 1993, proposed several amendments which were taken on board by the Commission.

---

[12] Statement of the Continental Members of Eurelectric concerning the Internal Electricity Market. *Eurelectric,* 23 September 1991. (The UK, with its pro-liberalization views, was not a party to this statement.)

[13] European Parliament Doc EN/PR/225/225592 A3–0000/93, 14 April 1993.

In addition, the Council submitted six principles to be fulfilled in opening up the gas and electricity sectors to competition. These included ensuring security of supply; environmental protection; protection of small consumers; transparency and non-discrimination; and recognition of differences between national systems. The Commission also took these principles into account when drafting the amended proposals.

Modified electricity and gas proposals: main amendments

Amended proposals, which took into account the opinion of the European Parliament and the views of the Member States, were eventually published in May 1994[14]. The key amendments to the proposals were in the following areas:

Public service obligations

In both the electricity and gas proposals, the Commission accepted Parliament's amendments and emphasized the concept of public service obligations.

Third party access (TPA)

Mandatory TPA was replaced by *negotiated* TPA. The Commission's amended proposals provide as follows:

> Third party access, which was regulated in the Commission's initial proposal, is a right open to autoproducers for the supply of their own establishments and branches and to producers and transmitters for the supply of large industrial consumers and distribution companies. *The conditions for exercising this right have to be negotiated both with the transmission system operator and with the distribution system operator*, except where the supply of electricity arises out of a tendering procedure for the construction of production capacity; in this case access to the network is guaranteed. (*emphasis added*)

The Commission did not fully follow Parliament's recommendations on the conditions for TPA to networks. Parliament's preference was for a long-term contract with large consumers; the Commission rejected this as an unjustifiable restriction on two grounds. First, that it would always be possible to refuse access to third parties where the performance of the service of general economic interest assigned to the system operator is compromised; and, secondly, that distribution companies should be able to avail themselves of the same possibilities as large consumers to enable small and medium-sized consumers to benefit indirectly from the opening up of the market.

---

[14] Amended proposal for a European Parliament and Council Directive concerning common rules for the internal market in electricity, OJ 1994 C123/1; amended proposal for a European Parliament and Council Directive concerning common rules for the internal market in gas, OJ 1994 C123/26 and COM(93) 643 final – COD 384, COD 385 1993.

Unbundling

The amended proposals removed the earlier provisions for the unbundling of management, replacing it by a requirement for the independence of the system operator to be guaranteed at an administrative level. The unbundling of accounts remained, which, together with accompanying provisions, permitted access by the authorities to internal company accounts.

Tendering system

The Commission acceded to Parliament's recommendation for the introduction of a tendering system for authorising all new production and transport capacities. However, a licensing system would be applied to all autoproducers and independent producers. The amended proposals also simplified the rules for operation of transmission and supply networks.

Legal base unchanged

The legal base for the amended Directive remained as Articles 57(2), 67 and 100a of the Treaty.

Application

The principles discussed in the draft Electricity Directive were also applied to the Natural Gas Directive; only the specificities differed.

Amended proposals deadlocked: introduction of new compromises

The Energy Council of May 1994[15] brought a new dimension to the liberalization debate. The majority of Member States were unable to reach agreement on key issues of the Commission's amended electricity proposal, to the extent that they had introduced almost 200 comments and reservations on the text.

'Single buyer' system to parallel third party access

Third party access was particularly difficult. The French proposed a new alternative concept to negotiated third party access, namely a single buyer or sole purchaser model, under which a single buyer would buy and sell electricity, manage the network, guarantee security of supply and plan energy strategy. The sole buyer could be the existing state monopoly. The German Presidency conclusions of the May 1994 Energy Council invited the Committee of Permanent Representatives to analyse whether this concept could co-exist with TPA in a way that would permit the comparable opening of the electricity market of the Member States and whether such co-existence would be compatible with the Treaty.

Despite the opposition of the Competition Commissioner, Mr Van Miert, the Commission nevertheless thought that third party access and the single buyer system might be able to co-exist, provided that certain conditions were

[15] Council of the European Union Press Release 6889/94, 25 May 1994.

met. The role of the single buyer would have to be clearly defined, including its public service obligations; it would have to be genuinely independent of commercial or economic interests in the sector; an appeals procedure against decisions of the single buyer would have to be available; there must be no import or export monopoly and the single buyer should act only as an intermediary between consumers and producers.

The May 1994 Energy Council coincided with the *Almelo* judgment[16] which had been given one month earlier. *Almelo* had established that an exclusive purchasing obligation on a distributor to buy electricity from a regional electricity company could be justified provided that such restriction on competition was necessary to fulfil a mission of general economic interest. Definitions of service of general economic interest and public service obligations were therefore important issues for the Council to determine.

Agreement reached on key liberalization principles

The driving force behind the single buyer concept has been France, or, more accurately, the French electricity monopoly, Electricité de France (EDF), which has been instrumental in influencing the French government to embrace the single buyer concept.

In March 1995 the Commission adopted a working paper on the organization of the internal market for electricity[17]. It concluded that negotiated third party access *could* co-exist with the single buyer system provided that precautions were put in place to ensure objectivity, transparency and non-discrimination.

These conditions were laid down under six headings:

(1) *Consumer choice for eligible consumers.* Under the single buyer system, eligible customers should be able to contract electricity supplies with external producers under the same conditions as those made with national independent electricity producers.

(2) *Import and export of electricity.* Compatibility between both systems could be achieved if the import regime under the single buyer system is governed by an obligation to buy unlimited quantities of imported electricity under established conditions. There should also be transparency of tariffs for use of the transmission system and for prices to be paid by the single buyer for imported electricity. An additional condition should be that electricity imports are only subject to objective and justified conditions; examples of such conditions being the lack of interconnected capacity or reasons of public security.

---

[16] Case C-393/92 *Commune D'Almelo* v. *NV Energiebedrijf Ijsselmij* [1994] ECR I-1477.

[17] European Commission Press Release: IP/95/276, 22 March 1995. Commission Adopts Paper on Electricity Market Organization of the Internal Market for Electricity. Working Paper on the organization of the Internal Energy Market of 22 March 1995, SEC(95) 464, 1995; Working Paper on small and very small electricity systems in the internal market of 10 May 1995, SEC(95) 685, 1995.

(3) *Transparency and distortion of competition.* Unbundling of production and supply operations by the single buyer are necessary to guarantee objectivity, transparency and non-discrimination, to avoid distortions of competition and to ensure neutral and independent treatment.

(4) *Competition at production level.* Tendering procedures for new and additional production capacities, which are more restrictive in competition terms than authorization systems, should only be organized and decided by public authorities or other independent bodies appointed for this purpose.

(5) *Parallel authorizations of independent producers.* Independent producers should, even under tendering systems, benefit from parallel authorization to strengthen competitiveness. A transparent definition for independent producers in single buyer systems should be introduced on the basis of quantitative capacity thresholds. Furthermore, other producers, such as autoproducers, should be able to benefit from parallel authorizations.

(6) *Direct lines.* In the single buyer system, all eligible consumers should have the freedom to construct and use direct lines to external or national producers, and vice versa for producers to supply eligible customers within the framework of Article 7 of the Draft Amended Electricity Directive.

The Commission working paper explains what it understands by negotiated third party access and the single buyer system.

> [Third Party Access] The negotiated TPA approach forms a system in which electricity producers can sell supplies directly to eligible consumers by means of negotiating access to the network. Negotiations with the network operator will deal with transport tariffs and conditions, and will be subject to a dispute settlement mechanism. Consumers can shop around inside and outside the system for competitive electricity supplies, while the network operator will ensure respect of system security.
>
> [Single Buyer System] The single buyer model, as originally put forward, is a system in which only a single entity will buy and sell electricity. All producers will sell to the SB on a competitive basis; all consumers will buy from the SB against optimized prices. The SB will manage the network, undertake long-term planning and optimization of investments, and will ensure respect of services of general economic interest. Direct contract negotiations are only foreseen, for electricity imports, managed via the Single Buyer.

On the basis of the Commission's report, the Energy Council of 1 June 1995[18] drafted a set of conclusions specifying the elements of agreement among the Member States, and, more particularly, listing the points for

---

[18] EU Council of Ministers Conclusions. Energy Council, 7565/95, 1 June 1995.

further negotiation where agreement was absent, with a view to achieving a common position by the end of 1995.

Underlying the points of agreement was the acceptance that the single buyer system and negotiated third party access *can* co-exist subject to conditions which ensure the equivalence and reciprocity of both systems. These were as follows:

- the single buyer must purchase the electricity in objective conditions which guarantee price transparency and total non-discrimination;
- in the area covered by the single buyer system, a system of parallel authorization procedures should be introduced for independent producers;
- within the single buyer system, eligible consumers should be able to negotiate foreign supply contracts;
- transparent conditions should be defined in the two systems for transport and distribution in order to avoid discrimination;
- appropriate mechanisms should be introduced to regulate conflicts between the two systems in order to avoid abuse of dominant positions;
- in the single buyer system, producers not bound by contract to the single buyer should be able to export their electricity via the single buyer network provided that the single buyer network has sufficient transport capacities and that it is technically feasible.

A Directive was finally adopted in December 1996.

Main provisions of the Electricity Directive

Briefly, the Directive will initially open up national markets at 22 per cent, to be calculated on the basis of the Community share of electricity consumed by final consumers using more than forty gigawatt hours (GWh) per year. The degree of opening will be progressively extended over the six-year period following implementation of the directive, to reach 32 to 33 per cent of opening up of national markets. This increase is to be calculated by reducing the Community consumption threshold of forty GWh to a level of twenty GWh of annual electricity consumption after three years and to a level of 9 GWh of annual electricity consumption after six years.

The Directive provides for a Commission review of its application and a report on its operation to enable the Council to assess the possibility of a further opening of the market after nine years of the date of entry into force of the Directive. The report would examine in particular the co-existence of negotiated third party access and the single buyer system. The Directive stresses that the operation of the two systems must lead to equivalent economic results and that public service obligations imposed by Member States must be clearly defined, transparent, non-discriminatory and verifiable and must comply with the competition rules and with Article 90.

The objective of the Directive is to establish common rules for the production, transmission and distribution of electricity (Article 1). Member States, taking into account the principle of subsidiarity, should ensure that their electricity undertakings operate in accordance with the principles of the Directive to achieve a competitive market in electricity. They should prohibit discrimination between undertakings, and ensure that the two systems of negotiated third party access and the single buyer system 'lead to equivalent economic results and hence to a directly comparable level of opening up of markets and to a directly comparable degree of access to electricity markets' (Article 3).

Public service obligations

The concept of public service obligation, as laid down in Article 3, remains an important component of the Directive. The Member States, in accordance with Treaty provisions, particularly Article 90(2) of the Treaty, may, in the general economic interest, impose public service obligations which may relate to security, including security of supply, regularity, quality and price of supplies and environmental protection. However these obligations 'must be clearly defined, transparent, non-discriminatory and verifiable'. They must be published, including any revisions, and notified by the Member States to the Commission without delay (Article 3(2)). Importantly, the Article expressly provides that long-term planning by Member States might be one way of carrying out public service obligations.

Member States are, thus, allowed for reasons of *general economic interest*, not to apply key provisions of the Directive relating, *inter alia*, to authorization criteria, tender procedure requirements, negotiated access to the system or the single buyer procedure. The Directive, however, clearly says that the application of the public service derogation should not be contrary to the 'interest of the Community'. In doing so, the Directive lays particular emphasis on the last phrase of Article 90(2) of the Treaty which also refers to the interest of the Community. The Directive explains that it includes, in particular, the need to ensure competition in respect of eligible customers. The derogation is thus permitted but only subject to compliance with Article 90, including notably the last phrase of Article 90(2) and the other provisions of the Directive.

Electricity production

For the construction of new generating capacity, Member States are allowed to choose between an authorization procedure and/or a tendering procedure. Both procedures are subject to objective, transparent and non-discriminatory criteria.

The criteria to be used under the authorization procedure may relate to safety and security of the electricity system, installations and associated equipment; protection of the environment; land use and siting; use of public ground; energy efficiency; the nature of the primary services; characteristics specific to the applicant, e.g. technical, economic and financial capacity. Reasons for refusal to grant an authorization should be 'well founded and substantive', objective and non-discriminatory, and should be given to the applicant and sent to the Commission for information purposes. An appeals procedure should be made available.

The Directive specifies the scope of the tendering procedure. Member States or competent bodies designated by them, should compile an inventory of new means of production, including replacement capacity, on the basis of a 'regular estimate'. The inventory should take account of the need for interconnection of systems. The 'regular estimate' of the generating and transmission capacity which is likely to be connected to the system; the need for interconnectors with other systems; potential transmission capacity and the demand for electricity, must be compiled and published at least every two years. There is also a requirement to publish in the *Official Journal of the European Communities* at least six months in advance of the closing date for tenders.

The tender document must contain a description of the procedure to be followed by all tenderers and an exhaustive list of the criteria governing the selection of tenderers and the award of the contract. The tender procedure must be organized, monitored and controlled by a public authority or by an independent entity appointed for the purpose. Autoproducers (legal or natural persons generating electricity essentially for their own use) and independent producers (producers who do not carry out electricity transmission or distribution functions in the territory covered by the system where they are established (Article 2(4)(a)), are also to be considered within this framework.

Transmission

Member States must designate, or require companies which own transmission systems to designate a system operator to operate, ensure the maintenance and, if necessary, the development of the transmission system in a given area and its interconnectors with other systems.

Member States are responsible for ensuring that the technical aspects of the system guarantee the interoperability of systems, on an objective and non-discriminatory basis. The system operator is responsible for managing energy flows on the system, taking into account exchanges with other interconnected systems, and for ensuring a secure, reliable and efficient system including the availability of all ancillary services. The system operator must provide the operator of any other system interconnected with its system, sufficient operation, co-ordinated development and interoperability of the interconnected system. Article 7(5) explicitly prohibits the system operator from discriminating between system users or classes of users, particularly in favour of its subsidiaries or shareholders. The system operator must also be independent, at least in management terms, from other activities not relating to the transmission system.

Distribution

Member States may impose on distribution companies an obligation to supply customers located in a given area. In such cases, the tariff for such supplies may be regulated, for example, to ensure equal treatment of customers. Member States are required to designate, or require undertakings which own or are responsible for distribution systems to designate, a system operator with responsibility for operating, maintaining and, if necessary, developing the distribution system in a given area and its interconnectors with other systems. The responsibilities of the distribution system operator include the maintenance of a secure, reliable and efficient distribution system; non-discrimination between users or classes of users of the system; and where a Member State requires it, the giving of priority to generating installations using renewable energy, waste or combined heat and power.

Unbundling and transparency of accounts

Arrangements must be made by the Member States to ensure that the accounts of undertakings fulfil the requirements of the Directive. Electricity undertakings, irrespective of their system of ownership or legal form, must compile, audit and publish annual accounts in compliance with their national legislation implementing Directive 78/660/EEC on the annual accounts of limited liability companies[19]. Companies not legally obliged to publish accounts must, nevertheless, make them available to the public in their head office (Article 14(2)).

Integrated electricity undertakings (i.e. vertically or horizontally integrated undertakings) must keep separate accounts for their generation, transmission and distribution activities, and, where appropriate, consolidated accounts for other non-electricity activities, as if the activities were carried out by separate companies. This is designed to prevent cross-subsidization and a profit and loss account must be included for each activity. Member States, which designate as a single buyer a vertically integrated electricity undertaking, or part thereof, must impose provisions requiring the single buyer to operate separately from the generation and distribution activities of the integrated undertaking. Member States must ensure there is no information flow between the single buyer activities or vertically integrated electricity undertakings and their generation and distribution activities, excepting information necessary to carry out the single buyer responsibilities.

---

[19] OJ L222/11; 14 August 1978 as amended.

Access to the system

Article 16 permits the Member States to choose between negotiated access to the system and/or the single buyer procedure. Both procedures must operate according to objective, transparent and non-discriminatory criteria.

Negotiated or regulated access

Article 17 requires Member States opting for *negotiated access* to take the necessary measures to enable electricity producers and where Member States authorize their existence, suppliers of electricity, as well as eligible customers, to negotiate access to the system so as to conclude supply contracts with each other on the basis of voluntary commercial agreements (Article 17(1)). Article 17(2) states that 'where an eligible customer is connected to the distribution system, access to the system must be the subject of negotiation with the relevant distribution system operator and, if necessary, with the transmission system operator concerned'.

In the interest of transparency and to facilitate negotiations for access, system operators must publish an *indicative range of prices* for use of the transmission and distribution systems in the year following the Directive's implementation; and indicative prices published for subsequent years should be based on the average price agreed in negotiations during the previous twelve-month period.

Member States may opt for a *regulated system of access*, on the basis of *published tariffs* for use of transmission and distribution systems.

Access to the transmission or distribution system may be refused by the system operator where there is a lack of the necessary capacity. However, in such circumstances, 'duly substantiated reasons' must be given for the refusal, having particular regard to the need to safeguard competition, non-discrimination and public service obligations.

Single buyer procedure

Member States should designate a legal person to be the single buyer within the territory covered by the system operator (Article 18(1)). They must take measures to ensure (i) that a non-discriminatory tariff for the use of the transmission and distribution system is published; (ii) that eligible customers are free to conclude supply contracts with national and foreign producers, and in certain cases with distributors established outside their territory; and (iii) that independent producers can negotiate access to the system with the transmission and distribution systems operators so as to conclude supply contracts with eligible customers outside the system, on the basis of a voluntary commercial agreement.

The single buyer may be *obliged* to purchase the electricity contracted by an eligible customer from a producer inside or outside the territory covered by the system at a price equal to the sale price offered by the single buyer to eligible customers less the price of the published tariff for access.

Article 18(3) provides that where the obligation to purchase is not imposed on a single buyer, Member States should take the necessary measures to ensure that the supply contracts with eligible customers and, in certain cases, distributors are either implemented through access to the system on the basis of the published tariff mentioned or through negotiated access to the system. In the latter case, the single buyer would not be obliged to publish a non-discriminatory tariff for use of the transmission and distribution system.

Article 18(4) permits the single buyer to refuse access to the system and to refuse to purchase electricity from eligible customers where the single buyer lacks the necessary transmission or distribution capacity. In such circumstances, 'duly substantiated reasons' must be given for the refusal, having particular regard to Article 3 on competition, non-discrimination and public service obligations.

Other provisions

Article 19, dealing with the modalities of the opening up of the market, was the subject of difficult negotiations in the lead up to the agreed text. It requires Member States to take the necessary measures to ensure the opening of their electricity markets in order to enable contracts based on negotiated access and the single buyer procedure to be concluded 'at least up to a significant level' and notified annually to the Commission.

The national market share is to be calculated on the basis of the Community share of electricity consumed by final consumers consuming more than forty GWh per year. Calculation of the average Community share is to be conducted on the basis of information regularly provided by Member States to the Commission. The Commission is to publish this average Community share defining the degree of market opening in the *Official Journal of the European Communities* by November each year together with all information appropriate to clarifying the calculation (Article 19(1)).

Member States must specify customers within their territory having the legal capacity to contract for electricity, in accordance with the Directive's provisions, through negotiated access or the simple buyer procedure, given that all customers using more than one hundred GWh per year must be included in the above category (Article 19(3)). Distribution companies, if not already specified as eligible customers, would have the legal capacity to contract under provisions for negotiated access or the simple buyer procedure for the volume of electricity being consumed by their customers designated as eligible within their distribution system, to supply those customers (Article 19(3)).

Member States are required to publish annually (by 30 January each year) the criteria for the definition of eligible customers able to conclude contracts under provisions on negotiated access or the single buyer system.

Article 19(5) contains provisions designed to avoid imbalance in the opening of electricity markets during the nine-year period following the date of entry into force of the Directive. Under these provisions, contracts for supply of electricity through the negotiated access or single buyer procedures with an eligible customer in the system of another Member State should not be prohibited if the customer is considered as eligible in both systems involved (Article 19(5)(a)); and, in cases where transactions are refused because the customer is eligible only in one of the two systems, the Commission may impose an obligation (taking into account the market situation and the Common interest) upon the refusing party to execute the requested electricity supply at the request of the Member State where the eligible customer is located (Article 19(5)(b)).

Other Articles deal with the need to ensure that independent producers and autoproducers are able to negotiate access to the system; the appointment of an independent competent party to deal with dispute resolution over contracts, negotiations and refusals of access or purchase; the regulation of dispute settlement on a cross-border basis.

The Directive contains provisions to deal with sudden crises on the market requiring safeguard measures. In such cases, safeguard measures must be notified to the Commission by the Member States without delay.

Finally, the Directive already foresees that harmonization measures are likely to be necessary. It provides that the Commission must report on the need for such measures before the end of the first year of entry into force of the Directive. Member States must implement the Directive by 19 February 1999.

The co-existence of the two systems and the concept of the single buyer itself has been the subject of criticism. The single buyer system has been condemned[20] as being discriminatory in favour of national suppliers in the case of the single buyer, contrary to the objectives of the single market, and that the system implies a suspension of the Treaty rules on free movement and free competition. The electricity Directive (and, in due course, the gas Directive) will inevitably be subject to substantial scrutiny in its implementation and operation, not only by the Commission, which has a mandatory duty to monitor and report, but from key interests in the sector as a whole. The impartiality of the role of former national monopolies will undoubtedly be closely examined, not only by the Commission, but by electricity operators throughout the EU, and it is conceivable that the body of cases before the Commission and the Court on the application of competition enforcement under Article 90 in the energy sector might increase as future complaints on the application of the Directive are made and investigated.

[20] Study conducted by the National Economic Research Associates (NERA) on behalf of British electricity companies. Quoted in Agence Europe, 3 March 1995; European Report, 25 February 1995.

Application of competition enforcement under Article 90 and other treaty Articles to the energy sector

Use of Treaty articles to enforce competition

The inability of the Member States to reach agreement in opening up the energy market to competition means that control and removal of monopolies that are contrary to the creation of an internal energy market have had to be considered largely under the Treaty rules. The competition rules, state aid rules and Articles 30, 37, 59 and 90 are the key instruments which can be used by the Commission and the Court to break up energy monopolies. For example, the formal proceedings referred to the Court by the Commission against several Member States holding exclusive import and export rights for electricity and gas, have as their legal base Articles 30 to 37 and 90 of the Treaty[21].

Until *Almelo* there was no case-law which dealt directly with the compatibility of energy monopolies with EC law. There had been, nonetheless, important rulings of the European Court of Justice in several cases concerning the anti-competitive effects of monopoly restrictions. The Court had confirmed the Commission's powers under Article 90(3) in the *Terminal Equipment* case[22] and the *Höfner* case[23]. In *Terminal Equipment* the Court confirmed the Commission's powers to use Article 90(3) to ensure compliance with the provisions of Articles 90 and 30. *Höfner* further extended the possibility to challenge monopolies by ruling that a state monopoly which is manifestly unable to satisfy market demand and which excludes private companies from offering similar competitive services is abusing its dominant position. *Terminal Equipment* clarified the Commission's power to challenge Member States' monopolies in respect of goods. In *Telecom Services*[24], the Court confirmed the Commission's power to challenge monopolies over services as well. It was not entirely clear whether the provision of electricity and gas would fall under the rules relating to goods or services but the issue was finally resolved in the *Almelo* and *Rendo*[25] cases where the Court firmly held that electricity should be treated as goods. It would appear logical to extend this case-law to gas.

At the time of writing, the only judgment of the Court which deals directly with the compatibility of energy monopolies with EC law was given in April

---

[21] Cases C-156 to 160/94, *Commission* v. *Ireland; the Netherlands; Italy; France; Spain* (judgment pending).

[22] Case C-202/88 [1991] ECR 1223.

[23] Case C-41/90 [1991] ECR 1979.

[24] Joined cases C-271/90, C-281/90 and C-289/90, *Spain, Belgium and Italy* v. *Commission* [1992] ECR I-5833.

[25] Case C-393/92 (unreported), OJ 1991 L28/32.

1995 in an important case, *Commune D'Almelo* v. *Energiebedrijf Ijsselmij*. In addition to the *Almelo* case, there is an existing body of case-law of the Court and the Commission based on Articles 85 and 86 which deal with a wide range of competition issues such as exclusive distribution agreements and long-term contracts which are examined below.

Exclusive distributorship arrangements

Ijsselcentrale decision

The key case in this area concerns *Ijsselcentrale*[26], a Commission Decision of January 1991 which was appealed before the Court of First Instance in the *Rendo* case, but without success. The parties have further appealed to the Court of Justice (Case C-19/93 – *Rendo II*).

The *Ijsselcentrale* decision concerns the Dutch electricity distribution system. It is based on a complaint by local electricity distributors, including Rendo and Almelo, against an exclusive purchase obligation imposed on local distributors by their regional distributor, Ijsselcentrale. The case also raised the question of the compatibility with EC law of an equalization charge imposed on the local distributors.

The Commission made a distinction between public and non-public supply. It held that as far as non-public supply was concerned the exclusive purchasing obligation was tantamount to an import ban. The Commission found that distributors were also banned from exporting electricity. It held that both the import and export bans could not be justified by Article 90(2). The Commission also held that the ban on export in the field of public supply could not be justified by Article 90(2). It, however, refused to examine the compatibility of the import ban with Treaty rules in the field of public supply because this ban was consistent with a Dutch electricity law and that the compatibility of the law with Treaty rules should be examined in priority. Finally, it held that the 'equalization' charge, a small extra charge imposed by Ijsselcentrale to eliminate the difference between the cost of distribution to large and small consumers, did not have an appreciable effect on trade between Member States. This issue became the subject of the *Commune D'Almelo*[27] case, where the Court was asked to examine whether the import ban, which is linked to the equalization charge, is compatible with Treaty rules.

Commune D'Almelo judgment

The Court gave its judgment in the *Commune D'Almelo* case in April 1994. The Advocate General's opinion was given in April 1994. Generally, the judgment closely follows the Advocate General's opinion although there are some differences on points of detail.

---

[26] Case T-16/91 [1992] ECR 2417.

[27] Case C-393/92, *Commune d'Almelo* v. *NV Energiebedrijf Ijsselmij* [1994] ECR I-1477.

The starting point of this case is the equalization charged imposed by Ijsselcentrale (now Ijsselmij). The plaintiff argued that if there had not been an import ban he would not have been subject to the equalization charge. The Dutch appeal court agreed that this was at least a possibility and decided to ask the Court whether the import ban was compatible with Articles 37, 85, 86 and 90 of the Treaty.

On Article 37, the Court followed the Advocate General and held that it did not apply because the exclusive purchasing obligation and the ban on import were not imposed by law but resulted from an agreement between enterprises.

On Article 85, the Court held that the conditions imposed on local distributors by Ijsselmij were standard conditions set by the association of distributors of electricity in the Netherlands. The cumulative effect of these standard conditions had to be taken into account. They resulted in the partitioning of the market contrary to Article 85.

On Article 86, the Court held that Ijsselmij alone could not be said to hold a dominant position in a substantial part of the common market. Ijsselmij, however, could be in a collective dominant position with other regional distributors. It was for the national judge to determine whether such collective dominance existed. This should be determined by reference to the following:

(1) Does Ijsselmij *belong* to a group of enterprises which collectively hold a dominant position on the market?
(2) Are the enterprises so linked that they will adopt a similar line of conduct in the market?

The Court seems to place the emphasis on the second condition.

The Court held that it is for the national judge to determine whether ijsselmij is in a collective dominant position with the other regional distributions in the Netherlands. If so, then the restriction on imports constitutes an abuse contrary to Article 86.

The key point is whether the infringement of Articles 85 and 86 can be exempted by Article 90(2).

The Court referred to the *Corbeau* case[28] which provides that enterprises which are entrusted with tasks of general economic interest may be exempt from the application of the competition rules in so far as restriction of competition and even the elimination of competition is necessary to enable them to carry out their task. The Court held that the obligation of an enterprise such as an energy distributor involves:

- the uninterrupted supply of electricity throughout the territory;
- supply to all consumers (local distributors and/or users);
- supply in the quantity demanded at all times; and

[28] Case C-320/91, *Paul Corbean* [1993] ECR I-2533.

- supply under uniform tariffs which can only vary in accordance with objective criteria applicable to all clients.

The Court held that restrictions of competition are permitted if they are necessary to enable an enterprise entrusted with a task of general public interest to carry out its task. In determining this question, account should be taken of the economic situation under which the enterprise operates, including the cost it must bear and the regulations, in particular in the field of environment, which it must observe.

In summary, the test laid down by the Court is thus close to that suggested by its Advocate General, although less precise. For example, the Court did not even refer to the last paragraph in Article 90(2) which provides that the development of trade must not be affected to such an extent as would be contrary to the interest of the Community.

The Court leaves it open to Member States or to enterprises to which they grant exclusive rights, to prove that such rights are necessary to enable them to carry out the specific task with which they have been entrusted in the general public interest.

The national judge will have the final say. Faced with expert evidence from a monopoly supplier or a beneficiary of exclusive rights, it may be difficult for the national judge to rule against the maintenance of the monopoly or exclusive rights. The Court recognizes that conditions may differ between Member States depending on the costs borne by the beneficiary of exclusive rights and the specific regulations which it must observe.

Altogether, following *Almelo*, it seems that pending further harmonization of legislation in the energy sector, it may be possible, in many cases, for Member States or beneficiaries of exclusive rights to prove before their national judges that restrictions of competition are necessary to provide electricity supply in the general public interest.

Further guidance on the compatibility with Treaty rules of exclusive rights, and of import and export bans in the field of electricity and gas should be given by the Court in the important case currently pending before the Court and involving four Member States[29].

Rendo judgment[30]

The complainants in the *Ijsselcentrale* case were concerned that the Commission had only dealt with part of the complaint. They therefore appealed the decision before the Court of First Instance on the ground that the Commission had failed to rule on whether the ban on import and export of electricity in the field of public supply was contrary to the competition rules. The Court, however, rejected the appeal.

---

[29] Cases C-156 to 160/94.

[30] *Rendo NV and others* v. *Commission*. Case T-16/91, 18 November 1992 confirmed to a large extent on appeal in *Rendo NV* v. *Commission*. Case C-19/93 P, 19 October 1995.

The Court found that the Commission was entitled to consider that the examination for compatibility with EC rules of the Dutch law, which contained an import ban, took priority over the examination of the import ban in the Ijsselcentrale agreement. It recognized that, had the Commission done otherwise, it would have prejudged the validity of the law with EC rules. It concluded that the Commission had properly exercised its discretion.

Long-term supply contracts

There are two important cases on long-term supply: the Commission Decision in *Jahrhundertvertrag*[31] and the Court's judgment in the *Campus Oil*[32] case.

Jahrhundertvertrag decision

The Commission decision in *Jahrhundertvertrag* was made in December 1992. It concerned an agreement in Germany between coal mining companies, electricity utility generators and electricity autogenerators. It was a long-term agreement covering the years 1980 to 1995 for a large quantity of coal (40.9 million tons of coal equivalent), and imposed an exclusive purchasing obligation.

Following the intervention of the Commission, the agreement was notified for negative clearance or exemption. The Commission eventually exempted the agreement following some small changes.

There are three main lessons from the Commission's decision:

(1) if energy generators enter into a long-term exclusive supply agreement for primary energy sources they should notify it to the Commission;

(2) the Commission may exempt the agreement if the parties show that it is limited to what is necessary to safeguard basic electricity supplies. It is likely that, if the Article 90(2) conditions are satisfied, the tests under Article 85(3) will also be satisfied; the Commission applied the same concept of proportionality (i.e. the minimum necessary to safeguard basic supplies);

(3) it may be open to the parties to discuss and argue with the Commission what the minimum level of supply should be.

Campus Oil judgment

Campus Oil and other traders in petroleum products established in Ireland started an action against the Irish Minister for Industry and Energy in relation to an order made by him which required all importers of petroleum oils to take approximately 35 per cent of their requirements from the Irish Whitegate refinery owned by the State.

---

[31] OJ 1993 L50/14.

[32] Case C-72/83 [1984] ECR 2727.

Campus Oil argued that the order constituted a restriction to the free movement of goods in breach of Article 30 of the Treaty and was not saved by the exemption in Article 36.

The Court's judgment is based on examination of Article 30, with a reference to Article 90 and a thorough analysis of Article 36. The Court held that the obligation to buy a proportion of requirements from a national supplier limits the possibility to import. It protects the national market to a certain extent from competition from other Member States. This is a measure having an equivalent effect to a quantitative restriction on import contrary to Article 30.

The Court stated that Article 90(1) read together with Article 90(2) did not allow a Member State to take measures restricting imports in favour of an undertaking carrying out a task of general economic interest.

It is uncertain, in the light of the latest case-law of the Court on Article 90 and in particular, the *Corbeau* case, whether the Court would now take the same view on the application of Article 90. It is arguable that there may be cases where a restriction falling under Article 30 could be exempted not only by Article 36 but also by Article 90(2).

The Court held that an order to buy specific quantities from the national refinery could be allowed on grounds of public policy and public security provided that the set quantities were strictly proportional to what was necessary to ensure public security.

Conditions of contract: price, discriminatory pricing

Grangevielle[33]

This case concerns an independent producer of electricity in France, a hydroelectric company, Grangevielle. This producer, situated in the Alps, was cut off from the French grid owner, EDF, by a large mountain. It was therefore linked to the Italian grid. EDF however sought to rely on its nation-wide monopoly to force Grangevielle to sell its electricity to it upon its usual prices; EDF would then sell on to ENEL, the Italian grid owner. If Grangevielle, which was physically linked to the Italian grid, had been able to contract directly with ENEL, it would have been paid twice as much.

Grangevielle complained to the Commission. The Commission held that EDF could not exercise its exclusive rights over Grangevielle since it had no access to its production facility. It followed that the terms imposed by EDF on Grangevielle, by reason of its monopoly, were not justified. The case was settled: EDF agreed to apply terms similar to those applied by ENEL to independent Italian producers, thus taking into account the specific circumstances of Grangevielle.

---

[33] *EDF* v. *ENEL*. Commission Press Release IP/92/6680, 14 August 1992.

The lesson of this case is that a utility may need to look at the specific circumstances of producers and customers. If a producer/customer is situated near a border and for some reason the owner of the grid cannot have access to it or provide it with a service similar to that provided by others, the owner of the grid may not be able to impose its monopoly position on that producer/customer.

Transit restrictions: EDF transit to Portugal[34]

The French grid owner, EDF, wished to export electricity to Portugal via Spain. The Spanish electricity company agreed but only on condition that it should itself supply a significant part of the electricity contracted for. The Commission considered that this amounted to an abusive refusal by Spain to transmit electricity. The case was eventually settled. It is thought that this case served as a model for the Transit Directive.

The case shows that if transit facility is refused, it may be possible to rely on Article 86. It may be possible to show that the refusal to allow transit is abusive or that conditions imposed in return for transit are abusive. This case could be helpful to utilities or other parties who would wish to rely on the concept of access to obtain transit rights from a grid owner.

Other cases on energy

Apart from the important cases examined above, the Commission has issued several Article 19(3) notices reporting on various projects in the field of energy. These notices provide further useful insight into the Commission's approach to agreements in that sector. The Commission has thus given its views on various joint ventures[35] involving several electricity generators and on projects for privatizations[36].

Further Commission Decisions or Notices may be expected in the years to come following implementation of the electricity Directive, even in advance of implementation. They are likely to cover access problems, pricing issues or joint venture agreements between existing or ex-monopoly enterprises and private operators. The outcome of the electricity and gas cases[37] currently pending before the Court should also provide important guidance to the Commission and all interested parties on the application of the competition rules and Article 90(2) to energy.

---

[34] This case was settled and did not come before the court for a ruling. Dr Kurt Ritter, formerly of the Commission, confirmed this to be the case in a paper to the Fordham Conference, 1993, 'EC Antitrust Law and Energy'.
[35] Independent power generators, OJ 1992 C92/4; Twinning program engineering, OJ 1992 C148/8; Electricidad de Portugal (the *Pego* case) OJ 1993 C265/3; International Energy Agency, OJ 1993 C300/8; Interconnector between UK and Eire, UK and Belgium; Britannia field.
[36] Electricity in England and Wales, OJ 1990 C191/9; Electricity in Scotland, OJ 1990 C191/9; Electricity in Northern Ireland, OJ 1992 C92/5.
[37] Cases C-156 to 60/94.

Conclusion

It is clear that there is an irreversible trend towards the gradual opening up of the Community energy sector. This is due to the initiatives taken by the Member States themselves, in particular, the UK, the Netherlands, Portugal and certain of the Scandinavian countries, as well as to pressure by the Commission. A key point to bear in mind is that, irrespective of the legislative progress by way of EC Directives or Regulations, it is open to companies to rely on articles of the Treaty if they feel that the conduct adopted by certain energy enterprises, whether or not as a result of a national law, are contrary to the rules of the Treaty. Thus, even if legislative harmonization is not forthcoming, companies and individuals may be able to force the pace of change by relying directly on fundamental principles of the Treaty and bringing actions before the Commission or their national courts on issues such as third party access (there have been many recent cases on the principle of access to an essential facility); state aids/transparency; and non-discrimination in terms and conditions offered to customers.

In other utilities, notably the railways, DG IV, the Competition Directorate responsible for enforcing the competition rules, has been prepared to take decisions going beyond the scope of existing or proposed legislation based on a strict interpretation of Treaty articles.

It follows that developments in the case-law of the Court and the Commission in the field of energy and other sectors benefiting from exclusive rights should be monitored simultaneously with the development of proposed legislation designed to deregulate the energy market. There may be new opportunities for all economic players to operate outside their Member States of origin or to obtain services from these Member States.

Chapter 10

COMPETITION AND EXCLUSIVE RIGHTS IN THE TELECOMMUNICATIONS SECTOR

Introduction

Telecommunications is one of the few areas where the Commission has addressed Article 90 Directives to Member States. This may be due to the fact that telecoms is one of the sectors which have been subject to the greatest technological changes over the past years and that it soon became absolutely clear that Member States' monopolistic rights in that sector constituted a serious obstacle to competition with the rest of the world. Urgent action became necessary and Article 90 was seen as an efficient tool to introduce rapid liberalization.

The Commission has used a three-pronged approach involving liberalization, harmonization and the application of the competition rules:

- *liberalization* consisted of the lifting of restrictions through several Article 90 Directives which led to the opening up of the telecoms market;
- *harmonization* was introduced to complement Member States' liberalization and ensure a coherent EU wide telecoms sector. It took the form of *Article 100a Directives*;
- *the application of competition rules* was reinforced through the issue of *guidelines* and *notices*.

This chapter first sets out the background to the considerable legislative changes which have occurred in the telecoms sector and then deals with these three main policies. It explains that liberalization has occurred by stages: first equipment and services, thus creating reserved and non-reserved areas in the short term. Particular sectors have then been liberalized, for example satellite and mobile. Use of alternative infrastructure and full competition of the sector are in preparation for the final stage of liberalization (see the selective list of significant EU documents in the telecommunications sector at the end of this chapter).

The key harmonization measures are then examined and consideration is given to the interaction of the liberalization and harmonization measures.

Finally, the application of EU competition provisions, as set out in the Commission's general guidelines and the draft Notice on access agreements, is briefly reviewed.

There is cause for optimism that an open telecoms market is well under way although implementation is an issue which will have to be addressed. Liberalization of the telecoms sector in the EU is part of a worldwide action which is in progress through the liberalization talks in the context of the World Trade Organization (WTO). This should help to secure progress towards global liberalization.

Background to Commission measures

Case-law and legislation

Community law and policy have evolved in this sector through legislation and case-law, from the key first recognition in the *British Telecom*[1] case that telecommunications organizations are economic operators with a business activity and therefore fall within the scope of the competition rules, to the landmark telecommunications services and equipment judgments which curtailed the scope of special and exclusive rights held by telecommunications organizations.

Importance of the sector

The importance of the telecommunications sector is recognized in a series of policy documents and statements issued by the Commission (see the selective list at the end of this chapter). Telecommunications policy has developed through Green Papers and consultations since 1984.

The wider implications of telecommunications were recognized in the White Paper on growth, competitiveness and employment (Delors White Paper[2]). This was followed by the Bangemann Group Report on Europe and the Information Society and Commission Communication on Europe's way to the Information Society[3], and by amendments to the Treaty through the introduction of trans-European networks, including telecoms, into Article 129b[4].

---

[1] Case 41/83 *Italian Republic* v. *Commission of the European Communities* [1985] ECR 873.

[2] COM (93) 700, 1993.

[3] COM (94) 347, 1994.

[4] Article 129b states, in the context of the internal market, that 'the Community shall contribute to the establishment and development of trans-European networks in the areas of transport, telecommunications and energy infrastructures', and that 'within the framework of a system of open and competitive markets, action by the Community shall aim at promoting the interconnection and interoperability of national networks as well as access to such networks. It shall take account in particular of the need to link island, landlocked and peripheral regions with the central regions of the Community'.

The early 1980s saw significant technical developments in the sector, and a recognition by the Community of the importance of telecoms to the economy. There was also an increasing awareness that liberalization would be beneficial to both industry and consumers and to the economies of Member States.

The preparatory work: the various Green Papers

The main principles deriving from consultation and policy documents particularly in the form of green papers were gradually formulated during the 1980s. A Telecommunications Action Programme based on a Commission Report[5] was adopted by the Council of Ministers in December 1984. Its objectives included the establishment of a Common Market for telecommunications equipment through a co-ordinated Community strategy for standardization, mutual recognition of type approval for terminals, and the opening up of access to public telecommunications contracts and the development of advanced telecommunications services and networks. Measures giving effect to these objectives were adopted by the Council during the following three years. These included a range of Council recommendations, Decisions, Directives and resolutions on harmonization at EU level on opening access to public telecommunications contracts; mutual recognition of type approval for terminal equipment; and the co-ordinated introduction of integrated services digital network (ISDN). Next came the publication of the Commission Green Paper on the development of a Common Market for Telecommunications Services and Equipment ('the Green Paper'[6]) in June 1987, which was adopted by the Council in a resolution one year later.

Green Paper on a Common Market for telecommunications services and equipment

Objectives

The Green Paper stressed the need for Community policy to promote an advanced and competitive European telecommunications infrastructure with a Community-wide market for services and equipment.

The Green Paper underpinned the Commission's perception that the inherent difficulty in the sector was that the telecommunications monopoly had been bound up with the idea of national policy and public service, and that new regulations had to be developed in each Member State to provide consistency throughout the Community. It pointed out that all Member

[5] Commission Communication: Progress Report on the thinking and work done in the field of telecommunications and initial proposals for an Action Programme.

[6] Communication from the Commission: Towards a Dynamic Economy. Green Paper on the Development of the Common Market for Telecommunications Services and Equipment, COM(87) 290, 1987.

States had granted exclusive rights for the provision of their network infrastructure to their telecommunications administrations.

Principles for a Community telecommunications policy

The Green Paper proposed that a Community telecommunications policy should establish rules to give effect to the following principles with a view to complete liberalization of the telecommunications market by 1998:

(1) phased but complete opening of the terminal market to competition;

(2) gradual opening of the services market to competition, with the exception of voice telephony. This would require telecommunications administrations to interconnect their own network infrastructures in order to maximize efficiency and to set criteria for access by competitors to the networks;

(3) separation of the regulatory and operational functions of the telecommunications administrations. Thus the telecommunications administrations could no longer be 'both regulator and market participant, i.e. referee and player'. Regulatory functions should, in particular, concern licensing, control of type approval and mandatory specifications, frequency allocation and surveillance of usage conditions.

The Green Paper stressed the need to redefine the future interface between telecommunications administrations and competitors in key areas:

(1) *universal service provision*. This was identified as the only area where telecommunications administrations might retain some exclusive or special rights. Value-added services and other services should be opened up to competition;

(2) *leased lines and resale of capacity*. These could involve 'a legitimate level of protection of financial viability' but should not involve misuse of a dominant position;

(3) *tariff principles*. The Commission accepted a growing recognition that telecommunications tariffs should follow cost trends and the inevitability of a certain rebalancing of tariffs. It took the pragmatic view that a 'fair trade-off between cost orientation and the aim of universal service on reasonably the same terms for all' would have to be developed;

(4) *adjacent infrastructures/services*. Exclusive provision would have to be narrowly defined.

During the 1980s Commission policy on telecoms, as elaborated in the Green Paper and other statements, was paralleled by two significant rulings of the Court, in which the Court clarified, first, that the EC competition rules applied to telecoms (the *British Telecom* case referred to in the text to footnote 1 above) and, second, that monopoly activities by national telecommunications administrations could not automatically be extended beyond the monopoly to include additional value added services (the RTT[7] case). These important judgments are referred to below.

EC competition rules apply: British Telecom case

In a key European Court of Justice ruling in the *British Telecom* case, the Court confirmed, *inter alia*, that the competition rules of the EC Treaty are applicable to telecommunications administrations.

The case arose from a complaint lodged with the Commission by a private UK message forwarding agency against the UK telecommunications authorities concerning restrictions imposed by British Telecom, pursuant to UK legislation, on the transit of telexes between Europe and the US and on telex or telefax message forwarding services through computers.

A Commission Decision found these restrictions to be in breach of Article 86 on abuse of a dominant position. The Italian Government brought an action to the Court challenging the Commission Decision.

The subsequent judgment, which dismissed the Italian Government's challenge, is regarded by the Commission to be the 'cornerstone' of Community case-law in the telecommunications field. The judgment lays the foundation for future jurisprudence of the Court concerning liberalization of the telecommunications market. The judgment is significant for several reasons:

(1) the competition rules of the Treaty were held to apply to telecommunications administrations;
(2) the Court confirmed that, on the facts, British Telecom monopoly power was limited to certain sectors, pursuant to English law, and did not extend to all ancillary services;
(3) an undertaking holding a statutory monopoly on the management of telecommunications networks infringes Article 86 by prohibiting the activities of certain private operators unless there is evidence that such operators are abusing the public networks;
(4) the Court upheld the right of a private operator to provide value-added services via the public telecommunications network as long as it was not proven that this right jeopardized the network.

The *British Telecom* judgment thus provided clarification that national telecommunications administrations are subject to the competition rules in operating their monopoly activities and that such activities are circumscribed and should not automatically extend to value-added services.

Extension of a monopoly: RTT case

The Court deliberated further on the issue of objective justification for extension of a monopoly, in a case involving the Régie de Télégraphe et de Téléphone (RTT) in Belgium.

[7] Case C-18/88 *Régie de Télégraphie et de Téléphone (RTT)* v. *GB-INNO-BM SA*. [1991] ECR I-5941.

Under Belgian law, the RTT held a monopoly to establish and operate the public telecommunications network. In addition, only equipment supplied or authorized by the RTT could be connected to the network. The RTT had the power, therefore, to grant or to withhold authorization for connection of equipment to the network; to lay down the technical standards such equipment must meet; and to determine whether equipment not manufactured by it conformed with the technical specifications it laid down.

The Court recognized that the RTT monopoly constituted a service of general economic interest within the meaning of Article 90(2). It pointed out, however, that 'the exclusion or the restriction of competition on the market in telephone equipment cannot be regarded as justified by a task of a public service of general economic interest within the meaning of Article 90(2) of the Treaty'[8]. It reasoned that any undertaking should be free to manufacture and sell terminals, particularly telephones, subject only to their meeting essential requirements relating, in particular, to safety to users and to network operators and protection of public telecommunications networks against harm of any kind. It was sufficient, however, to ensure such compliance by laying down specifications to be met by the equipment and to establish a procedure for type approval to check that those specifications are met.

The RTT argued that Article 90(1) could be infringed only if the Member States have condoned an abuse that the RTT itself had in fact committed, for example, by applying type approval provisions in a discriminatory manner. The Court rejected this argument. It said that Article 86 prohibited the *extension of the monopoly* to establish and operate the telephone network to the market in telephone equipment, *in the absence of any objective justification*. Where the extension results from a measure adopted by a State, it is in breach of Article 90(1) in conjunction with Article 86. The Court also put forward an additional criterion, namely, equality of opportunity between operators. It stated that 'a system of undistorted competition, as laid down in the Treaty, can be guaranteed only if *equality of opportunity is secured as between the various economic operators*' (emphasis added). Accordingly, the Court rejected the extension of the RTT monopoly in setting and checking standards for telephone equipment when it is a competitor with operators on the market for that equipment.

Satellite Green Paper

In 1990 the Commission issued a Green Paper on Satellite Communications[9] which proposed regulatory changes to the satellite sector in line with general

[8] Case C-18/88 at paragraph 22, page 5980.

[9] Commission Green Paper on a Common Approach in the Field of Satellite Communications in the European Community, COM(90) 490, 1990.

Community policy on the liberalization of the telecommunications sector. These gave rise to the Satellite Directive[10] examined later in the chapter.

Mobile Green Paper[11]

Following a detailed analysis of the mobile telephony sector, the Commission issued a Green Paper which paved the way for the Directive on mobile telecommunications examined later in the chapter.

Infrastructure Green Paper

Published during 1994 as two documents, the Green Paper on the liberalization of telecommunications infrastructure and cable television networks, comprises Part I, which establishes the principle and timetable, and Part II, which examines a common approach to the provision of telecoms infrastructure in the EU.

Providers of telecommunications services which are open to competition should, in accordance with the Green Paper, be given a free choice of the underlying infrastructure to deliver these services. However, this should be subject to safeguards which guarantee, *inter alia*, in a competitive environment, universal service provision, licensing, interconnection and interoperability.

The Green Paper (Part II) addressed these issues:

Universal service

(a) Elements comprising universal service

The general principles were established in Council Resolution 94/C48/01 as being 'access to a defined minimum service of specified quality to all users', provided at an affordable price, on the basis of the principles of universality, equality and continuity. The specific components for open network provision (ONP) have been considered in the ONP legislation[12]. The main elements for voice telephony are specified in the Green Paper (Part II) to include, in particular, the provision of the basic telephone service, guaranteed quality of service and a tariff policy that would involve tariff flexibility for 'socially desirable purposes'.

(b) Costing of universal service

The Green Paper underpinned a two-step approach. The first step is an assessment of the uneconomic cost of universal service obligations on the basis of common principles; the second step is to establish practical financing mechanisms for uneconomic universal service provisions. In the Commis-

---

[10] Council Resolution of 19 December 1991 on the development of the common market for satellite communications services and equipment, OJ 1992 C8/1.

[11] Commission Green Paper on a common approach to mobile and personal communications in the European Union. COM(94) 145, 1994.

[12] Directives 90/387/EEC, OJ 1990 L192/1; 92/44/EEC, OJ 1992 L165/27; Council Recommendations 92/382/EEC, OJ 1992 L200/1, and 92/383/EEC, OJ 1992 L200/10.

sion's view, national regulatory authorities should support tariff rebalancing and set up targeted schemes for 'needy users' and 'uneconomic customers', rather than providing general subsidies for access.

The Green Paper recommends that 'only telecommunications-related public service obligations should be treated as universal service obligations in voice telephony'. It suggests that the national regulatory authorities should 'calculate the cost of uneconomic universal service obligations as the net cost of providing service to uneconomic customers, plus any costs associated with the uneconomic provision of public telephones, emergency services and other social obligations such as targeted schemes'[13].

Licensing

The Infrastructure Green Paper warned that licences should not be granted in a manner which would lead to the creation of special rights[14] and that they 'must be consistent with the overall framework provided by Community law'.

Future developments would involve the withdrawal of special or exclusive rights over infrastructure. The new approach to licensing must ensure two things:

- an appropriate level of regulatory supervision of independent service providers, including their contribution levels towards the cost of universal service; and
- adequate flexibility to handle the convergence between broadcasting, information technologies and telecommunications: 'increasingly, communications infrastructure will be the vehicle over which both telecommunications services and non-telecommunications services are delivered'.

The Green Paper set out common principles for the future granting of licences by the Member States or their national regulatory authorities. These are:

- procedures must separate regulatory and operational functions;
- award procedures must be transparent, non-discriminatory and open;
- numerical limitations on licences should be based on the essential requirements;
- award of national licences for infrastructure in support of trans-European telecommunications networks and services should be co-ordinated to ensure that they become operational at the same time in the Member States they cross.

Interconnection and interoperability

Interconnection of networks and their interoperability is a major goal of European Union policy for the information society. The Commission is

[13] Infrastructure Green Paper, Part II, page 83.
[14] Green Paper, Part II, page 59.

aware that there are potential problems associated with interconnection and interoperability from a competition law standpoint. The Green Paper identifies, in particular, examples of situations which could result in abuse of a dominant position in breach of Article 86. For example, the existence of a competitive relationship between organizations, where one demands the use of the other's facilities to carry its telecommunications; or, similarly, in a complementary relationship between a long-distance and local operator. In such cases, the competition rules must safeguard against abuse of one party by another. The Green Paper specified the Commission's intention to 'consider further measures setting out the conditions under which interconnection agreements are compatible with the principle of undistorted competition'[15], in respect both of Article 85 on restrictions of competition and Article 86 on abuse of a dominant position.

It is clear that monopoly holders and holders of special and exclusive rights have greater potential ability to breach the competition rules. The Green Paper emphasizes the need to safeguard against network holders with dominant market positions for infrastructure provision from preventing third parties from offering services by refusing to interconnect; and to prevent infrastructure providers holding dominant positions from placing unreasonable restrictions on consumer choice of service provision or from unreasonable charges.

Subsequent legislative measures have been taken on interconnection in the context of ONP. These are referred to later in the chapter.

Article 90(3) Liberalization Directives

Introduction

In 1988 the Commission adopted a Directive on competition in the *terminal equipment market*[16] using its powers under Article 90(3) on the grounds that the special or exclusive rights held by national telecommunications monopolies in the Member States (where telecoms were either wholly or partly state monopolies), were exercised in such a way as to disadvantage equipment from other Member States. In particular, use of such rights blocked the freedom of users to select equipment best suited to their needs in terms of price and quality, irrespective of its origin.

This Directive was followed two years later by the adoption of a Commission Directive on competition in the *markets for telecommunications services*[17]

[15] Infrastructure Green Paper, Part II, page 71.

[16] Commission Directive 88/301/EEC of 16 May 1988 on competition in the markets for telecommunications terminal equipment, OJ 1988 L131/73.

[17] Commission Directive 90/388/EEC of 28 June 1990 on competition in the markets for telecommunications services, OJ 1900 L192/10, as amended by Commission Directive 96/19/EC, OJ 1996 L73/13.

also based on Article 90(3) of the Treaty. Its objective was to open up the telecoms market to competition, including access to networks. Its main provisions are similar in many respects to the Terminal Equipment Directive.

In July 1988, France, supported by Belgium, Germany, Greece and Italy, challenged part of the Terminal Equipment Directive on the basis of Article 90(3)[18]. In general, France challenged the Commission's right to issue a Directive under Article 90(3) which had a regulatory effect and required Member States to abolish certain special or exclusive rights granted to their telecommunications bodies, in order to introduce competition to the telecoms terminal equipment market. Much to the surprise of practising lawyers the Court upheld the Commission's right to issue Directives based on Article 90(3) to curtail certain monopolistic rights of State companies in the telecoms sector. The Court later confirmed the right of the Commission to limit State companies' monopolistic rights not only in the field of equipment but also in the field of services[19]. These two cases encouraged the Commission to issue further Article 90(3) Directives to limit monopolies. The Directives on Terminal Equipment and Services were thus followed by Directives dealing with satellites, alternative infrastructure, mobile telephones and full competition.

Terminal equipment

The Commission's basic premise for adopting the Directive was stated in the preamble:

> The market in terminal equipment is still as a rule governed by a system which allows competition in the common market to be distorted: this situation continues to produce infringements of the competition rules laid down by the Treaty and to affect adversely the development of trade to such an extent as would be contrary to the interests of the Community.

The Commission justified its use of an Article 90 Directive on the grounds that:

> the treaty entrusts the Commission with very clear tasks and gives it special powers with regard to the monitoring of relations between the Member States and their public undertakings and enterprises to which they have delegated special or exclusive rights, in particular as regards the elimination of quantitative restrictions and measures having equivalent effect, discrimination between nationals of Member States, and competition. The only instrument, therefore, by which the Commission can efficiently carry out the tasks and powers assigned to it, is a Directive based on Article 90(3).

---

[18] Case C-202/88 *France* v. *Commission* [1991] ECR I-1259.

[19] Joined cases C-271/90, C-281/90 and C-289/90, *Spain, Belgium and Italy* v *Commission*.

The Directive included telecommunications bodies as falling within the scope of Article 90(1), on the basis that: 'They are either public undertakings or private enterprises to which the Member States have granted special or exclusive rights for the importation, marketing, connection, bringing into service of telecommunications terminal equipment and/or maintenance of such equipment'.

The Directive provides that Member States which grant to certain undertakings special or exclusive rights to import, market, connect, bring into service and maintain terminal equipment, must withdraw those rights and submit to the Commission all measures taken for this purpose. The Court upheld the abolition of exclusive rights but not the abolition of special rights on the ground that the Commission had failed to define special rights clearly and to explain why they were contrary to the competition rules. Member States must ensure that users have access to new public service termination points and publish the physical characteristics of these points. Member States must provide the Commission with a list of all technical specifications and type approval procedures used for terminal equipment together with the publication references.

By 1 July 1989 Member States must ensure that responsibility for technical specifications and for type approval for terminal equipment is assigned to a body independent of public or private undertakings in the telecommunications sector.

The Directive also required Member States to ensure that contracts for the rental or maintenance of terminal equipment, which at the time of conclusion were subject to special or exclusive rights, could be terminated by customers, subject to maximum notice of one year. This provision was annulled by the Court on the ground that the Directive was addressed to *Member States* only and could not deal with contracts entered into by *undertakings*.

Services

The Telecoms Services Directive[20] has as its objective the opening up of the telecoms market to competition, including access to markets. The main principles established by the Directive relate to liberalization of services, access and 'essential requirements' and independence of the regulator.

The Directive obliges Member States to abolish all special or exclusive rights granted to undertakings for the supply of telecoms services, except voice telephony, and to take the measures necessary to ensure that any operator is entitled to supply such services. It goes on to specify that Member States which grant licences for telecoms services should ensure that the

[20] Commission Directive 90/388/EEC of 28 June 1990 on competition in the markets for telecommunications services, OJ 1990, L 192/10. Now substantially amended by Commission Directive 96/19/EC of 13 March 1996 with regard to the implementation of full completion in telecommunications markets, OJ 1996, L 74/23.

conditions under which they are granted are objective, non-discriminatory, transparent, with reasons given for refusal, and an appeals procedure against refusal. The Court upheld the abolition of exclusive rights but not the abolition of special rights for the same reason as in the *Terminal Equipment* case (see above).

The Directive requires Member States which maintain special or exclusive rights for the provision and operation of public telecommunications networks to take the necessary measures to ensure that conditions governing *access* to the networks are objective and non-discriminatory and that they are published. They should particularly ensure that requests by operators for leased lines should be granted within a limited period and that there are no restrictions placed on their use (except in limited circumstances). Member States must also inform the Commission each time the charges for leased lines are increased and provide justifiable reasons for such increases.

Member States must abolish restrictions on the processing of signals in advance of their transmission through the public network, or after their reception; unless the necessity of such restrictions for compliance with public policy or *essential requirements* can be shown. Article 1(1) defines 'essential requirement' as:

> the non-economic reasons in the general interest which cause a Member State to restrict access to the public telecommunications network or public telecommunications services. These reasons are security of network operations, maintenance of network integrity, and, in justified cases, interoperability of services and data protection.

Member States must establish a *regulatory* structure, a body independent of the telecommunications organizations that are responsible for granting operating licences, controlling type approval and technical specifications, allocation of frequencies and surveillance of usage conditions.

The Services Directive excluded specific services from its scope. These are telex, mobile radio telephony, paging and satellite services. Since the adoption of the Services Directive the Commission has adopted other Article 90(3) Directives which include satellite services, mobile, cable television networks and personal voice telephony within the scope of the Services Directive[21].

Review of the situation in the telecoms services sector

Under the terms of the Terminal Equipment and Services Directives, the Commission issued a mandatory review of the situation in the telecoms services sector in October 1992[22], which set out the progress made in liberalization to date, and identified restrictions preventing the future development of

---

[21] See head 4 of the selective list of documents at the end of this chapter (liberalization measures).

[22] 1992 Review of the situation in the telecommunications services sector. SEC(92) 1048, final, 1992.

the sector in the Community. It examined four policy options to resolve future problems. Option 1 would halt the liberalization process and maintain the status quo. Option 2 would introduce extensive regulation of tariffs and investment. Option 3 would liberalize all voice telephony. Option 4 would open voice telephony to competition between Member States.

On the basis of the review, the Commission launched a wide-ranging consultation process to discuss future options for the development of the sector. In its communication on the consultation[23], the Commission recommended full liberalization of public voice telephony services by 1 January 1998. A Council Resolution followed which endorsed the liberalization of voice telephony services by that date[24].

The Resolution laid down major goals for telecoms policy in the short and longer terms. Longer term goals were specified as:

- liberalization of all public telecoms services, whilst maintaining universal service;
- maintaining a balance between liberalization and harmonization;
- examining progress on structural, particularly tariff, adjustment in advance of full liberalization of voice telephony, to take account of difficulties of remote regions with less developed and small networks;
- establishing a future policy for telecoms infrastructure based on the result of the consultation process linked to the infrastructure Green Paper.

These goals and principles have formed, and continue to form, the basis of EU telecoms policy.

Satellite

In October 1994 the Commission adopted a Directive[25], using its powers under Article 90(3) to open up to competition the satellite communications market. This followed a Commission Green Paper on a common approach in the satellite field which called for complete liberalization of the satellite services and equipment sectors, including the abolition of all special or exclusive rights[26].

The Directive specifically addresses the defect pointed out by the Court in the *Terminal Equipment* case[27] where the Court declared provisions on *special rights* in the Commission's Terminal Equipment Directive to be void on the

---

[23] Communication to the Council and European Parliament on the consultation on the review of the situation in the telecommunications services sector. COM(93), 59 final, 1993.
[24] Council Resolution of 23 July 1993 on the review of the situation in the telecommunications sector and the need for further development in that market. OJ L1993, C-213/1.
[25] Commission Directive 94/46/EC amending Directive 88/301/EEC and Directive 90/388/EEC, in particular with regard to satellite communications, OJ 1994 L 268/15.
[26] Commission Green Paper on a Common Approach in the field of Satellite Communications in the European Community, COM(90) 490, 1990.
[27] Case C-202/88 *France* v. *Commission* [1991] ECR I-1259.

grounds of lack of specificity on the nature of such rights and the respect in which their existence was contrary to the Treaty.

This Directive defines special rights both in the preamble and in Article 2(ii). Special rights are rights granted by a Member State to a limited number of undertakings within a geographical area which, in accordance with criteria which are *not* objective, proportional and non-discriminatory, limit to two or more, the number of undertakings authorized to provide a service/engage in an activity; designate several competing undertakings as authorized to provide a service/engage in an activity; or confer on any undertaking(s) legal or regulatory advantages which substantially affect the ability of any other undertaking to provide the same telecommunications service or to undertake the same activity in the same geographical area under substantially equivalent conditions (Article 2(ii)).

The objective of the Directive is to abolish special or exclusive rights granted by Member States in connection with the supply telecommunications services except for voice telephony.

The Member States, moreover, must inform the Commission of their criteria for granting authorizations, together with any conditions attached, and the declaration procedures for the operation of transmitting earth stations. Any proposed changes to licensing procedures by the Member States must also be communicated to the Commission. Furthermore, fees imposed by Member States on service providers as part of authorization procedures, must be based on objective, transparent and non-discriminatory criteria and such information should be published.

The Directive also requires the Member States to abolish any regulatory prohibitions or restrictions on the offer of space-segment capacity to any authorized satellite earth station.

The effect of this Directive is to amend the Telecommunications Terminal Equipment Directive and Telecommunications Services Directive to include the abolition of all special or exclusive rights in connection with satellite services.

The Satellite Services Directive, which entered into force in December 1994, by its application of the mutual recognition of licences, represents a significant step towards removing the problems inherent in a fragmented market and has provided a regulatory basis for further competition in the sector. For example, in November 1995 the Commission submitted a proposal for a Council and European Parliament Decision on an action at European Union level in the field of satellite personal communications services in the European Union[28]. This proposes Community action over a three-year period, *inter alia*, to establish a selection procedure for space segment operators; to adopt common and harmonized conditions for authorizations; to co-operate with the European Conference of Postal and Telecommunications Administrations (CEPT) and the European standardization

---

[28] COM(95) 529, 1995.

bodies; to establish an advisory committee and a regulatory committee; and to consult and possibly take action at international level on the co-ordinated introduction of satellite personal communications on a global scale.

Alternative infrastructure

An Article 90(3) Directive was probably not what the Council Resolution of December 1994[29] on the principles and timetable for the liberalization of telecoms infrastructure had in mind when it requested the Commission to propose amendments to the Community regulatory framework, to be presented to the European Parliament and the Council. However, the Commission, in December 1994, adopted a draft Article 90(3) Directive. Whilst the Commission is empowered under its own initiative to adopt a Directive pursuant to Article 90 without consulting the Council or European Parliament, it, in fact, issued the text in draft form with the aim of consulting the Member States, European Parliament and other interested parties in advance of formally adopting a decision. (This procedure was also used in the adoption process for the Article 90 Directive amending the Services Directive (Directive 90/388/EEC) to include abolition of special or exclusive rights in the satellite services field.)

The Directive[30], which was adopted in October 1995, amends the Telecommunications Services Directive (Directive 90/388/EEC) in the following respects:

- it incorporates into the Services Directive in an amended Article 4 the provision that Member States 'shall withdraw all restrictions for the supply of transmission capacity on cable TV networks and allow operators to use the cable networks to deliver their services';
- Member States are required to 'ensure that interconnection of cable TV networks with the public telecommunications network is authorized for such purpose, in particular interconnection with leased lines, and that the restrictions on direct interconnection of cable TV networks are abolished';
- an amendment of Article 2 of the Services Directive provides that Member States must take the necessary measures to ensure the separation of financial accounts, as well as transparency and non-discrimination, where a single operator provides both telecommunications and cable television networks. Furthermore, where cable television networks are used for telecommunications purposes, the Member States must ensure that the cable television operators keep separate financial accounts for

[29] Council Resolution of 22 December 1994 on the principles and timetable for the liberalization of telecommunications infrastructure, OJ 1994 C379/4.

[30] Commission Directive 95/51/EC amending Commission Directive 90/388/EEC regarding the abolition of the restrictions on the use of cable television networks for the provision of telecommunications services, OJ 1995 L 256/49.

both their network capacity provider activities and for telecommunications operations. Where a single operator provides both telecommunications and cable television networks, the Commission mandates itself to engage in an impact assessment of such joint provision in meeting the aims of the Directive by 1 January 1998;
- the Directive entered into force in January 1996.

The Commission Directive elaborates, in an accompanying explanatory memorandum, its justification for proposing a Commission Directive. It specifically indicts regulatory restrictions by the Member States in preventing the use of alternative infrastructure, particularly cable television networks, as constituting the main cause of bottlenecks in the provision of liberalized telecommunications services.

The memorandum states that current restrictions imposed by Member States on the use of cable television networks for services provision, other than in television broadcasting, are intended to prevent the provision of public voice telephony on alternative networks to the public switched telephone network, in order to protect the main source of revenues of the telecommunications organizations. Since such restrictions arise from State measures, and have the object in each of the national markets of favouring telecommunications organizations either owned by the Member States, or to which Member States have granted special or exclusive rights, these restrictions must be examined under Article 90(1) of the Treaty. The memorandum states:

> This Article requires Member States not to adopt or maintain measures regarding such undertakings which remove the useful effect of Treaty provisions, and in particular of the competition rules. It includes a prohibition on maintaining measures regarding telecommunications organizations which result in limiting the free provision of services within the Union or lead to abuses of dominant position to the detriment of the users of a given service.

The Commission memorandum highlights in particular the restrictions on the provision of *new services* (pay television, interactive television, video on demand and multimedia services) and the resultant limitation on their intra-Community provision. According to the Commission, 'such regulatory restrictions cannot be justified for public policy reasons or essential requirements'.

The Commission thus concludes that State measures granting exclusive rights to telecommunications organizations to provide transmission capacity, and the regulatory restrictions on the use of cable television to provide liberalized telecommunications services, constitute a breach of Article 90 in conjunction with Article 59 (on free movement of services).

The Commission also examined the application of Article 86 (on abuse of a dominant position). It concluded, *inter alia,* that the structural dominance of the telecommunications organizations constituted a potential barrier to entry; and the fact that the telecommunications organizations held exclusive rights for the provision of voice telephony contributed to their dominance. Whilst acknowledging that the mere creation of a dominant position in a given market arising from the grant of an exclusive right is not as such incompatible with Article 86, nonetheless, a Member State is not allowed to maintain a legal monopoly where the undertaking concerned is compelled or encouraged to abuse its dominant position in a way that is liable to affect trade between Member States.

The result of restrictions on use of cable television infrastructure for telecommunications services is reflected in high charges of the telecommunications organizations; in fact, infrastructure charges are stated to be on average, ten times higher in the European Union than their equivalent in the US. According to the Commission, these high prices are a direct result of the restrictions imposed by Member States on the use of infrastructure other than that provided by the telecommunications organizations.

The Commission memorandum refers to the Court's jurisprudence in the *Terminal Equipment*[31] case quoting from paragraph 51 of the judgment which states that:

> a system of undistorted competition, as laid down in the Treaty, can be guaranteed only if equality of opportunity is secured as between the various economic operators. To entrust an undertaking which markets terminal equipment with the task of drawing up the specifications for such equipment, monitoring their application and granting type-approval in respect thereof is tantamount to conferring upon it the power to determine at will which terminal equipment may be connected to the public network, and thereby placing that undertaking at an obvious advantage over its competitors.

Similarly, by reserving to one undertaking which markets telecommunications services the task of supplying the indispensable 'raw material', i.e. transmission capacity, to all companies offering telecommunications services, gives it virtually unfettered power to determine the service to be offered by its competitors. Thus the Commission concluded that 'the exclusive rights granted to the telecommunications organizations to promote transmission capacity for telecommunications services to the public and the resulting restrictions on the use of cable TV networks for the provision of liberalized services are therefore incompatible with Article 90(1) in conjunction with Article 86 of the Treaty'.

The Commission examined the possible application of the exception contained in Article 90(2) i.e. where the application of the competition rules

[31] Case C-202/88 *France* v. *Commission* [1991] ECR I-1259.

would obstruct the performance, in law or in fact, of the particular tasks assigned to the telecommunications organizations. It concluded that, since under the Telecommunications Services Directive (90/388/EEC) Member States were entitled to reserve voice telephony to their national telecommunications organizations in order to ensure sufficient revenues for the establishment of a universal service network, a temporary prohibition on the provision of voice telephony on cable television networks was justifiable under Article 90(2) for the same reason. However, Article 90(2) could not be used to justify other restrictions.

The Commission memorandum assessed the issue of possible abusive cross-subsidies between the broadcasting activities of cable television operators and their telecommunications network capacity operations. It took the view that it would be 'extremely difficult to determine the cross-subsidies within it between the reserved activities and the services provided under conditions of competition', and it thus proposed the separation of financial records. Separation of financial accounts would also apply in cases where Member States grant the right to establish both cable television and telecommunications networks to the same undertaking.

While the Commission Directive confined itself to the liberalization of cable television networks for the provision of telecommunications services, the Infrastructure Green Paper[32] envisaged a broader use of alternative infrastructure operated by third parties, to encompass networks owned by energy companies, railways or motorways for delivery of telecommunications services. The Telecommunications Council of November 1994[33], approved the Green Paper, Part I, on the principle and timetable for infrastructure liberalization, subject to a compromise that excluded liberalization of other alternative networks such as railways or energy networks due to the opposition of some Member States[34].

Moreover, the Telecommunications Council, while taking the decision to liberalize telecommunications infrastructure, set the date at 1 January 1998, and not January 1996 as envisaged by the Commission. The Commission therefore proceeded to adopt the Directive which establishes 1 January 1996 as the date for the opening up of cable networks for the transmission of telecommunications services. The Member States' right to maintain monopolies on voice telephony provision would be unaffected until 1998 as the Directive is applicable to non-reserved services only.

---

[32] Green Paper on the Liberalization of Telecommunications Infrastructure and Cable Television Networks. Part I COM(94) 440, 1994.

[33] Commission statement concerning Council Resolution on Universal Service in the Telecommunications Sector, 94/C/48/06, OJ 1994 C48/8.

[34] Proponents of the use of alternative infrastructures are France, Germany, the Netherlands, UK and new Member States. The cohesion countries – Greece, Ireland, Portugal and Spain – together with Belgium have been less enthusiastic over infrastructure liberalization.

Mobile

A Directive designed to abolish monopolies in the mobile telephone sector has been adopted by the Commission[35]. The Directive amends the Telecommunications Services Directive[36] to include mobile telephony.

The Commission's decision to issue an Article 90(3) Directive followed deliberations at the Telecommunications Council of 13 June 1995[37] at which the Council was unable to reach agreement on a deadline for abolishing state monopolies over mobile telecommunications. The Commission's specified deadline was 1 January 1996. While the Council adopted a resolution welcoming the Commission's work to date on mobile and personal communications which broadly agreed with the Commission's priorities on liberalization of this sector, it nonetheless fell short of specifying a deadline for liberalization, asking only that the Member States should use their best endeavours to open up the sector to competition.

Publication of a draft Commission Directive, accompanied by a published Commission notice[38] stated the Commission's intention to initiate a consultation procedure with interested third parties in advance of the final adoption in January 1996 by the Commission of the Article 90(3) Directive. Under the Directive's provisions, Member States would be required to abolish state monopolies over mobile and personal communications services and to permit mobile services operators to construct their own networks or to use alternative infrastructure. The terms of this Directive go beyond liberalization into frequency allocation and licensing of DCS-1800 (see text and footnote 39 below).

According to the Commission, the Directive should enable the European Union to take the lead in establishing the right regulatory conditions to encourage the development of mobile and personal communications into a vast mass market. The EU market would thus benefit from a combination of liberalization of services and networks, together with the use of harmonized digital standards. The Directive is based on the consultative procedure following the Green Paper on mobile and personal communications.

Mobile and personal communications services are defined as meaning 'services other than satellite services whose provision consists, wholly or partly, in the establishment of radiocommunications to a mobile user, and makes use wholly or partly of mobile and personal communications systems'. Mobile and personal communications systems refer to 'systems consisting of the establishment and operation of a mobile network infrastructure whether connected or not to public network termination points, to support the transmission and provision of radiocommunications services to mobile users'.

---

35 Commission Directive 96/2/EC of 16 January 1996 amending Directive 90/388/EEC with regard to mobile and personal communications, OJ 1996, L20/59.

36 Commission Directive 90/388/EEC, OJ 1990 L 192/10.

37 EU Council of Ministers conclusions: 7840/95; 13 June 1995.

38 Commission notice concerning a draft Directive amending Commission Directive 90/388/EEC with regard to mobile and personal communications, OJ 1995 C197/5, 1995.

The Directive requires the Member States to abolish all exclusive and special rights in the area of mobile communications.

The Directive also establishes licensing procedures to authorize the launch of digital services (GSM, DCS-1800 and DECT[39]). It prevents Member States from refusing licences for operating mobile systems according to the DCS-1800 standard by 1 January 1998, and the DECT standard by 1 January 1996. Member States must abolish existing restrictions on combinations of mobile technologies and systems, in particular, on multi-standard equipment under existing licences. Member States must ensure that their implementing measures contain the requirement of ensuring effective competition between operators competing in the relevant markets. This provision should be particularly applied in justification of the extension of existing licences.

Mobile network operators must ensure transparency and non-discrimination between fixed and mobile network operators in common ownership. Furthermore, licensing conditions must not include unjustified technical restrictions. Licences may be limited by Member States only on the basis of essential requirements, such as lack of available frequencies, but the principle of proportionality would apply to this restriction.

The Directive provides that the future designation of radio frequencies for specific communication services must be based on objective criteria. Designation procedures must be transparent and published.

The Directive deals with access to infrastructures. It abolishes restrictions on use of facilities for mobile networks, permitting mobile networks to use their own infrastructure as well as that provided by third parties.

The Directive regulates interconnection without prejudice to the future harmonization of interconnection rules in the context of open network provision (ONP). It removes restrictions on direct interconnection for mobile networks. Under its provisions, mobile communications systems for the public have the right to interconnect with the public telecommunications network. Member States would be required to guarantee access to a minimum number of interconnection points to the public fixed network in the licences for mobile services. Interconnection conditions with the public telecommunications network must be based on objective criteria, be transparent and non-discriminatory and comply with the principle of proportionality. Member States must ensure in case of appeal that the national regulators have full access to interconnection agreements and that such information is given to the Commission as requested.

The Directive emphasizes the use of infrastructure controlled independently of the telecommunications organizations as a means of providing new entrants to the mobile market greater control over costs. According to the

[39] These standards are global system for mobile communications (GSM), and digital communications standard for cellular communications (DCS-1800), the two frequencies available for digital mobile services, and digital European cordless communications (DECT) for digital cordless telephony within a fixed radius.

Commission, leasing capacity currently represents a cost factor for second operators of between 30 per cent and 50 per cent. In addition, the right to establish self-provision of networks and to choose alternative infrastructure and connections gives mobile operators a great deal more flexibility which should encourage further development and innovation in the market. The Commission points out that in most Member States the second mobile operator must pass calls through the fixed network of the incumbent national network for interconnection into another Member State. Removal of such restrictions through direct interconnection with a selected operator in the country of destination is considered by the Commission to be more technically logical and less expensive.

There will be derogations of up to five years for Member States with less developed networks in order to take account of their specific situations. Such derogations will affect Portugal, Spain, Greece and Ireland. Very small networks (Luxembourg) will also benefit from a two-year derogation. The Directive entered into force in February 1996.

The Commission issued a press release in September 1995 on the opening up to competition of the GSM mobile telecommunications market[40], in which it reinforced its intention to intensify its screening of the GSM licensing procedures in the Member States to ensure a level playing field for new market entrants. It noted that in nearly all cases, the incumbent operator in the Member States was the former monopoly operator or one of its subsidiaries.

Competition Commissioner Van Miert has held negotiations with several Member States, notably Belgium, Italy, Ireland and Spain, in connection with problems over the licensing of a second GSM operator. These problems were successfully resolved between the Commission and the Member States involved. For example, Italy[41] was alleged to have discriminated against Omnitel, a consortium led by Olivetti and in competition with the state-owned telecommunications administration, Telecom Italia, on the grounds that it was charged an excessively high fee. The Commission asked the Italian government to ensure that the state communications operator paid the same fee or that it compensated the second operator accordingly. The Commission indicated that it would take a case to the Court against Italy, should it fail to comply with the Commission's request.

The award of the second mobile telephone network in Belgium was put to auction in May 1995 by the Belgian government. The use of an auction system for the award of a licence for the second mobile telephone network is questionable given its potential for abuse. However, on grounds of subsidiarity, the Commission may consider it beyond its scope to lay down detailed licensing conditions for the Member States. It can probably only go as

[40] Commission Press Release, IP/96/791, 8/8/96.
[41] Commission Decision concerning the conditions imposed on the second GSM operator in Italy, OJ 1995 L 280/49.

far as to require licensing procedures to be objective, non-discriminatory and transparent. The Commission has raised specific concerns over the auction system, on the grounds that the second licence is awarded not only on the basis of a comparison of qualitative elements, but also on the basis of a financial bid above a certain set threshold. It considers that the auction system leads to the imposition of unfair competition conditions and has threatened to use both Article 90(3) provisions and references to the Court, to require Member States to end infringements where they are perceived to exist in selection procedures. The Commission recognizes its inability to prohibit the auction system, however much it dislikes it. It would prefer Member States either to request a comparable financial contribution from the incumbent operator or decide on compensation measures to balance the effect of what it regards as an 'entrance fee' under the auction system[42].

The Commission's Competition Directorate (DG IV) has also sent warning letters to GSM/DCS-1800 handset manufacturers and network operators in the European Economic Area (EEA) limiting the use of the 'SIMLock' feature in mobile handsets, on the basis that this service feature effectively ties the customer to one GSM operator or service provider. The handset must be able to be unlocked upon demand by the customer. This is designed to prevent anti-competitive effects contrary to Articles 85 and 86 in respect of existing or new operators, and to avoid reinforcing the division of the mobile phone market along national lines[43].

Full competition

Following on from the 1992 Review and subsequent consultation and Council Resolutions, the Commission, in 1996, adopted a Directive under Article 90(3) on the implementation of full competition in telecommunications markets[44]. This Directive substantially amends the Telecoms Services Directive by introducing complete liberalization of public voice telephony services by 1 January 1998 (subject to transitional periods for Member States with less developed or small networks). The Services Directive has, as mentioned elsewhere in this chapter, undergone previous amendments to include satellite communications, cable television networks, and mobile and personal telecommunications within the Directive's scope, pursuant to Commission Directives 94/46/EC, 95/51/EC and 96/2/EC respectively.

The new Directive abolishes the Article 90(2) exemption granted under the Services Directive in respect of special and exclusive rights for provision of voice telephony services, on the grounds that this is no longer justifiable in a liberalized, competitive environment. However, Member States will have

---

[42] BIO/96/173; 22/4/96.

[43] Commission Press Release IP/96/791, 8 August 1996.

[44] Commission Directive 96/19/EC of 13 March 1996 amending Directive 90/388/EEC with regard to the implementation of full competition in telecommunications markets, OJ 1996 L 74/13.

until 1 January 1998 to enable telecommunications organizations to make the necessary adjustments to the liberalized regime, particularly for the rebalancing of tariffs. Member States with less developed or small networks may request a temporary exemption of up to five years to adjust to the new system structurally. The exemption is only applicable to Greece, Ireland[45], Portugal and Spain, in respect of less developed networks (up to five years) and to Luxembourg, in respect of small networks (up to two years).

The removal of special and exclusive rights for voice telephony would enable the Member States' telecommunications organizations to provide new voice telephony services in other Member States in the absence of competition restrictions. It should enable new entrants to compete with dominant operators in the Member States, using the currently available infrastructure held by the operators in the Member States, or alternative infrastructure, subject to fair competition, through commercially negotiated arrangements.

This important Directive contains provisions which facilitate access of new competitors to the market; clarifies licensing criteria; establishes provisions on interconnection; provides for the phasing out of restrictions on tariff rebalancing and cross-subsidization; refers to universal service provision; and deals with numbering and access to rights of way.

Licensing

The new Directive obliges Member States to ensure that licensing conditions are objective, non-discriminatory, proportionate and transparent. Reasons for refusal must be given and an appeals procedure should be instated. Telecoms services other than voice telephony, public network establishment and provision and networks using radio frequencies, may only require a general authorization or declaration procedure.

Member States must inform the Commission of the conditions under which licences, general authorizations and declaration procedures are grounded, together with any conditions attached to them.

Member States may limit the number of licences only where related to the lack of available spectrum and where justified under the proportionality principle.

Packet or circuit-switched data services no longer benefit from the Article 90(2) exemption. Thus the adopted set of public service specifications is abolished and may be replaced by a general declaration or a general authorization procedure.

The draft Council Directive on licensing of telecommunications services[46] (based on Article 100a) has been drafted to take account of complete

---

[45] A formal decision approving the timetable for full liberalization has been issued in respect of Ireland. Commission Annual Competition Report for 1996.

[46] Proposal for a European Parliament and Council Directive on a common framework for general authorizations and individual licences in the field of telecommunications services, COM(95) 545, 1995, as amended by the Common Position, Council Document 9944/96. 18 September 1996.

liberalization of telecommunications infrastructure provision by 1998, and reflects the principles contained in the Full Competition Directive, by imposing minimal regulatory obligations in order that undertakings in the telecoms field can exercise freedom of establishment and freedom to provide services in the EU within the context of the competition rules.

Interconnection

New provisions on interconnection have been added to the Services Directive by the new Directive. A new Article obliges Member States to ensure that interconnection to voice telephony services and public switched telecommunications networks of the telecommunications organizations by other authorized operators is conducted on the basis of non-discrimination, proportionality and transparency.

The Commission recognizes that, in principle, interconnection should be a matter for commercial negotiations between the parties concerned, but expresses concern, nonetheless, that an imbalance in negotiating power between new market entrants and the telecommunications organizations, whose privileged monopoly stems from their special and exclusive rights, could lead to anti-competitive results.

Therefore, the Directive intervenes to regulate interconnection in the following ways. First, Member States must ensure that the telecommunications organizations publish (by 1 January 1997) the terms and conditions for interconnection to their basic voice telephony service and public switched telecommunications networks, to include interconnection points and interfaces offered according to market needs. Second, Member States should not prevent organizations providing telecoms services and/or networks that request interconnection from negotiating interconnection agreements with telecommunications organizations for access to the public switched network in respect of special network access and/or conditions meeting their particular requirements.

Provision is made for the establishment of a disputes procedure by Member States, in the event of commercial negotiations failing to lead to agreement within a reasonable time. Member States, at the request of either party, and within a reasonable time period (unspecified in the Directive), may adopt a reasoned decision which sets out operational and financial conditions and requirements for interconnection.

The Directive also requires that cost accounting systems implemented by telecommunications organizations concerning the provision of voice telephony services and public telecommunications networks identify the cost elements relevant for pricing interconnection offerings.

The Directive specifies that the measures on interconnection are applicable for a five-year period from the date of the effective abolition of special and exclusive rights for the provision of voice telephony granted to the telecommunications organizations. However, this will be reviewed if a harmonizing Interconnection Directive is adopted before the end of the period. A draft Interconnection Directive[47] pending adoption, has been proposed by the Commission under Article 100a, which echoes the principles on competition and contains a similar definition of interconnection to that contained in the Full Competition Directive.

Interconnection was identified by the Infrastructure Green Paper as a major objective of EU policy for the information society. The draft Article 100a Directive on interconnection in the context of ONP follows the principles outlined in the Infrastructure Green Paper and the Full Competition Directive. The explanatory memorandum to the draft Interconnection Directive states that 'interconnection is seen as a key element in the future competitive environment, allowing new market entrants access to existing end-users, on a basis which will encourage increased investment and market growth in the telecommunications services sector, within a predictable and stable regulatory environment'.

It should also be noted with regard to interconnection and interoperability that they are subject to specific Treaty considerations under Article 129b which states that 'within a framework of a system of open and competitive markets, actions by the Community shall aim at promoting interconnection and interoperability of national networks as well as access to such networks'. Trans-European networks in telecoms, transport and energy are Community priorities for the development of the single market.

Universal service obligation

A new Article in the Directive on full competition sets out specific obligations for Member States. It is designed to avoid the possibility of over-burdening new market entrants with national financing schemes for universal service provision and further entrenching the dominant positions held by the telecommunications organizations, in breach of Article 90 in conjunction with Article 86. National financing schemes connected with contributions towards universal service obligations harmonized in the framework of ONP, should apply only to undertakings providing public telecoms networks, and contributions should be allocated to each undertaking in accordance with objective and non-discriminatory criteria and in conformity with the principle of proportionality.

It is stated in Preamble 19 of the Directive that funding schemes relate to the financing of the universal service only and not to other activities not directly related to universal service provision.

The need for tariff rebalancing is highlighted as necessary in order to balance tariffs with real costs. Article 4c states that:

---

[47] Proposal for a European Parliament and Council Directive on interconnection in telecommunications with regard to ensuring universal service and interoperability through the application of the principles of open network provision (ONP), OJ 1995, C-313/7, as modified to take account of the European Parliament's amendments at first reading, OJ 1996, C-178/3.

> Member States shall allow their telecommunications organizations to re-balance tariffs taking account of specific market conditions and of the need to ensure the affordability of a universal service, and, in particular, Member States shall allow them to adapt current rates which are not in line with costs and which increase the burden of universal service provision, in order to achieve tariffs based on real costs.

Where rebalancing is not possible by the 1998 deadline, Member States in this position must inform the Commission of their timetable for phasing out tariff imbalances. The situation with regard to universal service financing and rebalancing will continue to be assessed by the Commission. A Commission review, to be presented by 1 January 2003, will assess whether the financing schemes in place in the Member States do not limit access to the relevant markets, and whether further proposals are necessary. An earlier report from the Member States must be presented to the Commission by 1 January 1998 on the future phasing out of remaining tariff imbalances.

These provisions are in line with the Infrastructure Green Paper principles (referred to in the chapter) and with other Commission Communications[48].

Lifting of other restrictions

The Directive also removes restrictions relating to exclusive rights in respect of directory services.

It introduces competitive conditions into the area of access and to rights of way in order to protect undertakings which do not have their own infrastructure. Member States must not discriminate between public telecommunications network providers and must facilitate access to existing facilities which cannot be duplicated under reasonable terms. These measures are designed to ensure that the principle of access to an essential facility is not breached by the telecommunications organizations through the exercise of their dominant position in restricting access, contrary to Article 90(1) in conjunction with Article 86.

An addition to Article 3b of the Services Directive is intended to help authorized voice telephony operators to compete with existing operators as regards numbers. The Commission's concern that telecommunications organizations might retain the best numbers and give competitors insufficient or less attractive numbers, has led to a provision specifying that the Member States must ensure by 1 July 1997 that adequate numbers are available for all telecoms services, and that they should be allocated on an objective, non-discriminatory, proportionate basis, in respect of individual application procedures.

The Services Directive is amended by a new Article 8 which legislates against potential anti-competitive cross-subsidization of telecommunications

---

[48] Commission Communication on universal service for telecommunications in the perspective of a fully liberalized environment – an essential element of the Information Society, COM(96) 419, 1996 and draft Directive on the application of ONP to voice telephony and on universal service for telecommunications in a competitive environment (replacing Directive 95/62/EC) OJ 1996, C-371/22 as amended by COM/97531 final – COD 96/226.

activities. The Article provides that in authorization schemes for voice telephony and public telecommunications networks provision, Member States must '*at least* ensure' (emphasis added) that where authorizations are granted to undertakings to which they also grant exclusive or special rights in areas other than telecommunications, there must be unbundling of financial accounts between these activities and the voice telephony and/or networks provision activities. The threshold that triggers this requirement is a turnover in excess of ECU 50 million in the relevant telecommunications market. Preamble 24 implies that this figure could be subject to future revision, stating that '*for the time being*', a turnover of more than ECU 50 million could be considered as a significant turnover' (emphasis added).

Whilst Council Directive 95/51/EC removed restrictions to the provision of liberalized telecommunications services over cable televisions networks, the Commission is concerned that some Member States continue to maintain restrictions in this area. Hence a new Article indicates the Commission's intention to conduct an overall assessment on the situation concerning remaining restrictions on the use of public telecommunications networks for the provision of cable network capacity.

The Directive, which entered into force in April 1996, imposes an obligation on the Member States to inform the Commission, not later than nine months after the Directive's date of entry into force, of the manner in which they have complied with its requirements.

Harmonization Directives

Introduction

The telecommunications sector in the EU is subject to both liberalization by way of Commission Directives adopted under Article 90(3) and to harmonization of the rules on access to networks by way of Directives adopted under Article 100a. The first harmonization Directive, which established the framework for what was termed open network provision (ONP) was adopted by the Council contemporaneously with the Telecoms Services Directive. Like the Telecoms Services Directive, the ONP Directive was predicated on the basis of access to the facilities of telecommunication organizations, providing for the development of harmonised principles and conditions for open network provision. The Directive set out a work programme for the development of ONP conditions for particular services including leased lines, packet and circuit switched data services and voice telephony services.

ONP Directives were adopted in relation to these services. As the pace of liberalization increased, further Directives were proposed and adopted to adapt the ONP Directives to a competitive environment.

Following the adoption of the Infrastructure Green Paper, further directives were adopted on interconnection of telecommunications so as to ensure universal service and interoperability and in relation to licensing and general authorisations. A number of the liberalization directives deal with interconnection, seeking to establish a framework for interconnection pending the adoption of harmonised rules based on ONP[49].

ONP Directives and liberalization directives provide the basic principles for regulation by national regulatory authorities of the telecommunications sector. The Commission has provided helpful guidance on how the Directives affect the telecommunications sector in practice in the Draft Notice on Access Agreements (considered later in this chapter). This section provides an overview of the basic principles and requirements of ONP and the proposed adaptations in light of the increasingly competitive market and considers in brief some of the specific ONP directives.

ONP Framework Directive

Council Directive 90/387/EEC on the Establishment of the Internal Market for Telecommunications Services Through ONP the ONP Framework Directive, deals with the harmonization of conditions for open and efficient access to and use of the public telecommunications network and public telecommunications services. ONP conditions are intended to enable the provision of services between networks, both within and between Member States.

The principles underlying the ONP Framework Directive establish that conditions for ONP must be on objective criteria; be transparent and appropriately published; and ensure equality of access and non-discrimination in accordance with Community law.

ONP conditions cannot restrict access to the public telecommunications network and public telecommunications services (defined as telecommunications services whose supply Member States have specifically entrusted to one or more telecommunications organizations) other than on the basis of essential requirements. These requirements must be applied within the framework of Community law and cover: security of network operations; maintenance of network integrity; interoperability of services in justified cases and the protection of data as appropriate[50]. ONP conditions cannot impose additional restrictions on use of public networks or services unless such restrictions arise from the exercise of special or exclusive rights granted by Member States and which comply with Community law.

ONP conditions could be applied to telecommunications organizations operating public telecommunications infrastructure and to telecommunications organizations entrusted by Member States to provide particular

[49] See Article 3d of the Mobile Directive and Article 4a of the Full Competition Directive.

[50] The Commission determines the rules for uniform application of the essential requirements under the procedure established by Article 10. To date, the only justified case is that of voice telephony.

services. Telecommunications organizations are defined in the ONP Framework Directive as public or private bodies, to which a Member State grants special or exclusive rights for the provision of a public telecommunications network and where applicable, public telecommunications services. The Telecommunications Services Directive has an identical definition. Member States must notify the Commission of the bodies to which they have granted this status.

Annex II of the ONP Framework Directive sets out a reference framework for drawing up proposals on ONP conditions. This framework provides for common principles which should enable telecommunications organizations to operate the network whilst allowing service providers and users freedom of action: These principles cover:

- harmonized technical interfaces and/or other technical services features which should as far as possible be based on existing interfaces, taking into account the work of the European standards bodies;
- harmonized supply and user conditions in relation to access and provision of services;
- harmonized tariff principles for services including access services which should be based on objective criteria, be cost-oriented, whilst remaining a matter for national regulation, be transparent and appropriately published. Tariffs should also be unbundled to comply with EC competition rules, non-discriminatory and must guarantee equality of treatment.

This Directive entered into force on 1 January 1991.

Revisions to ONP Framework Directive

With the liberalization of non-reserved services and the 1 January 1998 deadline for the lifting of restrictions on the provision of voice telephony, the adaptation of the ONP principles was an increasingly important policy goal for the EU. Adaptation of the ONP Framework Directive would need to address the organizations to whom the ONP Framework Directive applied and such issues as universal service, interconnection and access charges in addition to reflecting changes to the regulatory environment ushered in by the adoption of the Directive on licences and general authorisations.

The approach proposed by the Commission still rested on the harmonization of access to and use of the public telecommunications network but the emphasis would be on establishing obligations only where competition by itself was not sufficient to meet Community policy goals. Structural separation of the regulatory and operational functions of telecommunications organizations remained a key plank on which ONP was to be based.

The Common Position adopted in September 1996 (which also addresses amendments to the Leased Lines Directive[51], considered below) provides

[51] OJ C315/41, 24 October 1996.

that in addition to the basic purpose of ONP conditions, ONP conditions must seek:

- to ensure the availability of a minimum set of services;
- to secure access and interconnection to public telecoms networks and services; and,
- to encourage the provision of harmonized telecommmunications services by promoting voluntary standards adopted at EU level;
- to guarantee the provision of universal service, taking into account any future evolution.

The approach with regard to the possibility of restricting access to and provision of services is similar to that established by the ONP Framework Directive. In line with the proposed approach to standardization, the Common Position includes a number of detailed provisions in relation to the development and adoption of standards. The Common Position also replaces definitions, in order to reflect the opening up of the sector to full competition and the removal of special or exclusive rights, and a new Annex updates the reference framework for the application of ONP conditions.

National regulatory authorities are to be separate from any entities providing telecommunications facilities or services, with open procedures and the possibility of appeal from any decision established.

The Commission is mandated to examine and report on the functioning of the Directive no later than 31 December 1999.

Formal agreement has been reached on the text and official publication is awaited.

Specific measures adopted on the basis of the ONP Framework Directive

The ONP Framework Directive identified areas for specific applications of ONP extending beyond the general principles contained in the ONP Framework Directive. To date, two Council Directives and two Council recommendations have been adopted to give effect to sector-specific requirements. These measures are:

(1) *Council Directive 92/44/EEC on the application of open network provisions to leased lines (the Leased Lines Directive)*[52]. The Directive, *inter alia*, laid down requirements on information to be supplied to users; access conditions under which restrictions to access and use of leased lines should be based only on essential requirements; the provision of a minimum set of leased lines in accordance with harmonized technical characteristics, with the option of updating the minimum offering in accordance with the procedure set out in the ONP Framework

[52] OJ 1992, L165/27.

Directive; control by the national regulatory authority; tariff principles and cost accounting.

The Common Position setting out the adaptation of the ONP Framework Directive to a competitive environment also addressed the amendment of the Leased Lines Directive. The amendments aim to secure access to leased lines (defined to reflect the increasingly competitive market) for all users from at least one operator subject to harmonised conditions of access and use. Member States will determine on whom this obligation to supply will be placed. This will be any organisation with significant market power[53], or in the absence of such an organisation to whom Member States apply these requirements, another organisation. Tariff regulation may be set aside once no organisation has market power in a particular market on the basis that a competitive market does not require cost orientation mandates to ensure competitive pricing.

(2) *Council Recommendation 92/382/EEC on the harmonized provision of a minimum set of package-switched data services (PSDS) in accordance with open network provision (ONP) principles*[54]. This recommendation requests the Member States to guarantee the provision on their territory of a minimum set of packet-switched data services with harmonized technical characteristics, which takes market demand into account. Other issues addressed are transparency of information, harmonized tariff principles and quality of service.

(3) *Council Recommendation 92/383/EEC on the provision of harmonized integrated service digital network (ISDN) access arrangements and a minimum set of ISDN offerings in accordance with open network provision (ONP) principles*[55]. Similar to the PSDS Recommendation, this Recommendation urges the Member States to guarantee the provision on their territory of an ISDN with harmonized access arrangements under a minimum set of offerings in conformity with European Telecommunications Standards Institute (ETSI) standards, as well as adequate and efficient interoperability between ISDNs in order to ensure Community-wide operations. It also addresses transparency of information, harmonized tariff principles, quality of services and numbering.

(4) *European Parliament and Council Directive on the application of open network provision (ONP) to voice telephony*[56]. A draft Directive on the application

---

[53] An organisation will be presumed to have market power where it holds a 25% share or more of the relevant market, to be determined on the basis of the type of leased line offered in a particular area. There is flexibility both above and below this figure in terms of the market power test subject to specified factors to be taken into account such as an ability to influence market conditions and its turnover relative to the size of the market.

[54] OJ 1992 L200/1.

[55] OJ 1992 L200/10.

[56] Directive 95/62/EC of the European Parliament and of the Council of 13 December 1995 on the application of open network provision (ONP) to voice telephony.

of ONP to voice telephony was first proposed in August 1992 pursuant to the co-operation procedure. The Directive was finally adopted in December 1995. It is considered in more detail below.

ONP voice telephony

A Directive on the application of ONP voice telephony (the Voice Telephony Directive) was adopted in accordance with the ONP Framework Directive in December 1995. Almost as soon as the Voice Telephony Directive was adopted, a proposal was put forward for a new Directive on the application of ONP to voice telephony and on universal service for telecommunications in a competitive environment. The main provisions of both the existing Directive and the draft amending Directive are considered below.

ONP Directive on the application of ONP to voice telephony

The Directive lays down the rights of users of the telephone system; ensures open and non-discriminatory access for all users, including service providers offering competitive services; and establishes the provision of voice telephony services at a European level.

Member States are required to ensure that their telecommunications organizations provide a fixed public telephone network and a voice telephony service so as to guarantee a harmonized offering throughout the Community. Users must be able to obtain a connection to the fixed public telephone network and to connect and use approved terminal equipment. Usage-based restrictions are only permitted if they are based on essential requirements or are imposed pursuant to the grant of special or exclusive rights to provide a voice telephony service.

The basic features of a universal public telephone service, such as connection times, discounts, itemized billing, dispute settlement procedures for billing problems and directory services are set out. Advanced services should be provided where technically possible. Provisions on interconnection and special network access are included.

National regulatory authorities must ensure that reasonable requests for interconnection to the fixed public telephone network from telecommunications organizations providing fixed public telephone networks in other Member States, are met and that interconnection agreements are entered into on a non-discriminatory basis.

Special network access concerns connection to the network at points other than at the standard network termination points.

Tariff principles should, *inter alia*, follow the principles of transparency and cost orientation set out in Annex II of the ONP Framework Directive (for which see above). Provisions on cost accounting principles and the elements to be included in a cost accounting system are included.

Proposal for a European Parliament and Council Directive on the application of ONP to voice telephony and on universal service for telecommunications in a competitive environment (replacing Directive 95/62/EC)

The proposed Directive aims to define harmonized conditions for the provision of fixed public networks and public telephony services from 1 January 1998. It also aims to improve the access of all users to the fixed infrastructure of public telephone networks, and to ensure a basic service which would include special tariffs for certain social groups, detailed billing, dispute settlement procedures and the introduction of a European number for free calls.

The draft Directive is concerned with *conditions of access* and use and *essential requirements.* Under Article 13, national regulatory authorities would be obliged to establish procedures to decide, on a case-by-case basis, and in the shortest possible period, whether or not to permit organizations providing fixed public telephone networks/publicly available telephone services, to take measures such as refusal to provide access to the network, or interruption/reduction in availability of service, on the ground of a user's alleged failure to comply with the conditions of use.

Restrictions on access based on essential requirements must be specified. Article 13(2) sets out a list of essential requirements which provide the basis of such restrictions. Similar to Directive 95/62/EC these relate to security of network operations, maintenance of network integrity, interoperability of services, data protection and effective use of the frequency spectrum. Restrictions should be imposed through regulatory means and be able to be challenged in court. They should also be published.

A safeguard exists concerning conditions imposed on users on the basis of special or exclusive rights granted to Member States by organizations providing public networks/services referred to in the draft Directive. These must be imposed by regulatory means and with the agreement of the national regulatory authority (Article 13/3).

Special market access must be provided by organizations with 'significant market power', who are required to deal with reasonable requests from organizations providing telecoms services for access to the network and network termination points other than those commonly provided network termination points referred to in Annex II, point I (Article 16(1)). Organizations with 'significant market power' must adhere to the principles of non-discrimination when making use of special network access for providing publicly available telecommunications services.

Tariffs for use of the fixed public telephone network and publicly available telephone services should follow the principles of cost orientation contained in Directive 90/387/EC.

Tariffs for facilities additional to connection to the public telephone network and publicly available telephone services should be *unbundled* so that users are not requested to pay for facilities that are unnecessary for the service requested.

Cost accounting principles, setting specific criteria for the operation of networks and services by holders of special or exclusive rights, are also set out.

The draft Directive contains provisions on public payphones which should meet the reasonable needs of users, and which may, where they can only be served at a loss, or fall outside normal commercial standards, fall within a universal service financing scheme. Member States should also take specific measures, where appropriate, to ensure access to affordability of telephone services for disabled users and users with special needs.

This Directive, when adopted, will replace the Voice Telephony Directive (Directive 95/62//EC). The draft Directive provides for a review by the Commission on the functioning of the Directive, by 31 December 1999, which also takes into account a Report on Universal Service to be published by the Commission by 1 January 1998.

European Parliament and Council Directive on a common framework for general authorizations and individual licences in the telecommunications services field (the 'Licensing Directive')[57]

The Licensing Directive takes account of the liberalization of telecommunications infrastructure provision that is to occur by 1 January 1998. It aims to impose minimal regulatory obligations on telecoms operators, providing for the use of general authorizations as far as possible. Individual licences may be granted where assignment of frequencies, imposition of public service obligations or other specific competitive safeguards cannot be guaranteed under the general authorization system.

The Licensing Directive addresses both the procedure for the grant of licences and the substance of licences.

All licence conditions must be non-discriminatory, proportionate and transparent. Furthermore, authorizations may only contain conditions that comply with the Annex, which sets out a list of conditions that may be attached to authorizations. Any conditions attached to authorizations must conform to the competition rules. Conditions include conformity with essential requirements for all authorizations, as well as specific conditions applicable to general authorizations and to individual licences. Examples include conditions on user protection, financial contribution to universal service for interconnection and public interest requirements; numbering (e.g. compliance with national numbering schemes), radio frequencies, provision of universal service obligations on interconnection and applications of open network provision principles to voice telephony.

There are also conditions applicable to operators holding a significant market position; requirements connected with quality, availability, or

---

[57] Directive 97/13/EC, OJ L117, 7 May 1997.

permanence of the service or network, including financial, management and technical requirements and conditions on a minimum period of operation; interconnection and interoperability of services; and universal service obligations, which are imposed in accordance with the Directives on interconnection and on the application of ONP principles to voice telephony (for which see above).

Section II of the Licensing Directive deals with general authorizations, providing that conditions in general authorizations should be transparent and should reflect the least onerous approach. The procedures for granting general authorizations should not be used by Member States to prevent an undertaking that complies with the conditions set out in a general authorization from providing the intended telecommunications service and/or telecommunications network. Member States may, however, require the undertaking to notify the national regulatory authority of its intention to provide the service to ensure compliance with the conditions set out in a general authorization. A period of up to four weeks may be required before the undertaking can begin service provision under a general authorization. This section also deals with non-compliance and the fees that may be imposed for operation under general authorization.

Section III regulates individual licences. Under Article 7(1), Member States may only issue individual licences for four purposes: to allow the licensee access to radio frequencies or numbers; to give the licensee particular rights with regard to access to public or private land; to impose obligations and requirements on the licensee relating to the mandatory provision of publicly available telecommunications services and/or public telecommunications networks, including obligations under ONP legislation which require the licensee to provide universal service; and to impose specific obligations, in conformity with Community competition rules, where the licensee has significant market power (as defined in the Interconnection Directive (see below)) in relation to the provision of public telecommunications networks and telecommunications services publicly available.

Section IV regulates the provision of telecommunications services throughout the European Union. Member States are obliged to ensure that the categories of telecoms services listed in the Annex to the Licensing Directive can be provided either without authorization, or on the basis of a general authorization. This section also addresses the harmonization of procedures for the granting of general authorizations, which should aim to develop the least onerous regime possible consistent with ensuring compliance with the relevant essential requirements and other public interest requirements set out in the Directive. Harmonization should also aim to establish balanced sets of rights and obligations for the undertakings benefiting from authorizations.

The Directive establishes conditions for a one-stop shopping procedure for individual licences.

The Directive must be implemented by the Member States by 31 December 1997.

The Commission remains vigilant in relation to licensing. For example, it initiated an Article 90 proceeding against Germany for its refusal to grant Vebacom a licence to operate a broadband telecommunications network based on synchronous digital hierarchy (SDH) technology. Germany, however, agreed to grant the licence and the procedure was therefore not pursued.

Directive 97/33/EC of the European Parliament and of the Council of 30 June 1997 on interconnection in telecommunications with regard to ensuring universal service and interoperability through application of ONP[58]

The Interconnection Directive addresses:

- the rights and obligations of service providers and network operators with regard to interconnection, in accordance with the principle of proportionality;
- the principles for interconnection charges and accounting systems, including universal service contributions;
- transparency of agreements; and
- the role of the national regulatory authority[59].

The Directive gives priority to commercial negotiations between parties wishing to interconnect, while reserving some conditions to be established *a priori* by national telecommunications regulatory authorities and clarifies the responsibility of national regulatory authorities (taking account of subsidiarity), including mechanisms for dispute resolution.

The competition rules under the Treaty are also applicable to interconnection agreements.

The main principles for interconnection and interoperability at national and Community level are set out in Article 3. Member States are obliged to take the necessary measures to remove restrictions that prevent organizations authorized by Member States to provide public telecoms networks and services from negotiating interconnection agreements between themselves, either nationally or in different Member States. As long as such agreements comply with Community law, technical and commercial arrangements for interconnection are a matter for agreement between the parties. Member States are required to ensure 'adequate and efficient' interconnection of

[58] OJ L199/32, 26 July 1997 (the Interconnection Directive).

[59] Communication from the Commission to the Council and the European Parliament. The Consultation on the Green Paper on the Liberalization of Telecommunications Infrastructure and Cable Television Networks, COM(95) 158 final, 1995.

public telecoms networks and services (mainly leased lines and European-wide services) to the extent necessary to ensure universal provision on a European-wide basis.

Organizations authorized to provide public telecoms networks and/or public telecoms services (notably leased lines) have a right and an obligation to negotiate interconnection agreements with similarly authorized organizations save for certain exceptional circumstances, to be determined by the regulator, where there are technical and commercially viable alternatives to the requested interconnection. Organizations which have 'significant market power'[60], should, in particular, meet all reasonable requests for access to the network.

The Directive outlines the conditions for interconnection and *universal service*[61] contributions under which the Member States may establish mechanisms for sharing the net cost of universal service obligations with other organizations operating public telecoms networks. Annex III sets out a system for calculating the cost of universal service obligations for voice telephony. The Annex defines universal service obligations as referring to 'those obligations placed on an organization by a Member State which concern the provision of a network and service throughout a specified geographical area for the provision of that service'.

The Annex establishes that 'the cost of universal service obligations shall be calculated as the difference between the net cost for an organization of operating with the universal service obligations and operating without the universal service obligations'. The burden of universal service provision may be shared by either directly sharing the cost of the universal service obligations, or by making financial contributions, for example, by way of a supplementary charge added to the interconnection charge. The principles of transparency, non-discrimination and proportionality apply to calculating the net cost of universal service contributions.

Interconnection to the public telecoms network and services by organizations with 'significant market power' must conform to requirements for non-discrimination and transparency. Interconnection should be provided under '*similar* circumstances to interconnected organizations providing *similar* services, and shall provide interconnection facilities and information to others under the *same* conditions and of the *same* quality as they provide for their own services or those of their subsidiaries or partners' (emphasis added). Moreover, interconnection agreements involving organizations which have 'significant market power' must be made available on request to interested

---

[60] 'Significant market power' is presumed when an organization has a share of over 25% of a particular telecoms market in the geographical area in a Member State within which it is authorized to operate. In some cases the percentage of market share can be lower.

[61] Universal service is defined in the Directive as 'a defined minimum service or a set of services of specified quality which is available to all users and in the light of specific national conditions, at an affordable price'.

parties, detailing interconnection charges, terms and conditions, together with any contributions to universal service obligations.

The principles for *interconnection charges* and *cost accounting systems* to which organizations having 'significant market power' must adhere, are set out in the Directive. Interconnection charges should be transparent and cost-oriented. National regulators would have the power to request an organization to justify its interconnection charges and to adjust them where deemed necessary.

Interconnection charges should be sufficiently *unbundled*, so that the applicant is not required to pay for anything not strictly related to the service requested. If charges related to the sharing of the cost of universal service obligations exist, they should be unbundled and separately identified.

In order to avoid *cross-subsidization*, an obligation is imposed on organizations providing public telecoms networks and/or services that hold *special or exclusive rights* for the provision of services in other telecoms sectors to keep separate accounts for the different activities. All cost and revenue elements should be separately identified. Member States may choose not to apply these requirements to organizations with an annual turnover of less than ECU 50 million.

National regulatory authorities must apply the following criteria to ensure adequate interconnection: satisfactory end-to-end communications for users; stimulation for a competitive market; promotion and development of trans-European networks and services; and interconnection of national networks and interoperability of services as well as access to such networks and services. Non-discrimination, including equal access and proportionality, and the maintenance and development of universal service must also be ensured.

The national regulatory authorities have powers to intervene to require the fulfilment of certain conditions, and even, if necessary, to impose interconnection terms in the interests of effective competition and interoperability. A disputes procedure, to be operated by the national regulatory authorities, is established under the Directive.

The Directive also addresses the application of essential requirements to interconnection, on the sharing of facilities, numbering, technical standards, interconnection with third countries, publication of information and other procedural provisions.

The Directive contains a number of important annexes setting out details on: specific public telecoms networks and services; organizations with rights and obligations to negotiate interconnection with each other in order to ensure European-wide services; calculating the cost of universal service obligations for voice telephony; a list of elements for interconnection charges; cost accounting systems for interconnection; thresholds for telecommunications turnover; a framework for negotiation of interconnection agreements.

The Directive must be implemented by the Member States by 31 December 1997, and a Commission Report on the operation of the Directive is mandated for 31 December 1999.

Establishment of a coherent telecoms regime at community level

Introduction: ONP and Article 90

The Community has sought to adopt a coherent set of principles aiming at liberalization of the industry and the establishment of regulatory principles. The interrelationship between Article 90 and Article 100a[62] have been at the basis of this strategy. Article 90 Directives cannot be considered in isolation from the important legislative measures adopted in the context of harmonization. The adoption of the Terminal Equipment Directive under Article 90, together with the reinforcement by the Court of Justice of the legitimacy of the Commission in adopting it, opened a new dimension in the use of Commission powers to inject competition into the sector, and forced the pace of deregulation.

Harmonization legislation, adopted under Article 100a by qualified majority voting in the Council, however, has been an equally important instrument in opening up markets by obliging Member States to adapt their standards, regulatory and administrative provisions to conform with common principles, such as the application of essential requirements, at a Community-wide level.

An important issue in the establishment of a coherent telecoms regime at Community level is the effect of the balance created between *harmonization measures based on ONP principles* and *Article 90 measures* to open up the sector to competition. According to the definition in the ONP Framework Directive, adopted under Article 100a, ONP 'concerns the harmonization of conditions for open and efficient access to and use of public telecommunications networks, and, where applicable, public telecommunications services'[63]. The Directive also establishes the ONP conditions applicable – technical interfaces, usage conditions and tariff principles, as well as the basic principles within which ONP conditions operate – in accordance with objective criteria, transparency, equality of access and non-discrimination. Exceptions to the application of open network provision are based on essential requirements and on restrictions arising from the legitimate exercise of special or exclusive rights. The individual Directives adopted on the basis of the Framework Directive (also under Article 100a) follow the general

---

[62] Article 100a provides that the Council, acting by qualified majority on a proposal from the Commission in consultation with the European Parliament and after consulting the Economic and Social Committee, shall decide measures regarding approximation of the legislation, regulatory and administrative provisions of the Member States, whose aim is to ensure the establishment and correct functioning of the Internal Market.

[63] Council Directive 90/387/EEC on the establishment of the internal market for telecommunications services through the implementation of open network provision. OJ 1990 L192/1.

principles and conditions for the operation of ONP in the specific sectors, namely leased lines, packet-switched data services, ISDN, voice telephony and interconnection.

Harmonization measures under Article 100a, particularly through ONP legislation, have put in place a common system of technical and regulatory standards to facilitate and accelerate the introduction of competition into the market, complemented by the legislation adopted under Article 90 with the objective of abolishing special or exclusive rights in order to open up the market to full competition. Legislation adopted under Article 90 has opened up markets; Article 100a legislation, particularly on ONP, has ensured a level playing field.

Interconnection and interoperability

Interconnection and interoperability are identified as key issues in the creation of a coherent Community telecoms regime. Interconnection is considered to be the cornerstone of the liberalized telecoms market, together with the ONP Directives, licensing and universal service. Interconnection and interoperability are particularly relevant in the context of trans-European telecoms networks, and the amended proposal for guidelines for trans-European telecommunications networks[64] specifies that 'the Commission shall undertake actions to ensure interoperability of the networks'[65]; that it is 'necessary to promote networks, interconnection and interoperability at European scale'[66]; but that 'activities undertaken in the context of these guidelines are subject to the full application of the competition rules set out in the Treaty and implementing legislation'[67].

Licensing

A licensing system at European level, based on the principle of proportionality, is envisaged as an important strategy for creating a balanced set of rights and obligations in respect of different categories of service. This regime, when operational, is intended to result in less regulation on service providers and small entities, and correspondingly greater regulation on large operators. Restrictions should be based only on physical limitations and not on an economic assessment of the market's capacity by the national regulator[68].

---

[64] Amended proposal for a European Parliament and Council Decision on a series of guidelines for trans-European telecommunications networks. COM (96) 108, 1996.
[65] Preamble, point 20.
[66] Preamble, point 21.
[67] Preamble, point 22.
[68] European Commission Memo 95/158, 27 November 1995.

Universal service

Universal service obligation is referred to by the Commission[69] as 'a dynamic and evolving concept' which is 'one of the essential elements of this information society and the priority attached to it must be considered in the context of the Community's policy of fully opening telecommunications markets to competition from 1 January 1998'.

The concept of universal service has evolved substantially since the 1993 Council Resolution on the review of the situation in the telecommunications sector[70] established, *inter alia*, the policy goal of liberalization of all public voice telephony services, whilst maintaining universal service. The concept has developed not only through progressions in policy contained in the Council Resolution and the Infrastructure Green Paper, but also through legislation.

The ground has shifted rapidly from the question of *whether* private operators should be at all permitted to provide universal service and perform public service tasks, to *how* to establish mechanisms for universal service provision by all operators, both public and private.

Building on the Infrastructure Green Paper and a Council Resolution on universal services principles in the telecoms sector[71], and a Commission statement of response thereto[72] have been several significant communications and specific legislative measures, further clarifying and expanding the universal service concept.

The Voice Telephony Directive[73] specifies the scope of universal service at EU level and, along with the Interconnection Directive[74], lays down details on the services to be costed in respect of universal services, and the method of the cost calculation. The Interconnection Directive also refers to the principles to be applied on costing and financing of universal service at EU level.

The Commission Communication on universal service for telecommunications in the context of a fully liberalized market[75] has drawn upon the results of an extensive consultation process involving the major interests in the sector in the Member States.

---

[69] Communication on universal service for telecommunications in the perspective of a fully liberalized environment. COM (96) 419, 1996.

[70] Council Resolution of 22 July 1993 on the review of the situation in the telecommunications sector and the need for further development in that market. OJ 1994, C213/1.

[71] Council Resolution on universal services principles in the telecommunications sector, OJ 1994 C48/1.

[72] Commission statement on Council Resolution on universal service in the telecommunications sector, OJ 1994 C48/8.

[73] European Parliament and Council Directive on the application of open network provision (ONP) to voice telephony. OJ 1995 L321/6.

[74] Proposal for a European Parliament and Council Directive on interconnection in telecommunications with regard to ensuring universal service and interoperability through application of the principles of open network provision (ONP), OJ 1995 C313/7, as modified by OJ 1996, C178/3.

[75] Commission Communication on universal service for telecommunications in the perspective of a fully liberalized environment – an essential element of the Information Society. COM (96) 419, 1996.

The Communication, *inter alia*, recommended the clarification of the requirement of affordability; emphasized that in remote rural areas price increases (other than from those adjustments associated with cost orientation) should 'not be used to compensate losses in revenue resulting in price decreases elsewhere', and that any differences in pricing between high and low cost areas should not jeopardize the affordability of universal service. The Communication noted the general support in the consultation process for tariff rebalancing, accompanied by protection for particular groups of users, e.g. the elderly. It also recorded the Commission's intention to encourage actions in the less favoured regions to speed up network digitalization, thereby narrowing the technology gap between regions in the Community.

The current concept of universal service as examined in the Commission Communication on universal service for telecoms in the perspective of a fully liberalized environment poses questions about its future development. For example, how will universal service remain affordable in a liberalized environment; what will happen to uniform national prices post-liberalization; how can universal service be improved in less favoured regions; how can service quality be monitored and how will standards be set; and how will users with disabilities or special needs be adequately provided for? In examining these issues, the Communication paved the way for the subsequent draft Directive on the application of ONP to voice telephony and on universal services for telecommunications in a competitive environment[76]. This draft Directive refers to the key areas of affordability, cost-orientation, special tariffs for certain social groups, detailed billing, dispute settlement procedures and the introduction of a European number for free calls.

The Commission intends to issue a report by 1 January 1999 appraising the scope, level, quality and affordability of universal service. Clues as to the type of criteria envisaged in extending the universal service concept are listed in the conclusions: universal service 'must combine a market-based analysis for demand for and availability of the service with a political assessment of its social and economic desirability'; public–private initiatives are foreseen to play 'a major role' in the development of new services, especially in the less favoured areas; and public access is regarded as 'an important means of bringing the information society to every citizen'.

Key principles emerging from these communications and legislative measures are as follows:

- the recognized need to develop and maintain a public telecoms network and services based on universal provision which guarantees services of specific quality, ensures rights of access, and is provided at reasonable and affordable prices throughout the territory;

[76] This draft Directive is discussed in detail earlier in the chapter. It is to replace Directive 95/62/EC, also discussed above.

- the universal service should promote interconnection to public networks and facilitate interoperability between different national and Community operators;
- where universal service provision of basic voice telephony can only be provided at a loss, Member States may establish universal service schemes to finance its provision. This may be by means of access fees, or other compensation mechanism. A national universal service fund may be created. Whatever mechanism is used, financing of universal service should take account of the principles of transparency, non-discrimination and proportionality and should comply with the EU competition rules;
- universal service should be provided for social reasons, and should protect particular groups of users, such as the elderly and disabled, to ensure that they do not leave the network as a result of the costs involved;
- universal service should make special provision for remote and less favoured areas;
- tariff rebalancing should be taken into account;
- cross-subsidization should be avoided;
- universal service is an evolving concept and its scope should continue to be monitored.

Application of the EC competition rules

The Commission has been particularly active in ensuring that the competition rules are applied to the telecoms sector. It has broken down national monopolies and given private operators an opportunity to take an initiative in the market. The Terminal Equipment and Services Directives, and subsequent legislation based on Article 90, have their legal basis in the competition rules of the Treaty. The Commission has attempted, under Articles 85, 86 and 90 of the Treaty, as well as under the Merger Regulation, to prevent restrictive practices and control monopolies.

It has taken Decisions addressed to Member States based on Article 90(3) ordering them to change regulations which led to an infringement of Article 86 by their national telecoms operators. It has also taken Decisions under the Merger Regulation and under Articles 85 and 86 to deal with mergers and strategic alliances falling short of a merger. The Commission merger Decisions and its Decisions and Notices based on Articles 85 and 86 relating to major national telecoms operators are worth mentioning briefly even though they do not involve an examination of Article 90 because they help understand the application of key competition law principles to state-related undertakings.

Article 90(3) Decisions

The Commission has taken several Decisions based on Article 90(3) against Member States which made it more difficult for new GSM operators to enter the market by imposing on them a fee which did not apply and had never been applied to the incumbent telecoms operators. A decision was thus taken against Italy in the *Omnitel Pronto Italia* case[77], against Spain[78], as well as against Belgium and Ireland.

In all these cases, the Commission found that the Member States concerned were discriminating in favour of the incumbent operator by asking a substantial fee from the new entrant to the market without asking a similar fee from the incumbent. In all these cases the Commission ordered the Member States to take measures so as to restore the balance.

The Commission has also taken proceedings against Member States which unfairly refused to license a new operator. An example of this is *Vebacom*[79] – a German company which tried to obtain a licence for a broadband telecommunications network based on SDH (synchronous hierarchy) technology making it possible to transfer data between thirty-six different sites of the German public television broadcaster ARD. This was a new service based on new technology which the Commission considered beneficial to customers. Negotiations took place between the Commission and the German government and the case was eventually settled in favour of Vebacom: the German authorities agreed to grant a licence to Vebacom for the establishment and operation of an alternative telecommunications network.

Mergers

Several merger cases involving telecoms operators in the broadcasting sector have been scrutinized by the Commission and eventually prohibited on the ground that they created a risk of foreclosure of the market for potential new entrants. These cases are examined in chapter 11 on broadcasting. Various mergers affecting the telecoms market – as opposed to broadcasting – have been approved without involving a second phase investigation[80]. One significant project, however, has given rise to a full investigation by the Commission

[77] OJ 1995 L280/49.

[78] Commission Decision of 18 December 1996 concerning conditions imposed on the second operator of GSM radiotelephony service in Spain, OJ 1996 L 76/19.

[79] Commission Decision of 16 August 1995 declaring a concentration to be compatible with the Common Market (Case IV/M.683 *Hermes* Inc./HIT Rail BV according to Council Regulation (EEC) no 4064/89 OJ 1996 C-157/13).

[80] Examples include the acquisition by ATT of certain business units of Philips Electronics NV in the market for the provision of public telecommunications equipment (IP/96/129); the acquisition by Ameritech International, Tele Danemark and Singapore Telecom of a strategic interest in Belgacom (IP/96/192), the *Hermes* joint venture between the American company GTS and European National Railway Companies. Commission Decision of 18 December 1996 relating to a proceeding under Article 85 of the EC Treaty and Article 53 of the EEA Agreement (Case IV/35.518 Iridium) OJ 1997 L 16/87.

under the Merger Regulation, namely the *BT/MCI* case[81]. The Commission was concerned that the merger between British Telecoms and MCI – an American diversified communications company – might increase BT's already dominant position in particular on the UK–US route and might give BT a dominant position in the field of audio-conferencing in the UK. Following undertakings by BT and MCI the Commission finally cleared the proposed merger. The parties undertook to make capacity available to competitors on the key trans-Atlantic submarine cable between the US and the UK. In addition, MCI undertook to divest itself of its audio-conferencing business in the UK.

Strategic alliances between undertakings which did not merge their operations but remained independent have been examined by the Commission in the light of Articles 85 and 86 pursuant to Regulation 17/62.

Articles 85 and 86 Decisions

The following decisions and notices are of particular interest: *Atlas*[82], *Global One*[83], *Unisource*[84], *Uniworld*[85], *Iridium*[86], *Global European Network*[87] and *Belgacom/ITT Promedia*[88]. All these cases had the following features in common:

- they involve *major national telecoms operators in Europe*;
- the Commission imposed *strict obligations* on the operators;
- they illustrate the application of *key principles* including right of access by third parties, cost orientation, transparency, non-discrimination, absence of cross subsidization.

In addition in *Atlas, Global One, Unisource* and *Uniworld*, the Commission *entered into negotiations with Member States governments* and not just the telecoms operators concerned. This may seem surprising given that the decisions are based on Article 85(3) and are accordingly addressed to undertakings and not to governments, unlike decisions based on Article 90(3). Negotiations with Member States took place in parallel to negotiations with telecoms operators. In return for an approval by the Commission, Member States have effectively been led into quickening the pace of liberalization of their telecoms network in advance of the deadline set in the liberalization Directives. This demonstrates the close relationship between the application of the competition rules based on Articles 85 and 86 of the Treaty and the

---

81 *BT/MCI* (IP/97/406).

82 Commission Decision in Case IV/35.337 *Atlas*, OJ 1996 L239/23; Commission Decisions concerning the conditions imposed on the *second operator of GSM radiotelephony services in Spain*, OJ 1997 L76/19, and on *second GSM operator in Italy*, OJ 1995, L280/49.

83 Commission Decision in Case IV/35.617 *Phoenix* v. *Global One*, OJ 1996 L239/57.

84 Commission Notice in Case IV/35.830 *Unisource* v. *Téléfonica*, OJ 1997 C44/15.

85 Case IV/35.738 *Uniworld*, OJ 1997 C44/4.

86 Notice in Case IV/35.518 *Iridium*.

87 *The Global European Network* (IP/97/242).

88 *Belgacom/ITT Promedia* (IP/97/292).

legislative measures taken by the Commission and the Council. This relationship is consistent with the Commission policy expressed in the Commission guidelines on competition in the telecoms sector and the draft Notice on the application of the competition rules to access agreements examined below.

Atlas and Global One

These two cases involved joint ventures between major telecoms operators. Atlas was a joint venture between France Telecom (FT) and Deutsche Telekom (DT) for the provision of telecommunications services to large users in Europe. The type of services offered would include network services, outsourcing and very small aperture satellite (VSAT) services.

Global One was an alliance between Atlas and Sprint for the supply of these services world-wide. It would also provide services to other telecommunications organizations. The Commission was concerned that the joint venture might reinforce the dominant position of FT and DT in certain telecommunications sectors.

The Commission took a Decision exempting both joint ventures from the prohibition contained in Article 85(1) but imposed stringent conditions on the parties. Of particular significance is the fact that the Commission felt that it could only authorize the joint venture if the telecommunications market was more open in France and Germany. The Commission could not impose such a condition on the telecoms operators themselves since this was a matter for the Member States which would involve a change in their national laws. Nevertheless because it was clear that the Commission would not agree to the joint ventures in the absence of further liberalization, the Member States undertook to take the measures necessary to open up their market in advance of the deadline set out in the existing Directives. The Commission effectively made the *Atlas and Global One* joint ventures conditional on the granting of the first two infrastructure licences in France and Germany, which took place on 1 December 1996. The Commission also imposed several conditions on the parties themselves including an obligation not to discriminate and not to cross subsidize.

Unisource and Uniworld

These two cases also involved joint ventures between major national telecoms operators. Unisource is between the Dutch and Swiss national telecoms operators and Telia of Sweden. Uniworld is a joint venture between Unisource and AT&T.

The Commission has issued detailed notices in respect of the two operations indicating that it is prepared to take favourable decisions but subject to a number of conditions.

As in *Atlas* and *Global One*, the conditions include a faster liberalization of the telecommunications market than originally anticipated in Spain and Switzerland. They also include various undertakings regarding interconnection, access and accounting rates.

Iridium

This case involved a joint venture between several parties including Motorola of the United States, Stet of Italy and Vebacom of Germany. The purpose of the joint venture is to provide global digital wireless communications services using satellites. The Commission found that the joint venture fell outside the scope of Article 85(1) because none of the parties could alone have entered the field of activity of the joint venture on financial and technical grounds.

Nevertheless because of Stet's strong position in the provision of satellite services in Italy, the Commission asked the parties to confirm that Iridium would not affect the ability of other interested operators to gain access to the telecommunications infrastructure of Stet other than those parts specifically developed for Iridium.

Global European Network

This case involved an agreement to provide high quality digital links between Member States. The agreement was entered into by five major European telecommunications operators, namely British Telecom, Deutsche Telekom, France Telecom, Telecom Italia and Téléfonica de España.

The Commission agreed to clear Global European Network on condition that the parties refrained from entering into a collective pricing arrangement and that they negotiated access to their network on a bilateral basis. In addition they were to offer access to third parties on a similar and non-discriminatory basis.

Belgacom/ITT Promedia

This case related to the conditions applied by the Belgian telecoms operator, Belgacom, to the supply of data necessary for the compilation of telephone directories. The data consisted of information on subscribers of Belgacom's voice telephone services. Belgacom would only communicate the data to ITT Promedia and other third parties at very high prices. The Commission considered that the prices were excessive and constituted a *prima facie* abuse of a dominant position.

The Commission asked Belgacom to apply more cost oriented prices which would nevertheless enable Belgacom to recover the costs of collecting and processing the data obtained and to add a reasonable profit margin. The Commission indicated that this cost orientation approach should lead to a reduction of more than 90 per cent in the amount originally charged to telephone directory publishers. The case was eventually settled on the basis that Belgacom would adopt a cost oriented approach for the supply of data which were under its exclusive control. With regard to other data that directory publishers could obtain from alternative sources, Belgacom was entitled to adopt a market oriented approach.

Competition rules and the telecommunications sector: Guidelines and a draft Notice

The principles of non-discrimination, transparency, cost orientation, and access as applied in the cases cited above offer a good illustration of the principles set out by the Commission in two important documents:

- Guidelines on the application of the competition rules to the telecommunications sector; and
- Draft Notice on the application of the competition rules to access agreements. This second document constitutes a specific application of the guidelines.

Guidelines on the application of the competition rules in the telecommunications sector: the Guidelines and Article 90

In June 1991 the Commission published guidelines on the application of the competition rules in the telecommunications sector[89] ('the Guidelines'). The Guidelines are addressed to public telecommunications operators and other business interests in the sector. They set out the key competition rules based on the jurisprudence of the Court of Justice (notably the *Terminal Equipment* and *Telecommunications Services* cases).

The Guidelines deal essentially with the application of Articles 85 and 86 of the Treaty. These Articles apply to the conduct of undertakings and not Member States. However, the Guidelines also point out that under Article 5(2) of the Treaty, Member States must refrain from any measure that could jeopardize the effectiveness of these Articles. Thus, if a Member State were to encourage breaches of Articles 85 and 86, for example, by encouraging abuses by dominant undertakings, such as requiring illegal cartels by public undertakings or undertakings that have been granted special or exclusive rights, then Article 90 could apply.

The Guidelines distinguish between certain situations in connection with such undertakings:

(1) where the conduct is entirely autonomous. In such a case, the undertaking can only be caught by Articles 85 and 86;

(2) where the conduct is imposed by a State measure and the undertaking is deprived of choice. In such a case, Article 90 may apply to the State in conjunction with Articles 85 and 86. The latter articles might be applied to the conduct of undertakings, taking into account the constraints imposed by the mandatory State intervention;

---

[89] Commission Guidelines on the application of EEC competition rules in the telecommunications sector, OJ 1991 C233/2.

(3) where the conduct arises from the free choice of the undertaking, but where the State has taken measures to encourage or strengthen the effects of its behaviour. In such a case, the undertaking would be caught by Articles 85 and/or 86 whilst Article 90 may be applied to the State measure.

The Article 90(2) exception for undertakings entrusted with the operation of services of general economic interest or having the character of a revenue-producing monopoly, may apply to both State measures and to practices by companies. For example, the Guidelines refer to a provision in the Telecommunications Services Directive which provides for a Member State to impose specified restrictions on the licences it can grant for telecommunications services. In the Commission's view, these restrictions may be imposed under Article 90(2). It also notes, however, that where Article 90(2) is invoked concerning a telecommunications organization's conduct which brings about restrictions in competition which its Member State did not impose pursuant to the Telecommunications Services Directive, it would be very hard to prove that the Article 90(2) exception applied. The Guidelines provide that account should be taken of the fact that in such a case, the State, whose role is to protect the public and the general economic interest, did not consider it necessary to impose the restrictions in question. It would thus be difficult to establish proof that the Article 90(2) exemption could apply to a company's behaviour involving restrictions of this nature.

The Commission infers from previous case-law of the Court[90] that it has exclusive competence, subject to control by the Court, to decide that the Article 90(2) exception applies and that national authorities may only find that the exception does *not* apply. However, as has been demonstrated in the first part of this book, this interpretation of the Commission is not correct[91]. It is now clear that the Commission shares the right both to reject the application of the Article 90(2) exemption as well as the right to apply it with the national authorities. The Guidelines further correctly recommend that where national authorities are unable to assess this issue clearly, they should suspend the decision to enable the Commission to find that the conditions for the application of 90(2) have been fulfilled.

The Guidelines define the relevant markets both in terms of products/services and in term of geographic coverage. It then gives specific examples of Article 85 and Article 86 infringements in the telecoms sector. It deals in particular with agreements on pricing; technical standards; exchange of information; different types of joint ventures; the issue of access to the network; and cross subsidization.

---

90 Case 10/71 *Müller-Heim* [1971] ECR 723, Case 66/86 *Ahmed Saeed* [1989] ECR 803.
91 Refer to chapter 3 for a detailed analysis of the new interpretation of Article 90(2).

Draft Notice on access[92]

The draft Notice is essentially a formal clarification of the access principles established under EU competition law and mentioned in the Guidelines and of the relationship between competition law and specific legislation adopted under Article 100a, particularly as between the competition rules and ONP legislation. It is also an assertion that the Commission intends to apply the Commission rules with consistency across the converging sectors involved in the provision of new multimedia services.

The draft Notice builds on the competition Guidelines referred to above by examining the issues connected with access. It examines the relationship between access agreements and the competition rules contained, in particular, in Articles 85 and 86 and the Merger Regulation; and refers to Commission powers of investigation and imposition of penalties for breach of the competition rules. It examines the concepts underlying Articles 85 and 86 such as relevant markets, dominance and its abuse by dominant access providers, and the restrictions imposed by Article 85.

The Commission explains that the market should be determined by looking primarily at substitutability from the point of view of the demand rather than from the point of view of the offer. Two products or services are thus likely to form part of the same market if they are substitutable in the eyes of users or consumers. Dominance of a specific market should be established by looking primarily at the turnover of a telecommunications operator as well as at the number of its customers. The draft Notice lays the emphasis not just on market dominance but also on market power which will normally be presumed when an operator has a market share above 25 per cent. The Commission explains that the relevant geographic market for the purpose of assessing dominance or market power is likely to be the national market for so long as full liberalization has not taken place across Europe, with large players operating on a significant basis across several Member States. The draft Notice recognizes that the ending of legal monopolies will not, *per se*, result in an end of dominance, since the development of effective competition from alternative networks will take time.

On procedure, the draft Notice explains the complementary rules of all relevant competition authorities besides the Commission, namely the national regulatory authority, the national competition authorities and national courts. The Commission explains that in many cases, it may be more efficient for a complainant to seek redress before its national institutions and, in particular, national courts where urgent interim measures are necessary.

The draft Notice states that the ONP framework imposes certain obligations on national telecoms operators that extend beyond those normally imposed by Article 86, particularly concerning strict standards in respect of

---

[92] Draft Communication from the Commission on the application of the competition rules to access agreements in the telecommunications sector. COM(96) 649, 1996.

transparency, obligations to supply and pricing practices. Enforcement of these obligations lies with the national regulators; consequently the draft Notice highlights the significance of the role of the independent national regulatory authorities (NRA): 'Community competition rules are not sufficient to remedy the various problems in the telecommunications sector. NRAs therefore have a significantly wider ambit and a significant and far-reaching role in the regulation of the sector'.

An important principle underlying the draft Notice is that of network access provision to new market entrants by the established telecommunications organizations, and the Commission shows that it intends to scrutinize dominance arising from control of essential facilities (e.g. interconnection facilities). The Commission gives a definition of the concept of essential facility which is similar to that adopted in its case-law[93]. Whilst refusal of access will be abusive only if it affects competition, the draft Notice points out that in all cases of refusal, the justification for refusal will be closely examined to determine whether it is objective. According to the Notice, withdrawal of access will usually be an abuse of dominance.

The draft Notice gives various examples of abuse. One of them is predatory pricing. Interestingly, the Commission indicates that the *Akzo*[94] test on predatory pricing based primarily on average variable costs may not be adequate in the telecoms sector because the variable costs of providing access to an already existing network are insignificant. The Commission therefore proposes a new test which would not just be based on average variable costs but would also take into account, to some extent, average total costs.

Another form of pricing abuse is discriminatory pricing. The Commission generally considers that price differentiation based on the use which is made of access to the network will normally constitute an abuse by an access provider which occupies a dominant position on the market. However, the Commission seems prepared to accept that in some limited circumstances a dominant access provider may apply different prices based on the use which is made of the access, provided this does not restrict competition between users and it is justified by a better utilization of the infrastructure and a better development of certain markets[95].

The draft Notice thus accentuates the importance of access agreements to the sector:

> Access agreements can have significant pro-competitive effects as they can improve access to the downstream market. Access agreements in the context of interconnection are essential to interoperability of services and infrastructure, thus increasing

---

[93] See Commission definition of essential facilities in Commission Decision of 21 Dec. 1993 relating to a proceeding pursuant to Article 86, *Sea Containers* v. *Stena Sealink*, interim measures, OJ 1994 L15/8, and in Commission Decision of 21 Dec. 1993 concerning a refusal to grant access to the Port of Rødby (Denmark), OJ 1994 L55/52.

[94] *Akzo*, Case C-62/86 [1991] ECR III-359.

[95] Para. 93 and footnote 68 of the draft Notice.

> competition in the downstream market for services, which is likely to involve higher added value than local infrastructure.

Consequently the Commission intends to exercise vigilance in examining access agreements in order to safeguard competition:

> The Commission will therefore pay particular attention to scrutinizing access agreements in the context of their likely effects on the relevant markets in order to ensure that such agreements do not serve as a hidden and indirect means for fixing or co-ordinating end-prices for end-users, which constitutes one of the most serious infringements of Article 85 EC.

Access agreements will also be assessed for their potential anti-competitive effects in respect of market sharing, exclusivity, and exchange of commercially sensitive information between the parties.

The draft Notice also notes that the Commission intends to 'avoid unnecessary duplication of procedures', especially competition procedures and EU and national regulatory procedures established under the ONP framework.

Conclusion

This chapter has examined the major themes in Community telecommunications policy, many of which are also common to the other liberalizing sectors (broadcasting, energy, transport, postal services). Key communications from the institutions and legislative measures refer to these common concepts. For example, access to an essential facility (i.e. access to a facility or infrastructure without which competitors would be unable to provide services to customers) underlies the principle of access contained, for example, in the Infrastructure Green Paper, draft Interconnection Directive etc. Other significant general concepts to be applied to the sector are those laid down under ONP principles, namely proportionality, transparency, non-discrimination. The concept of unbundling – notably referring in the telecoms sector to the separation of the management of infrastructure and the provision of services based on infrastructure, as well as to the separation of accounts – is considered to be a crucial condition for a liberalized market. Universal service (i.e. an essential service provided to all consumers throughout a territory on similar terms and conditions) is a key concept applied to the telecommunications sector, but common to other liberalizing sectors.

These concepts are applied in Community telecoms law and policy to enable the sector to conform to a set of general principles at EU level, in order to facilitate the sector to compete globally as well as within the internal market.

Telecommunications is by far the most rapidly evolving sector of those sectors under liberalisation. European telecommunications are operating within a global market of intense competition, particularly with the US and

Japan. The sector must therefore be able to innovate, act and react speedily to technical advancement. The abolition of special and exclusive rights in the sector, at an earlier stage than for other sectors, reflects a need for the adaptability necessary to match global competition. This has also had the effect of precipitating strategic alliances between undertakings in Member States and between undertakings in Member States and third countries. Open competition is expected to have positive effects for consumers, giving them greater choice, better quality and value for money.

Liberalization, however, also poses potential problems. Unless the regulatory framework is clearly defined to clarify the rules for state bodies and private operators, there could be confusion over the responsibilities and obligations of both private and public sectors. In particular, issues of licensing, interoperability, interconnection, access and universal service principles have to be considered in a manner that permits their practical application in the new telecommunications market, where the demarcation lines between public and private operators have lost their rigidity, and to allow restructuring, not only between private operators but also between private and public operators to facilitate their proper functioning in a liberalized market. There will also have to be vigilance over access charges to avoid discrimination in the charging of entities competing with national telecommunications authorities.

The convergence between telecommunications, information technology and audio-visual technologies has blurred the distinction between monopoly operators and new competitive services. This was recognized as early as the 1987 Telecommunications Green Paper, which noted that this convergence 'is outdating traditional boundary lines between the telecommunications network and terminals sector, and between services traditionally provided under monopoly and those provided in a competitive environment'[96].

The effect of the abolition of special and exclusive rights in today's climate of privatization and market restructuring, will require careful monitoring of the application of the competition rules and the harmonization legislation adopted under Article 100a, to ensure that the liberalized regime does not create new problems for competition policy in the newly integrating markets. Commissioner Van Miert has emphasized that the Member States are committed, politically and legally, to the 1998 liberalization timetable. While this is considered to be a significant achievement in itself, the 'true impact' is expected to be felt only when there is 'full and effective implementation' of these commitments. The Commissioner states that he is 'determined to use the full potential of the competition rules of the Treaty to maintain a tough stance in this area. We cannot tolerate tardy or incomplete implementation at this stage and luckily we have the tools to impose this'. It is therefore important that the EC regulatory framework is clear, and that it facilitates a predictable regulatory climate in which market entry is not impeded, investment is encouraged and universal service is guaranteed.

---

[96] COM (87) 290, 1987.

SELECTIVE LIST OF SIGNIFICANT EU DOCUMENTS IN THE TELECOMMUNICATIONS SECTOR

1. Policy documents

- Green Paper on the development of the Common market for telecommunications services and equipment. COM (87) 290, 1987.
- Green Paper on a common approach in the field of satellite communications in the European Community. COM (90) 490, 1990.
- Commission guidelines on the application of EEC competition rules in the telecommunications sector. OJ 1991, C 233/2.
- White Paper on growth, competitiveness and employment (Delors, White Paper). COM (93) 700, 1993.
- Green Paper on a common approach to mobile and personal communications in the European Union. COM (94) 145, 1994.
- Bangemann Group Report on Europe and the Information Society – Recommendations to the European Council and Commission Communication on Europe's way to the Information Society. COM (94) 347, 1994.
- Green Paper on liberalization of telecommunications infrastructure and cable television networks:
  Part one: Principle and timetable, COM (94) 440, 1994.
  Part two: A common approach to the provision of infrastructure for telecommunications in the European Union. COM (94) 682, 1994.

2. Commission Communications

- Commission Communication on the 1992 review of the situation in the telecommunications services sector. SEC (92) 1048, 1992.
- Commission Communication to the Council and European Parliament on the consultation on the review of the situation in the telecommunications sector. COM (93) 159, 1993.

- Commission Communication: Towards cost orientation and the adjustment of pricing structures – telecommunications tariffs in the Community. SEC (92) 1992.
- Commission Communication to the Council and European Parliament on developing universal service for telecommunications in a competitive environment. COM (93) 543, 1993.
- Commission Communication on the consultation on the Green Paper on mobile and personal communications in the European Union. COM (94) 492, 1994.
- Commission Communication on the consultation on the Green Paper on the liberalization of telecommunications infrastructure and cable television networks. COM (95) 158, 1995.
- Commission Communication on the future development of the market for directories and other telecommunications information services in a competitive environment. COM (95) 431, 1995.
- Draft Commission Communication on the application of the competition rules to access agreements in the telecommunications sector. COM (96) 649, 1996.

3. Council and European Parliament Statements, Resolutions, Recommendations

- Council Recommendation concerning the implementation of harmonization in the field of telecommunications. OJ 1984, L298/49.
- Council Recommendation concerning the first phase of opening up access to public telecommunications contracts. OJ 1984, L298/51.
- Council Recommendation on the co-ordinated introduction of the integrated service digital network (ISDN) in the European Community. OJ 1986, L382/36.
- Council Recommendation on the co-ordinated introduction of public pan-European cellular digital land-based mobile communications in the Community. OJ 1987, L196/81.
- Council Resolution on the development of the common market for telecommunications services and equipment up to 1992. OJ 1988, C257/1.
- Council Resolution concerning standardization in the fields of information technology and telecommunications. OJ 1989, C117/1.
- Council Resolution on the strengthening of the co-ordination for the introduction of the integrated service digital network (ISDN) in the European Community. OJ 1989, C196/4.

- Council Resolution concerning trans-European networks. OJ 1990, C27/8.
- Council Resolution on the strengthening of the European-wide co-operation with radio frequencies, in particular with regard to services with a pan-European dimension. OJ 1990, C166/2.
- Council Recommendation on the co-ordinated introduction of pan-European land-based public radio paging in the Community. OJ 1990, L310/28.
- Council Resolution on the final stage of the co-ordinated introduction of pan-European land-based public digital mobile cellular communications in the Community (GSM). OJ 1990, C329/25.
- Council Recommendation on the co-ordinated introduction of digital European cordless telecommunications (DECT) into the Community. OJ 1991, L144/47.
- Council Resolution on the development of the common market for satellite communications services and equipment. OJ 1992, C8/1.
- Council Resolution on the development of the integrated services digital network (ISDN) in the Community as a European-wide telecommunications infrastructure for 1993 and beyond. OJ 1992, C158/1.
- Council Recommendation on the harmonized provision of a minimum set of packet-switched data services (PSDS) in accordance with ONP principles. OJ 1992, L200/1.
- Council Recommendation on the provision of harmonized integrated services network (ISDN) access arrangements and a minimum set of ISDN offerings in accordance with ONP principles. OJ 1992 L200/10.
- Council Resolution on the promotion of European-wide co-operation on numbering of telecommunications services. OJ 1992, C318/2.
- Council Resolution on the assessment of the situation in the telecommunications sector. OJ 1993, OJ C2/5.
- European Parliament Resolution on the Commission Communication 'Towards cost orientation and the adjustment of pricing structures – telecommunications tariffs in the Community'. OJ 1993, C150/37.
- European Parliament Resolution on the Commission's 1992 review of the situation in the telecommunications services sector. OJ 1993, C150/39.
- Council Resolution on the review of the situation in the telecommunications sector and the need for further development in that market. OJ 1993, C213/1.
- Council Resolution on the introduction of satellite personal communications services in the Community. OJ 1993, C399/1.

- Council Resolution on universal service principles in the telecommunications sector (OJ 1994, C48/1) and Commission statement concerning Council Resolution on universal service in the telecommunications sector (OJ 1994, C48/8).
- European Parliament Resolution on the communication from the Commission accompanied by the proposal for a Council Resolution on universal service principles in the telecommunications sector. OJ 1994, C205/551.
- Council Resolution on the principle and timetable for the liberalization of telecommunications infrastructure. OJ 1994, C379/4.
- European Parliament Resolution on the Commission Green Paper on the liberalization of telecommunications infrastructure and cable television networks (Part One). OJ 1995, C109/3.
- European Parliament Resolution on the Commission Green Paper on a common approach in the field of mobile personal communications in the European Union and on the consultation on the Green Paper. OJ 1995, C151/473.
- European Parliament Resolution on the Green Paper on the liberalization of telecommunications infrastructure and cable television networks (Part Two). OJ 1995, C151/479.
- Council Resolution on the further development of the mobile and personal communications sector in the European Union. OJ 1995, C188/3.
- Council Resolution on the implementation of the future regulatory framework for telecommunications. OJ 1995. C258/1.
- Council Resolution on the industrial aspects for the European Union in the development of the Information Society. OJ 1995, C341/3.

4. Liberalization measures

- Commission Directive 88/301/EEC on competition in the markets for telecommunications terminal equipment. OJ 1988, L131/73.
- Commission Directive 90/388/EEC on competition in the markets for telecommunications services. OJ 1990, L192/10.
- Commission Directive 94/46/EC, amending Directive 88/301/EEC and Directive 90/388/EEC in particular with regard to satellite communications. OJ 1994, L 268/15.
- Commission Directive 95/51/EC, amending Directive 90/388/EEC with regard to the abolition of the restrictions on the use of cable television networks for the provision of already liberalized telecommunications services. OJ 1995, L256/49.

- Commission Directive 96/2/EC, amending Directive 90/388/EEC with regard to mobile and personal communications. OJ 1996, L20/59.
- Commission Directive 96/19/EC, amending Directive 90/388/EEC with regard to the implementation of full competition in the telecommunications markets. OJ 1996, L74/13.

5. Harmonization measures

- Council Directive 90/387/EEC on the establishment of the internal market for telecommunications services through the implementation of open network provision. OJ 1990, L192/1.
- Council Directive 92/44/EEC on the application of open network provision to leased lines. OJ 1992, L165/27.
- European Parliament and Council Directive 95/62/EC on the application of open network provision to voice telephony. OJ 1995, L321/6.
- Draft European Parliament and Council Directive on interconnection in telecommunications with regard to ensuring universal service and interoperability through application of the principles of open network provision (ONP). OJ 1996, C220/13.
- Draft European Parliament and Council Directive amending Directives 90/387/EEC and 92/44/EEC for the purpose of adaptation to a competitive environment in telecommunications. COM (95) 543, 1995.

6. Cases

- Case 41/83 *Italian Republic* v. *Commission of the European Communities* (British Telecom) [1985] ECR 873.
- Case C-18/88 *Régie de Télégraphe et de Téléphone (RTT)* v. *GB-INNO-BM SA* [1991] ECR I-5941.
- Case C-202/88 *France* v. *Commission of the European Communities* (Terminal Equipment) [1991] ECR I-1259.
- Joined cases C-271/90, C-281/90 and C-289/90, *Spain, Belgium and Italy* v. *Commission of the European Communities* (Telecom services) [1992] ECR I-5833.

Chapter 11

COMPETITION AND EXCLUSIVE RIGHTS IN THE BROADCASTING SECTOR

Introduction

Article 90 is of specific application to the broadcasting sector through the Directives on satellite communications and telecommunications services adopted under the legal base of Article 90, and referred to in detail in Chapter 10 on telecommunications. Otherwise, broadcasting is regulated in the Community principally by the Television Without Frontiers Directive[1], which has as its legal base Treaty provisions on right of establishment and free movement of services. There is also a substantial body of case-law of the Court concerning infringements of competition law resulting from restrictions on free movement connected with the grant by Member States of special or exclusive rights to national broadcasting monopolies. Thus Article 90 has been considered in conjunction with Treaty provisions on free movement and other Treaty Articles in several important cases. Jurisprudence of the European Court of Justice has also developed in ancillary areas of broadcasting as will also become evident later in this chapter.

The broadcasting sector is developing at a phenomenally fast rate due in particular to unprecedented technological advances in the broadcasting, computing and telecommunications sectors, coupled with a ground swell of international competition. The European Union is under tremendous pressure to establish conditions for effective global competition by European broadcasters. This is no easy task, given the fragmented national broadcasting markets and friction by some Member States over Community policy in the area. The Commission is tackling the broadcasting sector on several fronts: infrastructure liberalization which is driven by the telecommunications sector (which will not be covered in this chapter); regulation through the broadcasting activities (principally under the Television Without

[1] Council Directive of 3 October 1989 on the co-ordination of certain provisions laid down by law, regulation or administrative action in Member States concerning the pursuit of television broadcasting activities (89/552/EEC), OJ 1989, L298/23.

Frontiers Directive); development of case-law in the sector arising from several key referrals to the Court; issues relating to conditional access, media ownership and the technical implications of digital broadcasting.

The Television Without Frontiers Directive was adopted in 1989 as part of the single market programme and it entered into force in 1991.

The interpretation and implementation of the Television Frontiers Directive has been controversial and there has long been wide agreement that it requires amendment. There has, however, been wide disagreement as to the form the amendment should take, the outcome of which will be discussed later. To promote a coherent European Union policy on broadcasting and related areas, the Commission has also issued a series of Green Papers to stimulate discussion through consultation processes involving interested parties in the Member States. In particular, Green Papers have been issued on:

- strategy options to strengthen the European programme industry[2];
- legal protection of encrypted services[3] plus a follow-up proposal for a Directive[4];
- pluralism and media concentration in the internal market plus a follow-up communication[5];
- digital video broadcasting[6], and;
- multimedia and new media services[7].

The chapter also refers to key Commission Decisions taken under the Merger Regulation because of their significance for the enforcement of competition law.

Regulation of broadcasting in the European Union

The Television Without Frontiers Directive[8]

The principal purpose of the Directive was to remove restrictions on the freedom to provide broadcasting services within the Community.

[2] Strategy Options to Strengthen the European Programme Industry in the Context of the Audiovisual Policy of the European Union, COM(94) 96 final; 6 April 1994.

[3] Legal protection for encrypted services in the internal market. Commission consultation on the need for Community action COM(96) 76, 1996.

[4] Draft proposal for a European Paliament and Council Directive on the legal protection of services based on, or consisting of, conditional access. Provisional version of 9 July 1997.

[5] Pluralism and Media Concentration in the Internal Market. An Assessment of the Need for Community Action, COM(92) 480 final; 23 Dec. 1992. Follow-up to the Consultation Process Relating to the Green Paper on Pluralism and Media Concentration in the Internal Market – an Assessment of the Need for Community Action, COM(94) 353 final; 5 Oct. 1994.

[6] Digital Video Broadcasting. A Framework for Community Policy, plus Draft Council Resolution on a Framework for Community Policy on Digital Video Broadcasting, COM(93) 557 final; 17 Nov. 1993.

[7] Commission Green Paper on the protection of minors and human dignity in audio-visual and information services COM(96) 483, 1996.

[8] See footnote 1 above.

Recognizing that the need to obtain licences or authorizations and to comply with applicable law in every Member State to which a television service is broadcast acted as a barrier to the free movement of broadcasting services, the Television Without Frontiers Directive established the principle that those services should be regulated by the laws of the country of origin (though as we will see later in this chapter, establishing which State has jurisdiction has not been without difficulty) and that other Member States could not restrict retransmission of those services. Thus a licence to broadcast granted by one Member State should act as a 'passport' for retransmission in other Member States.

The Television Without Frontiers Directive recognized that, hand in hand with the concept of the passport, there was a need to establish certain minimum standards with which all broadcasts must comply (in relation to the protection of minors, rules governing advertising and sponsorship and a requirement to establish a right to reply). The minimum standards were also seen as a way of stimulating the European audio-visual industry, in the form of the controversial European works and independent productions quotas. The Television Without Frontiers Directive applies to television broadcasting which is defined as the initial transmission by wire or over the air, including satellite, in unencoded or in encoded form, of television programmes intended for reception by the public.

As outlined above, the Member States are required under the Television Without Frontiers Directive to ensure the freedom of reception and must not restrict retransmission on their territory of television broadcasts from other Member States (Article 2). Exceptions to this requirement are strictly limited, for example, to the protection of minors in particular from pornography or excessive violence, or to incitement to hatred on grounds of race, sex, religion or nationality (Article 22).

Each Member State must ensure that television broadcasts transmitted by broadcasters under its jurisdiction, or broadcasters who, while not under the jurisdiction of any Member State, use frequency or satellite capacity granted by, or a satellite link situated in, that Member State, comply with the law applicable to broadcasts intended for the public in that Member State.

The Directive lays down specific requirements in the following main areas:

- Member States must ensure that 'where practicable' broadcasters reserve for *European works*, a 'majority proportion' of their transmission time, not including time allocated for news, sport, games, advertising and teletext services (Article 4). This effectively means that over 50 per cent of programmes shown by channels should be of European origin.

  'European works' refers to works originating from the Member States of the Union, or from European third countries party to the Council of

Europe's European Convention on Transfrontier Television, provided that they comply with certain conditions (Article 6)[9].

- At least 10 per cent of transmission time excluding that given over to news, sport, games, advertising and teletext, or alternatively, at the discretion of the Member State, at least 10 per cent of programming budgets must be reserved by broadcasters for European works created by producers who are independent of broadcasters (Article 5): again, the requirement is to do so 'where practicable'.
- Unless otherwise agreed between the rights holder and the broadcaster, there must be a two year window between the first showing of cinematographic works and their television broadcasting. A one year window is sufficient if such works are co-produced by the broadcaster (Article 7).
- Advertising is regulated under the Directive in terms of its duration; the point at which it should be inserted into programmes; the exclusion of advertising from certain types of programmes – religious services or religious programmes of less than thirty minutes, documentaries, news and current affairs, and children's programmes.

  Advertising should not prejudice respect for human dignity, it must be non-discriminatory on grounds of race, sex, nationality, religious views or political beliefs and should not encourage behaviour prejudicial to health, safety or environmental protection. Cigarette and tobacco advertising are prohibited; as is advertising for medical products and treatment which is only available on prescription. Alcohol advertising is circumscribed and must meet strict criteria. Advertising must not be detrimental to the protection of minors. Generally, advertising should not normally comprise more than 15% per cent of daily transmission time, though that percentage may be increased to 20 per cent in certain circumstances (Articles 10 to 16 and 18).
- Sponsorship of television programmes must correspond with the criteria laid down in the Directive; for example, the content and scheduling of

[9] Article 6 provides that European works must comply with one of the following requirements: they are made by one or more producers established in one or more of those States; or production of the work is supervised and controlled by one or more producers established in one or more of those States; or, that the combination of co-producers of those States to the total co-production costs is 'preponderant' and the co-production is not controlled by one or more producers established outside those States. Works originating from other European third countries may qualify as European works provided that they are made exclusively or in co-production with producers established in one or more Member States by producers established in one or more European third countries with which the Community would conclude agreements under the Treaty, if those works are mainly made with authors and workers residing in one or more European States. Works which are not European works, but which are made mainly with authors or workers residing in one or more Member States would be considered European works on a basis proportional to the contribution of Community co-producers to the total production costs.

sponsored programmes must not interfere with the responsibility and editorial independence of the broadcaster; sponsored programmes must be clearly identified, e.g. by name or logo; and news and current affairs may not be sponsored.

- Provisions for the protection of minors are set out in Article 22.
- Provision is made for a Commission report on the application of the Directive five years after its adoption and every two years thereafter, the purpose of which is to decide whether to amend the Directive in line with developments in the television broadcasting area (Article 26). The first such report was published in June 1995[10], and led to the Commission drafting a proposed Directive to amend the 1989 Directive and help resolve the disputes currently raging within and between the various institutions of the European Union.

Proposed amendment to Television Without Frontiers Directive

The Commission initally has published a proposal to amend Directive 89/552/EEC[11] in the light of experience gained from the operation of the 1989 Directive and technical developments in the television broadcasting field in 1996. The proposed amendments were extremely controversial. The Council of Ministers was for a long time unable to agree a common position, in particular because of differences of opinion in relation to the quotas. A Common Position was finally agreed and published in July 1996 (96/C-264/09). That Common Position did not find favour with the European Parliament which initially proposed a radically different set of amendments (see Recommendation for Second Reading on the Common Position, European Parliament Session Documents A4–0346/96).

Amendments to the Television Without Frontiers Directive were finally agreed under the co-operation procedure, and published in the Official Journal on 30 June 1997[12].

The question of which Member State has jurisdiction over a particular broadcaster has caused some difficulty, in particular because the standard of what is acceptable varies greatly between different Member States. Thus what is regarded as pornographic in one Member State will not be so regarded in another. Similarly certain Member States have very strict rules on the adver-

---

[10] Report on the application of Directive 89/552/EEC, COM(95) 86 final; 31 May 1995. This report was accompanied by the proposal for a European Parliament and Council Directive amending Directive 89/552/EEC, OJ 1995 C185/4. The Commission has also published a Communication to the Council and the European Parliament on the application of Articles 4 and 5 of Directive 89/552/EEC, Television Without Frontiers, COM(94) 57 final; 3 March 1994.

[11] COM(96) 200, 1996, amended by Common Position 49/96, OJ 1996 C264/52 and opinion of the Commission in COM(96) 626, 1996.

[12] Directive 97/36/EC of the European Commission and of the Council of 30 June 1997 amending the Television Without Frontiers Directive.

tising of certain products, for example, alcoholic beverages[13] which are not applied in other Member States. This has led to a number of cases before the European Court of Justice where Member States have sought to take action against broadcasts subject to the jurisdiction of other Member States. This problem has been compounded by the different interpretations on the basis of which the jurisdiction of a particular Member State is founded. The two differing views are, broadly, the views of those countries which found jurisdiction on the basis of the seat of establishment and the UK which founds it on the basis of the country of uplink. Again there have been a number of cases before the European Court of Justice on this point[14]. The Commission proposal seeks to clarify the concept of jurisdiction in relation to establishment. The preamble (paragraph 9) refers to the Court of Justice judgment in *Factortame*[15] in which the concept of establishment was set out as involving the actual pursuit of an economic activity through a fixed establishment for an indefinite period. A revised Article 2 provides that the broadcasters under the jurisdiction of a Member State are, broadly, those established in the territory of that Member State, in which they must have a head office and the editorial decisions about programme schedules are taken in that Member State. Article 2 sets out a number of criteria for determining jurisdiction if those two main criteria are not satisfied. The criteria for broadcasters established outside the territory of the Community falling under the jurisdiction of a Member State are clarified. Such broadcasters must satisfy one of three conditions, i.e. they must use a frequency or a satellite capacity appertaing to that Member State or a satellite link-up situated in that Member State (Article 2(4)).

Also, of particular concern is the so-called quota system which stipulates that 'where practicable', European works must account for a 'majority proportion' of transmission time. The 'where practicable' qualification has been more loosely interpreted in some Member States notably the UK. The Parliament was initially very insistent on the removal of the 'where practicable exemption'; it has, however, accepted that the exemption should remain and the European works quota has not been amended save that teleshopping services have been added to the types of programming excluded from the calculation (subject to the requirement that the proportion of European

---

[13] In connection with this, the Commission, in July 1996, decided to send a reasoned opinion to France concerning restrictions preventing the television broadcasting of certain sporting events. The Commission initiated an enquiry following complaints into restrictions imposed by the Conseil Supérieur de l'Audiovisuel (CSA), under a provision of French law which regulates the advertising of alcoholic beverages in France. The complaints arose when the CSA intervened to prevent the television broadcasting in France of certain sporting events that occurred in other Member States on the basis that there was advertising for French alcoholic drinks in the stadiums where the events took place. The Commission considered that the measures taken to restrict the provision of advertising services by, for example, organizers of sporting events, were in breach of EC Treaty rules on the freedom to provide services contained in Article 59. The French authorities were required to discontinue the restriction, and warned that failure to do so could result in a referral to the Court of Justice.

[14] In particular Case C-222/94, *Commission* v. *United Kingdom* [1996] ECR I-4025.

[15] Case C-221/89, *R* v. *Secretary of State for Transport, ex parte Factortame* [1991] ECR I-3905.

works should, in the case of the majority of the Member States, be not less than that achieved in 1988, or in the case of Greece and Portugal 1990).

Other key amendments include:

- a requirement that Member States take measures to ensure that broadcasters do not broadcast events which are of major importance for society on an exclusive basis so as to deprive a substantial proportion of the public of the opportunity to view such events (Article 3a);
- a refined definition of European works (Article 6(3) and (4));
- the elaboration of a set of rules governing teleshopping (Articles 10 to 20);
- clarification of the rules relating to advertising (Articles 10 to 20); and
- the establishment of a contact committee, comprised of representatives of the competent authorities of the Member States, charged with facilitating the implementation of the Television Without Frontiers Directive, amongst other things.

These amendments are to be implemented by Member States by 19 December 1998. A further review of the Television Without Frontiers Directive is to take place before 19 June 2002 (Article 25a).

Although the Parliament lobbied strongly for new media services to be included within the ambit of the Television Without Frontiers Directive the decision was eventually taken not to do so. The examination of the issues involved in new services has been initiated in the Green Paper on the protection of minors and human dignity in audio-visual and information services[16].

The interpretation of the Television Without Frontiers Directive by the Court of Justice

The Leclerc case

The interpretation of the Television Without Frontiers Directive and EC Treaty provisions was at issue in a European Court of Justice case[17] in which the Court issued a preliminary ruling in response to a French court in an action brought in France by an importer, Leclerc-Siplec, against the French television companies TF 1 and M6, for refusing to accept some of its advertisements. French law prohibits distributors from advertising on television. Defendants TF 1 and M6 joined the appellant, Leclerc-Siplec, in arguing that the French law provision was contrary to EC law.

On examining the Television Without Frontiers Directive and EC Treaty provisions, the Court ruled that they 'do not preclude Member States from prohibiting, by statute or by regulation, the broadcasting of advertisements

[16] See footnote 7 above.

[17] Case C-412/93, *Société d'Importation Edouard Leclerc-Siplec* v. *TFI Publicité SA and M6 Publicité SA* [1995] All ER (EC) 343. [1995] 3 CMLR 422.

for the distribution sector by television broadcasters established on their territory'. The ban was not contrary to EC law since it applied to all distributors, had no effect on intra-Community trade or competition, and did not prevent French companies from advertising by means of other media.

The Court has also ruled on the failure of Member States to implement the Television Without Frontiers Directive or to implement it correctly.

The Commission v. UK

The Court ruled in September 1996 in a case taken by the Commission against the UK[18], that the UK failed to fulfil its obligations under Articles 2(1), (2) and 3(2) of the Televison Without Frontiers Directive by incorrectly implementing the Directive in its national legislation contained in the Broadcasting Act 1990. In this case, the Commission particularly objected to the provisions of section 43 of the UK Broadcasting Act 1990 which it alleged did not comply with Article 2(1) of the Directive in four respects: i.e. that it applied criteria other than that of establishment for determining which broadcasters fall within the jurisdiction of the UK; that it applied a criterion which was not relevant for the purposes of such jurisdiction, that is, the criterion of reception; that it failed to make third country broadcasts falling under the UK's jurisdiction subject to UK law; and that it applied different regimes to 'non-domestic satellite service' and 'domestic satellite service'.

The Court examined the criteria for determining which television broadcasters fall within the jurisdiction of the UK, particularly through interpretation of Article 2(1) of the Directive and through the Council of Europe Convention on Transfrontier Television. The Court stated that Article 2(1) requires each Member State to ensure that television broadcasters under its jurisdiction, or in respect of which it is required, for the purposes of broadcasts, to exercise jurisdiction under the second indent of Article 2(1), comply with the law applicable to broadcasts intended for the public in that Member State. Furthermore, Member States are also required, under Article 3(2) of the Directive, to ensure that television broadcasters under their jurisdiction comply with the Directive's provisions.

The Court rejected the UK's interpretation of the term 'jurisdiction' of a Member State, used in the phrase 'broadcasters under its jurisdiction', contained in Article 2(1). The UK contended that the Member State having jurisdiction within the meaning of Article 2(1) is that from whose territory the broadcast is uplinked (that is, transmitted to the satellite). The Commission argued to the contrary, contending that broadcasters under the jurisdiction of a Member State within the meaning of Article 2(1) refer to those established in the Member State concerned.

The Court stated that from an analysis of Article 2(1), the concept of jurisdiction of a Member State must be understood as 'necessarily covering jurisdiction *"ratione personae"* over television broadcasts', on the basis that

---

[18] Case C-222/94, *Commission* v. *UK* [1996] ECR I-4025.

'this interpretation is borne out by the wording of the first indent of Article 2(1) of the Directive in that it refers to broadcasters as being under the jurisdiction of a Member State without referring, in that context, to the place from which they transmit their broadcasts'. Moreover, 'a Member State's jurisdiction *"ratione personae"* over a broadcaster can be based only on the broadcaster's connection to that State's legal system, which in substance overlaps with the concept of establishment as used in the first paragraph of Article 59 of the EC Treaty, the wording of which presupposes that the supplier and the recipient of a service are "established" in two different Member States'.

The Court also considered the difference between Article 2(1) of the Television Without Frontiers Directive and Article 5(2) of the Council of Europe Convention on Transfrontier Television. Under Article 5(2) of the Convention, and within the meaning of Article 3, the State having jurisdiction is that in which the satellite link-up is situated or the State which is not party to the Convention. The Court rejected the UK's contention that it would be absurd if the Directive sought to regulate intra-Community broadcasting in a radically different manner from the Convention. The Court argued that the difference reflects a difference between their respective aims: the Directive is designed to establish the internal market in television services, whereas the Convention is designed to facilitate the transfrontier transmission/retransmission of television programme services.

The Court found that the criteria set out in section 43 of the Broadcasting Act were not in conformity with Articles 2(1) and 3(2) of the Directive and that sections 43 and 44 of the Broadcasting Act, concerning licensing of domestic satellite services ('DSS') and non-domestic satellite services ('NDSS') did not conform with Article 2(2) of the Directive, in that the definition of DSS and NDSS in section 43 of the Broadcasting Act included broadcasters falling under the jurisdiction of other Member States, thus giving rise to the possibility of double control. The Court noted that the UK did not deny that section 43 extended to all broadcasters transmitting from its territory.

The Court accordingly held that:

> . . . by adopting, with respect to satellite broadcasts, the criteria set forth in section 43 of the Broadcasting Act 1990 for the purpose of determining which satellite broadcasters fall under the jurisdiction of the United Kingdom and, in the context of that jurisdiction, by applying different regimes to domestic satellite services and non-domestic satellite services, and by exercising control over broadcasts which are transmitted by broadcasters falling under the jurisdiction of other Member States when those broadcasts are transmitted by a non-domestic satellite service or conveyed to the public as a licensable programme service, the United Kingdom has failed to fulfil its obligations under Article 2(1) and (2) and Article 3(2) of Council Directive 89/552/EEC of 3 October 1989 on the co-ordination of certain provisions laid down by law, regulation or administrative action in Member States concerning the pursuit of television broadcasting activities.

The Commission, following what it termed 'less than accurate coverage' of the consequences of this Court case in 'some British media', gave a press briefing the day after the ruling, accompanied by a background note[19] clarifying that the Television Without Frontiers Directive does not authorize cross-border broadcasting of pornographic programmes. It stated that the Directive established the legal framework for broadcasting in Europe on the basis of the sole responsibility of the transmitting State, which has to ensure that its broadcasters comply with the rules laid down under the Directive.

It noted that the Directive expressly provides in Article 22 for an exception to this principle in the case of offensive programme material such as pornography or gratuitous violence. In cases of alleged violations of Article 22, the principle of the *transmitting* State is reversed and the *receiving* State is explicitly authorized to take measures against the offending broadcasters.

The Commission made it clear that the Court of Justice ruling did nothing to diminish the power of the receiving State in this regard; in fact, by clarifying complex issues of jurisdiction, it would make it easier for Member States to determine responsibility for broadcasters, thereby strengthening their powers to act against channels broadcasting offensive material. The Commission further pointed out that in the 'Red Hot Dutch' case of 1993 in which the UK sought to prohibit broadcasts of a pornographic service uplinked from other Member States, neither the UK nor the Netherlands were prepared to accept responsibility for this pornographic channel. However, had the UK complied correctly with the Directive in transposing its provisions into national legislation, it would have been able to act immediately and independently against 'Red Hot Dutch'.

Commenting on the judgment, Sir Leon Brittan[20] stated that the clarification given by the Court on who is responsible for granting licences to satellite broadcasters now that barriers to broadcasting are removed, closes a legal loophole through which Dutch television pornography had been screened in the UK. He said that 'as a result of the ruling British Satellite broadcasters will be freer to beam their clean programmes into the rest of Europe, while the government will be able to curb pornographic programmes from abroad more swiftly'. Following the adoption of the final text of the revised Television Without Frontiers Directive the UK government has now amended the relevant provisions of UK law.

Commission v. Belgium

Judgment was given on another case involving the interpretation of the Television Without Frontiers Directive on the same day as the UK ruling. In a case taken by the Commission against Belgium[21], the Court ruled that

---

[19] BIO/96/399; 11 September 1996.

[20] Speech by Sir Leon Brittan, 'Don't Shoot the Referee' to the European Union of Women, London, 12 September 1996.

[21] Case C-11/95, *EC Commission* v. *Belgium* (re Cable Television Broadcasts) [1997] 2 CMLR 289.

Belgium had failed to fulfil its obligations under the Directive in several respects. The Court examined the Commission's allegation that a provision of Belgian law, Article 22 of the Decree of the Council of the French Community on Audio-visual Communications, 1987 (the Decree), constituted a serious restriction on the retransmission in the French speaking region of Belgium of television broadcasts from other Member States, thereby infringing Article 2(2) of the Television Without Frontiers Directive.

Belgium contended that the Directive concerns only primary broadcasting and does not cover transmission by cable which is a secondary form of broadcasting. It produced three arguments to support this. First, that 'television broadcasting' as defined in Article 1(a) refers only to the initial broadcasting of television programmes, second, that the term 'retransmission' referred to in Article 2(2) does not cover cable television broadcasts; and, third, that cable television broadcasting was not covered by the Directive because it was not widespread at the time the Directive was adopted. The Court rejected these arguments on the grounds that the Directive (ninth recital) does not draw any distinction between primary and secondary television broadcasting, and, according to the tenth recital, all restrictions on freedom to provide broadcasting services within the Community which result from disparities between the Member States' laws, regulations and administrative measures concerning the pursuit of activities as television broadcasters and cable operators must be abolished. The Court stated that the concept of 'television broadcasting' contained in Article 1(a) cannot be interpreted as a restriction in its scope. It also stated that the fact that cable distribution was not widespread at the time of adoption of the Directive could not be relied on to support the argument that it should be excluded from the Directive's scope; in fact, the fourth recital of the Directive expressly refers to the adoption by the Council of Europe of the Convention on Transfrontier Television, and it was apparent from Article 3 thereof that the Convention also applies to television programmes retransmitted by cable.

The Court also rejected Belgium's argument that the Belgian Decree covered only the provision of services by cable operators established in the territory of the French Community, and consequently that it did not give rise to any restriction on the free movement of foreign broadcasts. The Court stated that according to its case-law, cable retransmission of foreign programmes constitutes a transfrontier service; and that furthermore, it was apparent from the sixth, seventh and ninth recitals to the Directive that the Directive is specifically aimed at removing restrictions on freedom to provide television broadcasting services arising from disparities between the Member States' laws.

Belgium argued that a television programme may circulate freely through the Community only if it complied with the applicable law of the originating State, including the provisions of the Directive. Belgium maintained that this necessarily followed from the principle that the receiving Member State must be able to verify whether foreign television broadcasters applying for authorization to retransmit their programmes in the territory of the French Community of Belgium comply with the law of the originating State.

The Court did not accept these arguments on the basis that the Television Without Frontiers Directive divides obligations between the Member States from which programmes emanate and the Member States receiving them. It stated that:

> it is solely for the Member State from which television broadcasts emanate to monitor the application of the law of the originating Member State applying to such broadcasts and to ensure compliance with Directive 89/552, and the receiving Member State is not authorized to exercise its own control in that regard.

Only subject to provisions contained in Article 2(2) may the receiving Member State exceptionally suspend retransmission of televised broadcasts. Furthermore, if a Member State considers that another Member State has failed in its obligations under the Directive, it may invoke infringement proceedings under Article 170 or request the Commission to take action against that Member State under Article 169. It is established case-law that a Member State cannot unilaterally adopt measures to obviate any infringement by another Member State of rules of Community law.

The Court also rejected the Belgian government's argument that under a provision in Article 10(1) of the European Convention for the Protection of Human Rights and Fundamental Freedoms there was express provision for the possible establishment of a compulsory authorization system applying to broadcasting undertakings and, *a fortiori*, to organizations engaged in the distribution of television programmes. The Court said that the fact that a compulsory authorization system applying to the television sector is not contrary to Article 10 of the Convention does not prevent such a system from being contrary to Community law.

Belgium further submitted that the system contained in the 1987 Decree, which was intended to secure respect for certain cultural objectives, was justified in so far as the Television Without Frontiers Directive, particularly Articles 4 and 5, must be construed in the light of the culture title of the Treaty contained in Article 128. The Court accepted that Article 128(1) provides for the Community to contribute to the flowering of the cultures of the Member States, whilst respecting their national and regional diversity and that Article 128(4) provides that cultural aspects must be taken into account in Community action under other Treaty provisions. It pointed out, however, that Article 128 in no way authorizes the receiving State, by way of derogation from the system established by Directive 89/552/EEC (the Television Without Frontiers Directive), to make programmes emanating from another Member State subject to further controls.

The Court also rejected an argument of the Belgian government on the basis of the subsidiarity principle and dismissed an argument based on the preservation of pluralism in the media.

Consequently, the Court held that Belgium had failed in its obligations under Directive 89/552/EEC, particularly concerning Articles 2, 14 and 15, in respect of:

(1) the French Community 'by maintaining, in the French speaking region, a system of prior authorization for the retransmission by cable of television broadcasting emanating from other Member States';

(2) the French Community 'by maintaining in the French speaking region, a system of express, conditional prior authorization for the retransmission by cable of television broadcasts emanating from other Member States which contain commercial advertising or teleshopping programmes especially for viewers in the French Community';

(3) the Flemish Community, 'by maintaining, in the Dutch speaking region, a system of prior authorization for the retransmission by cable of television broadcasts emanating from other Member States';

(4) the bilingual Brussels region, 'by failing to adapt the laws, regulations and administrative measure necessary to comply with Article 2(2) of Directive 89/552'; and,

(4) the French Community, 'by failing to adapt the laws, regulations and administrative measures necessary to comply fully with Articles 14 and 15 of Directive 89/552'.

The remainder of the application was dismissed and Belgium was ordered to pay the costs.

Application of competition rules under the merger regulation to broadcasting

European competition authorities are being faced with a growing number of merger cases in the media field (in part as a response to technological changes expanding the available capacity for distribution and the convergence between broadcasting, telecommunications and computing). Competition Commissioner Van Miert has stated that there is a need to intervene where effective competition is prevented. This is something the Commission has seen fit to do in the four instances below.

Merger cases in the media field

MSG[22]

This case concerned the creation of a joint venture, MSG Media Service, by Bertelsmann AG, Deutsche Bundespost Telekom and Taurus Beteiligungs (a

[22] Commission Decision of 9 November 1994 relating to a proceeding pursuant to Council Regulation MO 4064/89 (IV/M.469-MSG Media Service), OJ 1994 L364/1.

member of the Kirch group), to supply administrative and technical services to pay TV operators. The Commission considered that MSG would affect three distinct product markets:

(1) *Administrative and technical services for pay-TV*: Although the Commission recognized that there was as yet no market in Germany for the services provided by MSG, such a market was expected to develop in particular following the introduction of digital TV. The Commission stated that 'according to the conception underlying MSG, it must be assumed that there will be a single market for services relating to digital pay-TV and other digital interactive television communication services'. The Commission also considered it possible that a separate market for subscriber management by specialized firms might develop, as might programme packaging services;

(2) *Pay-TV*: Pay-TV constituted a distinct product market from advertising-financed TV because, in the case of pay-TV, there was a trade relationship between the programme supplier and the viewer as subscriber, as opposed to a trade relationship between the programme supplier and the advertising industry. This resulted in different conditions of competition – in the case of advertising-financed TV, the audience share and the advertising rates were the key factors, whereas in the case of pay-TV, the key factors were the programming and the level of subscriber prices. In addition, content would be different as digitalization would allow 200 or more television programmes to be available and therefore a variety of new, payment-financed special-interest programmes would be expected to be produced; and

(3) *Cable television networks*: The Commission considered that there were considerable differences between broadcasts via terrestrial frequencies, satellite or cable networks both in terms of technology and financing. The way in which the final consumer would pay for the different forms of transmission were different, as were the considerations of the customer when choosing to acquire a cable linkup or a satellite dish. As a result, the two forms of reception were not interchangeable. In addition, from the programme supplier's point of view, cable and satellite were not interchangeable in terms of cost.

The geographic market for all three markets was restricted to Germany. Although the Commission recognized that MSG would over time extend its service activities to other countries as well, it concluded that, in the case of pay-TV, the programmes offered in Germany were not interchangeable with programmes offered in other countries largely because of language barriers and regulatory differences. As regards the operation of cable television networks, the Commission considered that the conditions of competition in Germany were substantially different from those in other countries due to Telekom's statutory monopoly on laying and operating cable network in public roads.

The main factor influencing the Commission's analysis of the effect on competition of the joint venture was the strength of the parties. Telekom operated 90 per cent of the cable networks in Germany while the only pay-TV supplier in Germany was Première, jointly controlled by Bertelsmann, Kirch and Canal Plus.

In the Commission's view, without the creation of MSG, each of these enterprises would otherwise have had to install an infrastructure for digital pay-TV and provide the corresponding services. The most likely potential competition was therefore excluded already in the development phase of the market as a result of which MSG would acquire a long-term monopoly. Moreover, the Commission believed that holding a dominant position on the market for technical and administrative services would give Bertelsmann and Kirch a durable dominant position in the market for pay-TV. Bertelsmann and Kirch both had preferential access to software and considerable programme resources, in addition to which Bertelsmann had an important potential distribution network as it was the leading book club operator in Germany. Telekom in the meantime owned most of the cable network in Germany with direct access to more than four million cabled households and thereby controlled the return channels necessary for interactive services. Finally, the Commission was concerned that, by jointly operating the pay-TV structure together with the leading pay-TV suppliers, Telekom would also strengthen its position as cable network operator in such a way that, following liberalization, competition in the cable network market would be substantially impeded and Telekom's dominant position safeguarded. In particular, the Commission considered that there would be a risk that private cable operators would not obtain the programmes from the leading pay-TV suppliers, Bertelsmann and Kirch, or would obtain them on unfavourable terms.

The parties proposed giving certain undertakings in relation to the introduction of a common interface for decoders, ensuring non-disclosure to parent companies of programme or subscriber data of other pay-TV suppliers, non-discrimination in the presentation of Electronic Programming Guides, charging reasonable market rates and operating a transparent tariff policy. In addition, Telekom undertook to open up its networks for further digital transmission of programmes in order to avoid any shortage of channels. This was not sufficient for the Commission which regarded the undertakings either as too difficult to enforce or merely statements of compliance with the legal obligations of a dominant operator.

The submission by the parties that the rapid acceptance of digital television would be promoted by the services offered by MSG was rejected by the Commission on the basis that, in its view, potential suppliers of digital pay-TV would not enter the market as a result of the joint venture being created and therefore the successful spread of digital television would be hindered rather than promoted.

HMG[23]

The Commission was asked by the Dutch government to examine a proposed venture, Holland Media Group (HMG), between RTL4 SA (RTL), Vereniging Veronica Omroep. Organisatie (Veronica) and Endemol Entertainment Holding BV (Endemol). RTL was a broadcaster of Dutch-speaking television and radio programmes operating under the responsibility of Compagnie Luxembourgeoise de Télédiffusion (CLT), a Luxembourg broadcasting company involved in televison, radio and publishing. RTL broadcasts two free Dutch-speaking commercial channels, RTL4 and RTL5. Veronica in the past had been a public broadcaster in the Netherlands but had in 1995 become a commercial broadcaster. Finally, Endemol was an independent producer of television programmes mainly in the Netherlands. The purpose of the joint venture was to broadcast three general interest channels broadcast by itself, CLT, Veronica and others to the Netherlands and Luxembourg.

The Commission considered the joint venture affected three product markets in the Netherlands:

(1) *Television broadcasting*: The Commission noted that, but for the joint venture, Veronica would have been the strongest commercial competitor to RTL4 and RTL5, the main channels broadcast in Dutch by RTL. The creation of the joint venture, however, would mean that through the combination of the three channels, RTL4 and RTL5 and Veronica, HMG would be able to co-ordinate their programming schedules in order to attract a maximum of viewers and thereby react against any competing channel. The combination would also be strengthened by HMG's structural link to Endemol, by far the biggest Dutch independent producer of television programmes. In addition, a subsidiary of Veronica was a leading Dutch publisher and published a large number of family weekly magazines incorporating television programme information. In the Commission's view therefore, public broadcasters would not be able to react to the competition created by HMG, nor would the new market entrant, SBS, a new Dutch commercial broadcaster in the Netherlands, be likely to be successful as a result of the creation of HMG;

(2) *Television advertising*: The Commission believed that, given that HMG's market share in the television advertising market would be at least 60 per cent and that HMG would be able to offer package deals for advertisers which went beyond the normal practice by relating advertising on one channel with advertising on one or more of its other channels. HMG would be the clear market leader in the Dutch television advertising market. It would be in a position to counteract all active attempts to compete by existing players and would render the entrance of newcomers to the market very difficult;

---

[23] Commission Decision of 20 September 1995 relating to a proceeding pursuant to Council Regulation 4064/89 (IV/M.553-*RTL/Veronica, Endemol*), 96/346/EC.

(3) *The market for Dutch independent television production*: It was clear that Endemol had an overwhelming position in the market. It owned a large number of the most popular Dutch formats and also had preferential access to foreign formats which were then adapted to the Dutch audience. The Commission concluded that given these resources, despite the fact that prior to the creation of the joint venture, Endemol had been the main programme supplier for RTL4 and RTL5 and Veronica, the creation of a structural link between Endemol and a national broadcaster through HMG would mean that the market for Dutch independent television production would not be capable of being attacked by a competitor.

Rejecting the parties' submission of proposals for undertakings, the Commission declared the merger incompatible with the common market. However, in July 1996, the Commission approved the joint venture following various modifications. These included the withdrawal of Endemol from the project and the transformation of RTL5 into a news channel which meant that HMG would operate only two general interest channels and thereby provide for more competition in general interest channels. Endemol challenged the fidelity of the Commission's original decision and this is still pending before the Court of First Instance.

Nordic Satellite Distribution[24]

This case involved the creation of a joint venture, Nordic Satellite Distribution (NSD), between Norsk Telekom AS (NT), TeleDanmark AS (TD) and Industrifoervaltnings AB Kinnevik (Kinnevik), for the provision of transponder capacity and the transmission and distribution of satellite television channels via cable networks of direct to home broadcasts in the Nordic region (Denmark, Sweden, Norway and Finland). NT was controlled by Telenor, the principal provider of telephone services in Norway which owned and/or leased transponder capacity. NT also had a subsidiary which provided distribution services to the direct-to-home market in Norway, Sweden, Finland and Denmark. TD was the monopoly Danish public telecom operator with a national broadband network used for transmitting television signals to local distribution networks. Kinnevik was a Swedish company active in television and media and telecommunications.

The Commission considered the operation involved three product markets:

(1) *Satellite television transponder capacity and related services*: The Commission confirmed the decision in MSG that distribution of television signals via satellite was a distinct market from television distribution by

[24] Commission Decision on 19 July 1995 declaring a concentration to be incompatible with the common market and the functioning of the EEA agreement (case IV/M.490-*Nordic Satellite Distribution*), OJ 1996 L53/20.

terrestrial links. Telenor controlled all three satellites in the position 1 degree West and Kinnevik and TD had leasing agreements with the Swedish operator which ensured that NSD controlled the majority of the transponder capacity situated at 5 degrees East. These satellite positions were necessary to broadcast to the Nordic region. The Commission rejected the parties' claim that competition was being provided from Astra and Eutelsat, since all channels on Astra and Eutelsat, except for Kinnevik's four channels and a pay-TV channel, were broadcast in foreign languages while national channels were by far the most popular. Moreover, due to Kinnevik's TV channels, NSD would be able to offer some very popular Nordic TV channels on an exclusive basis which would mean that the majority of Nordic direct-to-home households would direct their dishes towards NSD's satellites. Finally, because of NSD's links to TD and NT as major cable operators, in the Commission's view, broadcasters would have to anticipate not getting access to a large part of the Nordic cable networks if they transmitted from Astra or Eutelsat. The Commission concluded as a result that NSD would have a dominant position in the market for transponder capacity suitable for Nordic viewers. In the Commission's view, NSD would be in a position to bar other satellite operators from leasing transponders to broadcasters and therefore it was very unlikely that new satellite operators would be able to challenge NSD's dominant position in the medium and long-term.

(2) *Operation of cable television networks*: The Commission noted that TD controlled approximately 50 per cent of the cable connections in Denmark and had a dominant position on the Danish market due to the legal regime there. The Commission concluded that the creation of NSD would result in a strengthening of TD's dominant position because NSD would be able to discriminate in favour of TD when offering channels to cable operators. NSD's monopoly position as regards the provision of programming would mean that it could ensure that the terms offered to the cable operators would favour TD rather than other cable operators and that cable operators in competition with TD would have to negotiate with TD, their competitor, as an NSD partner. The situation was unlikely to change after the first step of liberalization of the telecommunications market in Denmark as TD would still retain many advantages over its competitors due to its past legal monopoly. With respect to the other Nordic countries, it was unlikely that NSD would hold a dominant position because of the relative strengths of the competitors in Norway and Sweden while in Finland, NSD had no control over or influence over the cable networks.

(3) *Distribution of pay-TV and other encrypted television channels to direct-to-home households*: The Commission noted that there were three major

distributors in this market, namely FilmNet, Telenor CTV and Viasat which was 100 per cent owned by Kinnevik. To be competitive, a distributor had to have a TV channel or package of TV channels on its smart card which a considerable number of viewers found attractive. Given that, measured by the number of smart cards sold, Viasat had a very strong position in the market and as the Viasat's smart cards would also contain the CTV package and include Kinnevik's channels sold exclusively by Viasat, the Commission concluded that the operation would create a dominant position for Viasat over the Nordic market. Moreover there would be little room for a new distributor in the Nordic market since by its control of Nordic transponder capacity and its link to Kinnevik as a broadcaster, NSD would be the dominant provider of TV channels to Nordic viewers and would, by means of its parental links to cable operators, to a large extent control access to the Nordic cable sector.

According to the parties, NSD would lead to economical and technical progress by improving the distribution of satellite TV in the Nordic region. The Commission rejected this claim, since in its view, NSD did not add any new transponder capacity and as a result, the number of satellite TV channels offered to Nordic viewers would not be affected by the operation. In addition, because of NSD's dominant position as provider of TV channels from Nordic transponders, the Commission considered that it was most likely that the majority of direct-to-home households and independent cable operators in the Nordic countries would be forced to use the encryption system used by NSD. In the long term, after the introduction of digital television, the parties claimed that NSD would be used to create an integrated infrastructure for the distribution of satellite TV and other related services. The Commission recognized the benefits to customers of having an integrated system for transmission of satellite TV but stipulated that such infrastructure would need to be accessible to all interested parties.

In the Commission's view, the vertically integrated nature of NSD was not necessary in order to create such an infrastructure. The Commission therefore declared the merger incompatible with the common market and the functioning of the EEA agreement.

Téléfonica de España/Sogecable/Cablevision[25]

Téléfonica de España and Sogecable SA notified the Spanish authorities that they proposed to acquire joint control of Cablevision SA, a subsidiary of Canal Plus in order to provide multimedia services to local cable operators. Téléfonica was involved in the construction and the operation of networks and the provision of telecommunication services in Spain while Sogecable provided satellite and terrestrial pay-TV services. Cablevision was involved in

[25] Case IV/M.0709 – *Téléfonica/Canal plus/Cablevision.*

the provision of technical, administrative and commercial services to cable operators. The Commission sent letters on 6 and 7 February 1996 to the parties and to the Spanish Court for the defence of competition in which the Commission concluded that the operation which took Cablevision under joint control constituted a concentration with a Community dimension. On 16 April 1996, Sogecable brought an interim action against the Commission for the suspension of the operation of the Commission's decision until the Court of First Instance had ruled on its validity. It claimed that the concentration had a national dimension and therefore that it was excluded from the scope of the Merger Regulation. On 31 May 1996, however, the parties notified the concentration to the Commission. On 12 July 1996, the Court of First Instance dismissed the interim application. The Commission then opened a second phase investigation. In its view the merger affected the market in the supply of services to cable operators, the market in pay-TV and the market in cable television and had an indirect effect on the telephony market. On 7 November 1996, Competition Commissioner Van Miert announced that the Commission's anti-trust services and the advisory committee on mergers (against the opinion of the Spanish member) had reached negative conclusions on the project on the basis that it would result in dominance on the cable television market in Spain. On 8 November 1996, Téléfonica and Sogecable renounced the Cablevision agreement.

The Commission has recognized that, in order for the European audiovisual industry to operate, a certain amount of convergence is necessary. Though if that convergence is likely to involve the possible foreclosure of markets, the Commission is unlikely to clear the relevant merger or joint venture. An example of a permitted joint venture is when the Commission cleared the joint acquisition of Telepiu, the Italian pay-TV operator, by the Richemont Group (through its shareholding in FilmNet), and the Kirch Group. In this case, the Commission took the view that the acquisition would be unlikely to create or strengthen a dominant position in the television market in Italy, since Telepiu, whilst being the only pay-TV operator in Italy, had a very limited market penetration compared with the other public and commercial channels (which together accounted for about 90 per cent of the Italian television audience). Moreover, neither Richemont nor Kirch had a presence in the television broadcasting market in Italy. Therefore the Commission did not oppose the acquisition and declared it compatible with the common market.

Similarly, the Commission declared the creation of the joint venture, CLT-UFA[26] compatible with the common market. In this case, CLT (the holding company of the CLT group), which has activities in television and radio broadcasting throughout Europe, and UFA (a subsidiary of Bertelsmann and holding company for the television, film and radio activities of Bertelsmann), planned to merge into a joint venture; this joint venture was to be controlled

---

[26] European Commission Press Release: IP/96/897; 8 Oct. 1996.

jointly by Bertelsmann and the holding company, Audiofina. The Commission noted that the concentration would create a leading European player in the audio-visual sector, which would focus mainly on free access television. However, it decided it would not lead to the creation of a dominant position in any of the affected markets.

In its analysis, the Commission pointed out that UFA jointly controls with Canal Plus and Kirch, Première, which was until recently the only German pay television channel. It considered, however, that the proposed concentration between UFA and CLT would have no significant impact on the position of Première on the German pay television market since CLT had no pay television activities in Germany.

In its press statement on approval of the joint venture, the Commission observed that 'although in certain niche markets there are already programmes broadcast through Europe, television broadcasting in the European Union generally takes place on national markets', and that this is 'mainly due to different regulatory regimes, existing language barriers and cultural factors'.

Development of EU policy in the broadcasting sector

Green Papers, Communications, Resolutions and Directives

The European programme industry: Green Paper on strategy options to strengthen the European programme industry[27]

The 1994 Commission Green Paper set out to consider the full implications for the future of the programme industry (cinema and television) in the European Union. In analysing the characteristics of the European programme industry and assessing future priorities, the Green Paper underlines priority themes that specifically include the need for a European policy that establishes and implements rules 'to stimulate private initiative with particular emphasis on competition, universal service, standardization and openness but with guarantees for data protection and privacy and the security of information and communication systems'. Such rules should include provision in the Community of basic trans-European telecommunications services with emphasis on the interoperability of networks (the other priorities relate to best practice, training and the harnessing of technologies).

The Green Paper discussed optimization of the potential of the broadcasting sector in many different contexts – technical, financial and

[27] See footnote 2 above.

international amongst others. It recognizes the significant role of competition policy in applying EC competition rules even-handedly throughout the European Union. The Green Paper perceives the development of infrastructure and liberalization of services in the telecommunications sector, such as open network provision and liberalization of voice telephony by 1998, as essential conditions for the creation of a favourable environment for the development of the broadcasting sector. The opening up of the telecommunications sector to free competition is seen as particularly important for the development of new services in broadcasting.

Conditional access[28] is also underlined as an essential element in the development of new broadcasting services. According to the Commission, the regulation of conditional access should create an environment sufficiently open to permit all potential pay television companies to have fair and non-discriminatory market access, but without consumers having to buy different decoders in order to benefit from a wide range of services.

Pluralism and media concentration: Green Paper on pluralism and media concentration in the internal market

In response to requests from the European Parliament, the Commission has examined the issue of safeguarding pluralism in the media in the light of mergers and acquisitions taking place within the sector, and taking account of the global dimension. In December 1992, the Commission published a Green Paper on the subject[29]. This produced options for discussion and invited interested parties to submit views. In October 1994 the Commission published a follow-up document[30], commenting on this consultation process. It suggests that action at Community level might be necessary to ensure the free movement of broadcasting and to prevent restrictions on competition arising from fragmentation of the internal market as a result of the disparity of national rules on pluralism in media ownership which gives rise to legal uncertainties.

The Commission reported that the response to the Green Paper produced a virtually unanimous conclusion that the current disparate legal framework is inappropriate to deal with high output digital technology offering a range of new media services on a cross-border basis. It said that disparities in Member States' legislation could result in operational difficulties. For example, broadcasting a bundle of specialized channels by digital compression to a small target audience could only be profitable if operated on a cross-border basis. While such broadcasting might be legal in some Member States, the legislation of other Member States could restrict the number of channels. The legal issue would inevitably arise as to whether a Member State can

[28] 'Conditional access' describes the mechanisms by which subscribers can be authorized to view channels or individual programmes or services to which they have subscribed.
[29] See footnote 4 above.
[30] See footnote 4 above.

legitimately restrict such broadcasts. Other matters which could create potential problems for cross-border operators include whether the definition of televised media covers electronic newspapers, and the issue of whether limits should be placed on shareholdings in television channels.

The consultation process has not yet given rise to any concrete proposals and Community powers to intervene in this field are ambivalent.

On the issue of pluralism in media ownership, there remains some doubt as to the legal competence of the Community to act in this area. When the European Parliament discussed media concentration in October 1994[31], there was severe criticism from MEPs of Commission performance in this area. The Commission Communication was criticized for not pushing for a European Union Directive on media concentration. MEPs also expressed frustration that two rounds of consultation seemed excessive. Replying for the Commission, the former Internal Market Commissioner, Vanni D'Archirafi, appeared to have considered the legal competency issue, stating that in his view the only suitable legal base for action was under Article 7a of the Treaty[32] since the Article in the Treaty on European Union on cultural co-operation specifically ruled out harmonization and EC competition rules were not appropriate in this case.

There are arguments to support the view that the Community does not have the legal competence to act in this area. The Treaty of Rome, as amended by the Single European Act and the Treaty on European Union, gives no specific power to legislate to harmonize the national rules on pluralism in the media.

Case-law of the European Court of Justice would tend to support this view. In *Veronica*[33] the Court of Justice has stated that 'such cultural policy objectives constitute objectives of public interest which a Member State can legitimately pursue in establishing in an appropriate manner the status of its own radio broadcasting entities'. Hence the cautious approach by the Commission of engaging in two rounds of consultation before submitting draft legislation.

In addition to the cases under the Merger Regulation referred to earlier, the Commission has intervened in a media ownership issue albeit based on restrictions of freedom of establishment under Article 52.

The Commission recently sent a reasoned opinion under formal infringement proceedings (Article 169 reference) to Belgium in connection with a measure requiring that at least 51 per cent of the capital of a private company broadcasting television to the whole of the Flemish community should be reserved for publishers of Dutch speaking daily and weekly newspapers

[31] European Parliament Session News Press Release, 26 Oct. 1994.

[32] Article 7a states that the 'Community shall adopt measures with the aim of progressively establishing the internal market over a period expiring on 31 December 1992, in accordance with the provisions of this Article and of Articles 7b, 7c, 28, 57(2), 70(1), 99, 100a and 100b and without prejudice to the other provisions of this Treaty'.

[33] Case C-148/91, *RTL, Vereniging Veronica Omroep Organisatie and Commissariat voor de Media* [1993] ECR I-0487.

and magazines[34]. The Commission considers these measures to be in breach of Treaty rules which prohibit unjustified restrictions on the freedom of establishment (Article 52). This is on the basis that it is unlikely that a publisher of Dutch speaking newspapers would be established other than in Belgium or the Netherlands. Thus, in effect, companies based in other Member States face restrictions if they wish to establish themselves as television broadcasters in Belgium. The Commission rejected the argument that the promotion of language and culture as a general interest objective was valid in this particular case, since the majority shareholding requirement did not appear to be appropriate to the objective of promoting the Dutch language and culture.

Legislation to regulate media ownership would also have to be examined on grounds of subsidiarity (i.e. that action should be taken at the most appropriate level) and proportionality (i.e. that action must be proportionate to the objectives sought to be achieved)[35]. Moreover, the legal basis for legislation on the harmonization of media laws would have to be carefully considered, since unanimous adoption by the Member States under Article 235[36] might, in fact, be the only appropriate means of action within the scope of the Treaty. Whether the required unanimity would be forthcoming is another matter.

The Commission has stated that it intends to present a proposal for a Council Directive on harmonizing the rules on the concentration of the media. However, should the Commission decide to consider the question of the Community's general legal competence in this area to be a problem in view of possible legal risks, it could then turn to its own powers to adopt legislation under Article 90(3). The Commission's powers to issue Directives under Article 90(3) to require broadcasting companies to which Member States have granted special or exclusive rights, to comply with the competition rules by withdrawing or amending the special or exclusive rights granted to them, is a power well recognized by the Court. It could, for example, use its powers under Article 90(3) to ensure access for existing competitors of a state broadcasting monopoly or for new market operators.

Digital video broadcasting

Commission Communication on digital video broadcasting

The Commission issued a Communication to the Council and European Parliament on a framework for Community policy in digital video broadcast-

[34] Commission Press Release. IP/96/716, 26 July 1996.

[35] Article 3b of the EC Treaty.

[36] Article 235 states that if 'action by the Community should prove necessary to attain, in the course of the operation of the common market, one of the objectives of the Community and this Treaty has not provided the necessary powers, the Council shall, acting unanimously on a proposal from the Commission and after consulting the European Parliament, take the appropriate measures'.

ing in November 1993[37]. This communication examines the technological evolution of the field and its global exploitation. It underlines that the proliferation of new television services through digital compression techniques, while holding enormous potential for new service provision, might also 'pose threats to the established broadcasting scene by introducing a powerful new element of competition which some economic actors may be able to exploit more rapidly than others'. It also recognizes that the scope for digital technology will extend beyond television to a wide range of existing and new audio-visual and other communications services.

The implications of digital technology, therefore, raise new kinds of political, economic and technical issues for Community policy. In the competition field, the Commission identifies the protection of the public interest through ensuring free and fair competition as an area for possible future Community action.

Competition issues arising from digital video broadcasting are identified by the Commission as including interoperability between services, flexibility in the provision of television services and cost considerations. Moreover, new technical processes and concepts such as source coding, multiplexing, channel coding and modulation and conditional access systems[38] introduce a series of interrelationships in the digital television system which will have strong competition elements.

The projected time scale of the transition from the current analogue system to digital television could be as long as twenty years. This adaption period, as well as requiring technical co-ordination, will also need to take competition aspects into account. In this context, the Commission refers to the duty of the Community and the Member States to ensure the defence of the public interest 'against threats from powerful industrial and commercial sources'. Ensuring fair competition and consumer protection are described as 'cornerstones of national and Community policy in general and are also specifically relevant in the field under discussion'.

The communication emphasizes the importance of access to the digital broadcasting market:

> at a time of increasing media concentration it is . . . intended to ensure that all broadcasters have access to digital broadcasting technologies and that no barriers to

---

[37] See footnote 6 above.

[38] 'Source coding' refers to the process of compressing picture information which can be reconstructed. 'Multiplexing' permits different components of a television programme (picture, several sound channels, associated data etc.), to be combined into a single bit-stream and likewise several different television programmes to be simultaneously carried in a single broadcast television channel. 'Channel coding' provides protection against loss or corruption of broadcast data by introducing additional data derived from it into the multiplex signal. The process of 'modulation' provides for the broadcast of multiplexed information. 'Conditional access systems' enable broadcasting companies to determine which potential viewers may receive their programmes. This is notably in connection with so-called pay-TV. A conditional access system involves a variety of processes: scrambling the signal; use of encryption techniques; operation of a subscriber management centre which deals with commercial aspects of the system; and subscriber interface equipment, e.g. use of 'smart cards'. Digital Video Broadcasting Communication, COM(93) 557 final, 17 Nov. 1993, page 5.

entry are erected for new market entrants. In order that the market may be kept open, media companies should be prohibited from using restrictive practices in order to impose their own technologies or to limit competition from competing television channels.

Access is becoming increasingly controversial as pay-TV broadcasters (such as BSkyB) defend their proprietary rights. Pay-TV broadcasters[39] have committed major investments to infrastructure development. They are reluctant to provide unrestricted access to other operators. Community policy in this area is developing towards a form of conditional access where, on the one hand, pay-TV broadcasters could own or control digital decoders but, on the other hand, they must give their competitors access to digital services on terms that are fair, reasonable and non-discriminatory. The pay TV-broadcasters want to ensure that access by other operators is negotiable on commercial terms. The public broadcasters' perspective centres on opposing the concentration of excessive power in a small number of powerful media operators, and they favour a prohibition on pay-TV broadcasters from owning or controlling digital decoders.

While it might be ironic that concerns over the creation of monopolies in this sector have emanated from public broadcasters which themselves have tended to be beneficiaries of monopoly privileges, it is nevertheless a legitimate concern.

Council Resolution on a framework for Community policy on digital video broadcasting

This Council Resolution[40] originally declared that whilst the Council's preferred means of achieving coherent market development in digital video broadcasting was through voluntary agreements by the operators themselves, if such a consensus approach failed and/or if the requirements of fair and open competition, consumer protection or other public interest so demand, it was prepared to introduce regulatory measures to achieve its objectives to protect those interests.

Conditional access: the Advanced Television Standards Directive[41]

Following the perceived failure of the consensus approach and the outcome of a widely based consultation process, the European Commission recognized the need to regulate the provision of conditional access services within the European Union, whilst allowing operators of those services to earn a

[39] Pay-TV is principally financed by subscribers. Traditional, non-pay-TV is financed by advertising and/or licence fees.

[40] Council Resolution of 27 June 1994 on a framework for Community policy on digital video broadcasting, OJ 1994 C181/3.

[41] European Parliament and Council Directive of 24 October 1995 on the use and standards for the transmission of signals, EC Directive 95/47/EC, OJ 1995 L281.

return on investments made in developing those services and the provision of those services to broadcasters. This regulation is contained in the 'Advanced Television Services Directive, which should have been implemented into the national laws of the Member States by 22 August 1996.

The main requirements of the Advanced Television Services Directive, which apply in relation to digitally transmitted television services, are that:

- set boxes and other consumer equipment capable of decoding digital television signals must be capable of decoding signals scrambled using a common European scrambling method;
- whilst retaining control of proprietary conditional access technology, holders of industrial property rights to conditional access products and systems must ensure that manufacturers are licensed on fair, reasonable and non-discriminatory terms;
- operators of conditional access systems, must offer to all broadcasters on a fair, reasonable and non-discriminatory basis conditional access services and operating accounting separation regarding their activity as conditional access providers; and
- conditional access systems must have the necessary technical capability for cost-effective transcontrol at cable head ends (allowing cable operators to take full control of the services using such conditional access systems).

Application of competition enforcement under Article 90 to the broadcasting sector and ancillary services

As in other utility sectors, such as postal services and energy, where EC legislation has remained relatively limited, Community law and policy have been to a significant extent dictated by the European Court of Justice's interpretation of relevant Treaty provisions in a number of key cases. Given that the national state broadcasting monopolies have been granted special or exclusive rights by their Member States, many of these rulings have interpreted and clarified the application of Article 90 to the specific circumstances of the broadcasting sector itself, as well as to the development of Article 90 generally.

Key cases

Sacchi[42]

The *Sacchi* case was a key judgment in establishing the Court's jurisprudence on the application of the EC competition rules to State monopolies.

[42] Case 155/73 [1974] ECR 409.

Judgment in the case was given in 1974, and, as discussed in Part I of this book, the case-law of the Court has superseded certain premises then established in relation to monopolies operating in the public interest. Nonetheless, the Court in its *ERT* judgment of 1991 concerning the Greek national broadcasting monopoly, relied extensively on the jurisprudence established in *Sacchi*.

The case revolves around the following facts. Mr Sacchi, a director of a television relay company, Telebiella, took issue with the Italian State television monopoly, RAI, after having been charged with being in possession of television relay equipment without holding the requisite licence prescribed under Italian law. In Italy, the State granted a television monopoly to RAI, which involved a monopoly both on televised commercial advertising and on the right to ban any person/enterprise from receiving for retransmission audio-visual signals transmitted from Italian territory or from foreign stations. The national Italian court made a referral to the Court under Article 177, questioning the compatibility of Italian law with EC law.

The Court of Justice established several principles in its ruling, including the following:

- Article 90(1) permits Member States to grant special or exclusive rights to undertakings. The Court stated that: 'nothing in the Treaty prevents Member States, for considerations of public interest, of a *non-economic nature*, from removing radio and television transmission, including cable transmission, from the field of competition by conferring on one or more establishments an exclusive right to conduct them' (emphasis added). The statement is qualified, however, and refutes Italy's reliance on Article 90(2) as a protection against the competition rules. The Court went on to state: 'However, for the performance of their tasks, these establishments remain subject to the prohibitions against discrimination and, to the extent that this performance comprises activities of an economic nature, fall under the provisions referred to in Article 90 relating to public undertakings and undertakings to which Member States grant special or exclusive rights'.
- Exclusive rights held by an enterprise to transmit television advertisements are not incompatible with the free movement of goods (such advertisements being designed to promote the goods in question). However, this would not be the case if the exclusive rights were used 'to favour within the Community particular trade channels or particular economic operators in relation to others'.
- The grant of exclusive rights to transmit television signals is not contrary to Article 7 of the Treaty (on discrimination on grounds of nationality). However, discrimination by undertakings holding such exclusive rights against nationals of Member States based on their nationality, is incompatible with Article 7.

- It would be an abuse of Article 86 if an undertaking holding a monopoly on television advertising imposed unfair charges or usage conditions for its services or if it discriminated between commercial operators or national products and those of other Member States in connection with access to television advertising.
- Article 90 has direct effect in the Member States, 'even within the framework of Article 90', thereby conferring on interested parties rights which must be safeguarded by the national courts.

ERT[43]

This significant case on broadcasting monopolies originated from a reference by a Greek court to the European Court of Justice in connection with proceedings brought by the Greek radio and television broadcasting monopoly ('ERT') against a private television channel set up by the Mayor of Thessaloniki. The Court was asked to give a preliminary ruling on the compatibility with Community law of the Greek statutory radio and television monopoly in connection with the television monopoly itself, free movement of goods and services, the competition rules under Articles 85, 86 and 90, as well as other human rights aspects of the case.

Under Greek law, ERT is a non-profit making organization which holds a monopoly of radio and television broadcasting and retransmission. The Greek law also prohibits any person from undertaking, without authorization by ERT, activities for which ERT has an exclusive right. The Court referred to its earlier jurisprudence in *Sacchi* where it held that there was nothing in the Treaty to prevent Member States from maintaining radio and television broadcasting outside the scope of competition for public interest reasons, by granting an exclusive right of broadcasting to one or more enterprises:

> Nothing in the Treaty prevents Member States, for considerations of public interest, *of a non-economic nature*, from removing radio and television transmissions, from the field of competition by conferring on one or more establishments an exclusive right to conduct them. However, it follows from Articles 90(1) and 90(2) EC Treaty that the way in which such a monopoly is organized or exercised should not infringe EC Treaty provisions relating to the free movement of goods, freedom to provide services and the competition rules. (paragraph 11)

Referring to jurisprudence established in the *Sacchi* case, the Court stated that television broadcasting falls within the scope of Treaty provisions relating to freedom to provide services (since a television monopoly is a monopoly in the provision of services) and on the free movement of goods (in that trade in products used for television broadcasting are subject to such rules). The Court ruled that the granting to a single enterprise of exclusive

---

[43] Case C-260/89, *Ellenike Radiophonia Tileorassi* v. *Dimotike Etairia Pliroforissis*, 18 June 1991.

television broadcasting rights which include the exclusive right to import, hire or distribute broadcasting material and products is not contrary to Article 30 of the Treaty, unless it results in direct or indirect discrimination between national and imported products to the detriment of the latter. However, it left the national courts to determine the presence of such discrimination.

On the freedom to provide services, the Court noted that Article 59 of the Treaty requires abolition of any form of discrimination against a provider of services established in a Member State other than that in which the service is provided. The Court said that although the existence of monopoly rights for the provision of services is not in itself incompatible with Community law, incompatibility may occur where the monopoly is organized in such a manner as to discriminate against national television broadcasts and those originating in other Member States, to the detriment of the latter. With regard to ERT, the possibility existed that in the absence of a guarantee concerning the retransmission of programmes from other Member States, there might have been discrimination resulting from ERT's statutory monopoly on both the broadcasting of its own programmes and the retransmission of other Member States' programmes. However, the question of whether the exclusive broadcasting and retransmission rights, taken together, actually leads to such discrimination, is a question of fact to be determined by the national court. Such discrimination, if found by the national court, could only be justified on one of the grounds specified in Article 56 of the Treaty, i.e. public order, public security or health.

On the question of compatibility of the Greek broadcasting monopoly with the EC competition rules, the Court observed that an enterprise benefiting from a statutory monopoly on the territory of one Member State may be held to have a dominant position in a substantial part of the common market for the purpose of Article 86. It said that although Article 86 does not prohibit monopolies as such, it nevertheless prohibits abuses by them. The Court recalled that under Article 90(2), undertakings entrusted with the operation of services of general economic interest are subject to the competition rules, provided it is not shown that the application of those rules is incompatible with the performance of the particular task. The Court again left it to the national court to determine whether the practices of ERT were compatible with Article 86 and to check in cases where such practices are contrary to Article 86, that they may be justified by the requirements of the particular task with which the undertaking may have been entrusted.

The Court pointed out that under Article 90(1) Member States may not maintain nor introduce measures which could undermine the provisions of Articles 85 and 86.

The Court ruled that Article 90(1) prohibits the granting of an exclusive right to retransmit television programmes to an undertaking which has an exclusive right to transmit broadcasts, where those rights are likely to create a situation in which that undertaking is led to infringe Article 86 as a result of a discriminatory broadcasting policy in favour of its own programmes, except where the application of Article 86 would obstruct the performance of the particular tasks assigned to it.

The ERT case has progressed the law by ruling that the granting of an exclusive right to retransmit television broadcasts to a single undertaking is an infringement of Article 90(1) 'where those rights are *liable* to create a situation in which that undertaking is led to infringe Article 86 . . .' (emphasis added). This suggests that an abuse of a dominant position under Article 86 may occur by the simple *existence* of such dominant position without an abuse actually having taken place.

Dutch broadcasting cases[44]

The first of these two cases was an infringement action initiated by the Commission against the Netherlands on the basis of Article 169 (Case C-353/89); the second case was a request for a preliminary ruling made by a Dutch court (Case C-288/89).

Both cases raised the issue of the compatibility with Article 59 of a provision of Dutch law imposing several conditions on the broadcasting of foreign radio and television programmes, including advertising specifically aimed at the Dutch public by advertising in Dutch or with Dutch subtitles. According to these provisions, the broadcasters must be non-profit making entities, should not procure benefit to third parties and must entrust the production of advertising to a distinct legal person. Dutch national broadcasting companies were thus obliged to entrust the broadcasting of any commercial advertising to the 'Stichting Sterreclame' (STER), a public body. Advertising messages should not exceed 5 per cent of broadcasting time, should be clearly separated from the rest of the programme and should not be broadcast on Sundays. In addition, foreign programmes could only be broadcast if they did not contain advertising in the Dutch language.

The Court ruled that these conditions constituted a double restriction on the free provision of services. Dutch broadcasters were prevented from broadcasting radio and television programmes offered by transmitters established in other Member States if they did not fulfil the conditions. Moreover, there was a limitation on the possibilities for foreign television and radio stations to broadcast (for the benefit of advertisers established in the Netherlands) commercials designed for the Dutch public. The Court could not find any valid ground for justifying such restrictions.

The Court rejected the Dutch government's argument that these conditions could be justified on the grounds of cultural policy in the audio-visual sector with a view to maintaining pluralism. The Court said that whilst maintaining pluralism in the audio-visual sector is a legitimate concern, it is not

---

[44] Case C-353/89, *Commission of the European Communities* v. *Kingdom of the Netherlands* [1991] ECR I-4069. Case C-288/89, *Stichting Collectieve Antenne Voorziening Gouda and Others* v. *Commissariat voor de Media* [1991] ECR I-4007.

an indispensable requirement that national legislation should impose on bodies established in other Member States the need to align themselves to the Dutch model, if they intended to broadcast programmes containing advertising messages for the Dutch public.

In respect of the conditions imposed by Dutch law on the advertising messages themselves, the Court ruled that restrictions concerning the transmission of advertising messages such as an advertising ban on certain days, a restriction as to the duration or frequency of the messages, or a requirement that the commercial be clearly distinguished from the rest of the programme, might in principle be justified on grounds of general interest such as consumer protection or cultural policy. However, the Court found, on the facts, that Dutch law was merely protecting the revenue of the Dutch monopoly, STER, for commercial advertising and protecting STER from competition. Provisions in Dutch law, therefore, concerning the transmission of advertising were in breach of Article 59 on the freedom to provide services.

In the infringement action brought by the Commission against the Netherlands[45], the Commission had also brought a complaint against a provision of Dutch law according to which Dutch radio and television stations, when producing their own programmes, must use the services and technical means of a Dutch company, namely Nederlands Omroepproduktie Bedrijf (NOB). The Court found that such obligation effectively restricted the free provision of services within the EC since it prevented Dutch broadcasters from using the production services of companies established in other Member States. In this respect, the Court emphasized that Article 59 was applicable even if only one national company and not all benefited from preferential treatment. Thus the Court used Article 59 to impose a control over the grant of exclusive rights. It rejected the Dutch government's contention that its media law did not produce discriminatory effects because it applied to Dutch and foreign broadcasting companies alike. The Court refused to accept this argument, holding that it was sufficient that the preferential system set up should benefit a national provider of services.

The Court dismissed the Dutch government's argument that the statutory monopoly held by NOB was justified in the public interest. According to the Court, the obligation for Dutch radio and television stations to use the services of NOB goes clearly beyond what is necessary to protect freedom of expression and was therefore disproportionate to the aim of the legislation. In the Court's view, pluralism in the Dutch audio-visual sector cannot be affected by the possibility which would be open to the various Dutch broadcasting bodies of addressing themselves to production services abroad.

The Court also relied on its judgment in *Terminal Equipment*[46] and confirmed that Article 90 does not imply that all exclusive or special rights

---

[45] See note 44 above.

[46] Case 202/88, *France* v. *Commission* [1991] ECR I-1223.

granted to companies such as those conferred on NOB are compatible with the Treaty.

Screensport[47]

In a significant Commission Decision in *Screensport*, the Commission held that a joint venture involving, in particular, the Eurosport consortium and Sky Television Plc, for the establishment of a dedicated European satellite sports channel, infringed Article 85(1) and could not be exempted under Article 85(3).

Eurosport was a joint venture set up by News International, the parent company of Sky Television plc, and members of the European Broadcasting Union (EBU). It was exclusively a satellite television sports channel which transmitted both through cable systems and direct via the Astra satellite to domestic satellite dishes. Screensport, a small sports television channel company, complained to the Commission about the joint venture, notably on the basis that Eurosport had access to the institutional exchange system for television programmes through the Eurovision network. All EBU members are eligible to participate in Eurovision. Screensport, as it did not provide a national broadcasting service, was not eligible to be an EBU member. It alleged that its exclusion from Eurovision was discriminatory and that Eurosport's access to Eurovision provided Eurosport with a competitive advantage, particularly as regards live sports coverage. The Commission examined the complaint under Articles 85 and 90(2).

The Commission held the agreements which established and operated Eurosport to be restrictive of competition on two grounds: first, between the partners to the joint venture in so far as they were actual and/or potential competitors on a market for sports broadcasting; and second, with regard to third parties seeking to broadcast sports events, particularly transnational dedicated sports channels. The Commission further rejected Eurosport's claims that it should be granted an exemption under Article 85(3), in so far as the setting up of a second transnational commercial satellite television channel dedicated to sport was not an 'improvement' within the meaning of Article 85(3) in the sense of promoting technical or economic progress. Its effect was, in practice, a disproportionate distortion of competition on the market in question.

The Commission examined Eurosport's claim that it was an undertaking entrusted within the meaning of Article 90(2) with the operation of services of general economic interest and should therefore be exempted from application of the competition rules in the performance of the particular tasks assigned to it.

The Commission pointed out that whilst it might be possible that the 'public mission' obligations imposed by Member States on their national

---

[47] Commission Decision of 19 February 1991 relating to a proceeding pursuant to Article 85 of the EC Treaty (IV/32.524 – *Screensport/EBU Members*), OJ 1991 L63/32.

broadcasting organizations made them undertakings entrusted with services of general economic interest, it was nevertheless highly doubtful that, given the national nature of these obligations, they could be interpreted as 'extending to transnational activities such as those of Eurosport'. Furthermore, there was nothing in the application of Article 85(1) to the joint venture agreement which would prevent them from fulfilling these tasks. In any case, the joint venture was of a commercial character, entered into for purely economic reasons.

EBU/Eurovision System[48]

The decision relates to the validity under EC law of the rules of the European Broadcasting Union (the EBU) governing the acquisition of television rights to sports events, the exchange of sports programmes within the framework of Eurovision and contractual access to such programmes for third parties. Members of the EBU are broadcasting organizations which have a 'public service mission' within a national territory.

The Commission granted the EBU an exemption under Article 85(3) to the provisions of Article 85(1). The exemption was subject to two conditions. Firstly, the EBU and its members were required to acquire television rights collectively to sports events only under agreements which offered a possibility of access to third parties or, subject to the approval of the EBU, on conditions more favourable to the non-members. The second condition required the EBU to inform the Commission of any amendments and additions to the rules notified, all arbitration procedures connected with disputes under the access scheme, and all decisions concerning applications for membership by third parties. The exemption was granted for a period of five years.

However, proceedings were taken to the Court of First Instance to seek an annulment of the Commission Decision. The Court upheld the applicant's case and annulled the decision in the *Métropole Télévision* ruling.

Métropole Télévision[49]

The Commission Decision in *EBU/Eurovision System*, was subsequently annulled on appeal to the Court of First Instance, on the grounds that the Commission committed an error of law in its interpretation and application of Article 85(3) on two grounds: first, in respect of the EBU's membership rules; and, second, with regard to the concept of 'particular public mission'.

---

[48] Commission Decision of 11 June 1993 relating to a proceeding pursuant to Article 85 of the EC Treaty (IV/32.150 – *EBU/Eurovision System*), OJ 1993 L179/23.

[49] *Métropole Télévision SA and others* v. *Commission*. Joined cases T-528/93, T-542/93, T-543/93, T-546/93; unreported.

The Court stated that the Commission had based its Decision on an erroneous interpretation of Article 85(3)(a)[50] by failing to examine first whether the membership rules of EBU 'were objective and sufficiently determinate and capable of uniform, non-discriminatory application in order next to assess whether they were indispensable within the meaning of Article 85(3)(a) of the Treaty'. The Court examined the concept of 'particular public mission', having regard to Article 85(3)(a). It considered whether or not the concept of 'particular public mission', as used in the Decision, constituted a relevant factor to be taken into account for the purposes of the application of Article 85(3), particularly Article 85(3)(a).

The Court's examination of the Commission Decision found that the concept, of 'particular public mission' used constituted a fundamental component of the statement of reasons of the Decision.

According to the Court, the Commission considered that the fulfilment of a particular public mission enabled broadcasters to become active members of the EBU and that the constraints connected with such a mission were such as to justify the EBU's having a special status with regard to the competition rules. Therefore, the concept of 'particular public mission' as defined by the Commission was 'the principal factor in defining the circle of beneficiaries of the exemption at issue'.

The Court referred to the Commission Decision, noting that 'particular public mission' was characterized by the obligation to provide varied programming including cultural, educational, scientific and minority programmes without any commercial appeal and to cover the entire national population irrespective of the costs. Thus, the Court reasoned, the decision essentially takes over the elements of the particular mission of operating services of general economic interest provided for in Article 90(2) of the Treaty, as it has been interpreted by the Community Court, namely and in particular, operations (on behalf of all users throughout the territory of the Member State concerned) irrespective of the specific situations or the degree of economic profitability of each individual operation: therefore, in order to hold that the condition relating to Article 85(3)(a) was fulfilled, the Commission took account of factors falling within the scope of Article 90(2).

The Court admitted that in the context of an overall assessment the Commission was entitled to base itself on considerations related to the pursuit of the public interest in order to grant an exemption under Article 85(3). But in this case 'it should have shown that exclusivity was indispensable in order to allow the EBU members a fair return on their investments and to enable

---

[50] Under Article 85(3), Article 85(1) provisions may be declared inapplicable in the case of any agreement or category of agreements between undertakings; any decision or category of decisions by associations of undertakings; any concerted practice or category of concerted practices; which contributes to improving the production or distribution of goods or to promoting technical or economic progress, while allowing consumers a fair share of the resulting benefit and which does not:

> *(a) impose on the undertakings concerned restrictions which are not indispensable to the attainment of these objectives.* (emphasis added)

them to carry out their public service mission'. The Court held that the Commission had failed to prove indispensability on the basis of a minimum amount of actual economic data.

Consequently, the Court concluded that:

> by using in this case as a criterion for granting exemption from the rules of Article 85(1) of the Treaty simply fulfilment of a particular public mission defined essentially by reference to the mission of operating services of general economic interest referred to in Article 90(2) of the Treaty, the Commission based its reasoning on a misinterpretation of Article 85(3) of the Treaty. That error of law is likely to have distorted the assessment which it made of the indispensable nature of the restrictions of competition for which it granted exemption.

The Court, therefore, annulled the Commission Decision.

Telemarketing[51]

In the *Telemarketing* case, the Court established that an abuse of a dominant position occurs where, without any objective necessity, an undertaking holding a dominant position in a particular market reserves to itself an *ancillary activity* which might be carried out by another undertaking as part of its activities on a neighbouring but separate market, with the possibility of eliminating all competition from such an undertaking.

In *Telemarketing*, the Court recognized that the existence of a legal monopoly does not in itself entail a dominant position, nevertheless, an undertaking holding exclusive rights granted under the law of a Member State, could, by virtue of its position of economic strength, restrict competition on the relevant market by being able to behave to a significant extent independently of its competitors: 'The fact that the absence of competition or its restriction on the relevant market is brought about or encouraged by provisions laid down by law in no way precludes the application of Article 86 . . .'[52].

In *Telemarketing* the Court stated that:

> Although it is true . . . that it is not incompatible with Article 86 for an undertaking to which a Member State has granted exclusive rights within the meaning of Article 90 of the Treaty, to enjoy a monopoly, it is nonetheless apparent from the same article that such undertakings remain subject to the Treaty rules on competition and in particular those contained in Article 86. In its aforesaid judgment of 30 April 1974 in the *Sacchi* case, the Court also stressed that, if certain Member States treat undertakings entrusted with the operation of television, even as regards their commercial activities and in particular advertising, as undertakings entrusted with the operation

—

[51] Case 311/84, *Centre Belge d'Etudes de marché – Télémarketing (CBEM) SA* v. *Compagnie luxembourgeoise de télediffusion SA and Information publicité Benelux SA* [1985] ECR 3261.

[52] On this point the Court referred to its jurisprudence in case 26/75, *General Motors* v. *Commission* [1975] ECR 1367; case 13/77. *INNO* v. *ATAB* [1977] ECR 2115; and case 41/83, *Tilly* v. *Commission* [1985] ECR 880.

of services of general economic interest, the prohibitions of Article 90 apply, as regards their behaviour within the market, by reason of Article 90(2), so long as it is not shown that the said prohibitions are incompatible with the performance of their tasks.

The Court held, therefore, that Article 86 should be interpreted as being applicable to an undertaking holding a dominant position on a particular market, even where that position is not due to the activities of the undertaking itself, but due to provisions laid down by law, where competition is eliminated or severely curtailed on that market.

Magill[53]

Whilst not decided on the basis of Article 90 (although Article 90(2) was mentioned in the original Commission Decision), the *Magill* judgment marks a landmark ruling of the Court on the balance between EC competition law and national intellectual property rights. Its potential impact extends not just to the broadcasting sector, to which the specific subject matter refers, but to any sector, for example, telecommunications, where the existence and exercise of intellectual property rights are important for commercial operations.

The original case goes back to a 1988 Commission Decision which held that Independent Television Publications Limited (ITP, publisher of independent television listings), the BBC and RTE (the Irish state broadcasting company) had abused their dominant position, contrary to Article 86, by refusing to license the advance weekly television and radio programmes listings to Magill TV Guide Limited for the purpose of compiling a comprehensive weekly television guide of programmes to be shown on ITV, BBC and RTE in the week ahead.

Following the Commission decision, the parties appealed to the Court of First Instance in 1991. The Court of First Instance upheld the Commission decision. Subsequently ITP and RTE appealed to the European Court of Justice. The Advocate General's opinion disagreed with the Court of First Instance on the basis that a copyright owner's exclusive right to refuse licences should only be denied on the basis of 'substantial' and 'weighty' grounds. The Court of Justice did not take on board the Advocate General's arguments but upheld the Court of First Instance ruling.

The Commission Decision observed, firstly, with reference to *Sacchi*[54], that the competition rules were applicable to these public broadcasting organizations. It considered the television guides market to be a separate but ancillary

[53] Commission Decision of 21 December 1988 in *Magill TV Guide/ITP, BBC and RTE*, OJ 1989 L78/43. Case T-69/89 *RTE* v. *Commission* [1991] ECR II-485. Case T-70/89 *BBC* v Commission [1991] ECR II-535. Case T-76/89 *ITP Limited* v. *Commission* [1991] ECR II-575. Joined cases 76, 77 and 91/89R, *Radio Telefis Eireann and Others* v. *Commission of the European Communities* [1989] ECR 1141.
[54] See footnote 42 above.

market to that of broadcasting services; that the broadcasting organizations held both a factual and legal monopoly in relation to their programme listings which were protected under copyright laws in the UK and/or Ireland; that the effect of the copyright protection restricted competition from third parties on the market in question; and that, consequently, ITP, BBC and RTE held a dominant position on the market. Moreover, by refusing to license or by limiting the licence conditions concerning the reproduction of the television listings, they had acted in an unduly restrictive manner. In doing so, they had protected their own individual television guides and had consequently abused their dominant position by preventing the introduction onto the market of a new product, i.e. a comprehensive weekly television guide.

Dismissing the parties' arguments in relation to copyright, the Commission went so far as to state that the parties 'in fact used copyright as an instrument of the abuse, in a manner which falls outside the scope of the specific subject matter of that intellectual property right'.

Considering the possible application of the exemption from the competition rules under Article 90(2), the Commission said that even if the three broadcasters were under a public or statutory duty to produce and publish their individual programme listings in the form of television guides, the application of the competition rules in this case would in no way obstruct the performance of that particular task within the meaning of Article 90(2), therefore this article was not applicable.

The Commission decision held, therefore, that the policies and practices of the three broadcasters infringed Article 86 by preventing the publication and sale of comprehensive weekly television guides in the particular markets in question, and the parties were instructed to end the infringement immediately by supplying each other and third parties on request with their television listings.

The *Court of First Instance*, on appeal, upheld the Commission decision. The Court of First Instance considered that the exercise of the national intellectual property rights should be restricted to the extent necessary to reconcile the respect to which intellectual property rights were entitled, with the fundamental principles of the Treaty concerning freedom of competition and the free movement of goods. While the exercise of an exclusive right to reproduce a protected work was not in itself an abuse, this was not so when the right was exercised in a manner that was clearly contrary to the objectives of Article 86.

The reservation of an exclusive right by the broadcasters prevented the emergence on the market of a new product, i.e. a general television guide, for which there was consumer demand on that ancillary market of weekly television guides. This had the effect of excluding all competition from that market simply to secure the broadcasters' respective monopolies. This went beyond what was necessary to fulfil the essential function of the copyright as permitted under EC law. The refusal to license by the broadcasters was arbitrary and not justified by the requirements pertaining to the activity of publishing television magazines. It was possible for the broadcasters to adapt to the conditions of a competitive television magazine market to ensure the commercial viability of their weekly publications. Therefore, their conduct could not be covered in Community law by their copyright protection in the programme listings. The Court of First Instance thus dismissed the appeal of the broadcasters on the basis of a breach of Article 86.

In contrast, the Advocate General's opinion took the view that 'substantial' and 'weighty' grounds must be submitted to demonstrate that intellectual property rights contained in the specific subject matter of copyright can be exercised in a way that does not comply with the Treaty. Interference with copyright by a compulsory grant of licences was recognized as a serious matter. In this specific case, he concluded that such interference could not be justified. He considered the refusal to license as justifiable in order to retain the market free from competition and that the broadcasters, as copyright owners, had a right to prevent their protected work from use by competitors.

The Court of Justice, in a short judgment, did not follow the opinion of the Advocate General, and reinforced the Court of First Instance ruling that the refusal to license amounted to an abuse of a dominant position and was contrary to Article 86.

The *Magill* judgment has caused substantial uncertainty at the margin between application of the competition rules and the legitimacy of national intellectual property protection in copyright. Broadcasters, or operators in any other sector which rely on national intellectual property protection rights, will have to give serious consideration to the issue of refusal to grant licences for intellectual property rights when issues of Community competition law are at stake. In fact, as a result of the *Magill* judgment, it would appear that intellectual property rights might now, in certain circumstances, be regarded as 'essential facilities' for which access must be given.

Conclusion

Technological developments affecting the broadcasting sector have forced the pace of Community policy development in this area. EC Treaty obligations, particularly in respect of the right of establishment and freedom to provide services, and the competition rules under Articles 85 and 86 and 90, form the legal framework for future policy.

The broadcasting sector now has to accommodate major infrastructure changes and a whole new range of services. As a result of new digital technology in broadcasting, the boundaries are extending beyond recognisable traditional broadcasting parameters. It is evident from examining recent Commission Decisions taken under the Merger Regulation and Article 85 that the trend towards the convergence of competition policy and the regulatory framework between the audio-visual sector and telecoms is shaping policy in the emerging information society.

The legal issues emerging from the broadcasting sector are similar to those arising from the other liberalising sectors (telecommunications, transport, energy and postal services). Access is a major issue. The new Article 90 Directive on removing restrictions on the use of cable television networks carrying liberalized telecommunications services will be instrumental in providing wide access to a range of new multimedia services.

An increasingly important area in the overall development of the information society, including broadcasting, is the provision of a satisfactory regulatory framework in the intellectual property field. The Commission has engaged in a two-year consultation process on the Green Paper on Copyright and Related Rights in the Information Society[55]. According to the Commission[56] the consultations underlined four main points: that existing copyright concepts continue to be valid in the new information society environment and no new structure of copyright and related rights is needed; that adaptations to the existing rights, however, are needed and should be applied in the framework of the single market; that the enforcement and management of rights merit particular attention; and that an appropriate international response in the form of international minimum standards for intellectual property protection in the new environment is needed in parallel to the 'domestic' Community initiatives.

On 9 July 1997 the Commission proposed a draft Directive on the legal production of television and radio broadcasting and information society services offered to the public where access is subject to payment (including pay-TV and video-on-demand).

In broadcasting, as in other sectors, the Commission's vigilance is continuing to extend into areas of Community law where restrictions are placed on free competition. The lack of current Community legislation in areas such as access is not preventing the Commission from using its powers under the Treaty or pursuant to EC competition legislation, to ensure that markets remain open to competition.

---

[55] COM(95)382, 19 July 1995.

[56] Commission Press Release; Memo/96/94, 7 Oct. 1996.

Chapter 12

COMPETITION AND EXCLUSIVE RIGHTS IN THE POSTAL SECTOR

INTRODUCTION

The postal services sector in the Community is a rapidly developing growth area. In addition to the traditional forms of post such as letter, express mail and parcel deliveries, there are new electronic services and ancillary services such as financial services. Traditionally, as a result of exclusive rights granted to public postal monopolies in the Member States, private operators deal mainly with express and parcel services, since letter services tend to fall within the 'reserved' sector allocated to the postal administrations. A Community postal policy has been envisaged since 1989, but over the years has proved difficult to achieve, given the diversity of postal services and the different views of Member States, particularly the lack of a common concept of universal service provision. Community efforts to reach a common approach to postal services in the European Union has led to a Green Paper on the development of the single market for postal services[1], Commission guidelines for the development of Community postal services[2] and a Council resolution on the development of Community postal services[3]. Following these preliminary works, the Commission prepared a draft Notice on the application of the competition rules to the postal sector and a draft Directive on common rules for the development of Community postal services which is on its way to adoption by the Council of Ministers.

Also crucial to the development of a Community strategy on postal services has been a range of cases before the Court concerning the use of privileged

[1] Green Paper on the Development of the Single Market for Postal Services, Communication from the Commission, COM(91) 476(final), 11 June 1992.
[2] Guidelines for the Development of Community Postal Services, Communication from the Commission to the Council and the European Parliament, COM(93) 247(final).
[3] Council Resolution of 7 February 1994 on the Development of Community Postal Services, OJ 1994 C48/3, 16 Feb. 1994.

monopoly positions by undertakings holding special or exclusive rights[4], the *Corbeau* case being of particular significance since it specifically addressed issues in the postal sector. In addition, there have been a number of Commission decisions addressing issues of liberalization in express courier services[5].

Green Paper on the development of the single market for postal services

Following a Council request in autumn 1989 to draft measures for a Community postal services policy, the Commission produced, in mid-1992, the Green Paper on the development of the single market for postal services. The Commission's analysis concluded that in determining future action for the postal sector:

> the guiding principle must be the maintenance and, if appropriate, the development of a universal postal service which would provide collection and delivery facilities throughout the Community, at prices affordable to all and with a satisfactory quality of service. Then, provided that the universal service is secured, there should be as much freedom of choice as possible.

The Green Paper identifies five areas of concern for the Community. These are:

(1) *Lack of harmonization.* Since universal service providers in each Member State have developed independently, this has created interoperability problems between Member States.

(2) *Single market implications.* The diversity by Member States has implications for the single market, where divergences in performance of universal service are capable of leading to market distortions. For example, 'it does not appear coincidental that one of the Member States with very poor service has a mail order sector only one-fifth the size of that of other Member States of comparative economic size'. Accordingly this not only harms the mail order companies' prospects, but reduces consumer choice. Such disruptions affect both senders and receivers of mail.

(3) *Cross-border service performance.* Cross-border services often take twice as long from collection to delivery as domestic services: 'Since effective

---

[4] Case C-41/90, *Höfner* [1991] ECR I-1979; Case C-202/88, *France v. Commission* [1991] ECR I-1223; Case C-260/89, *ERT* [1991] ECR I-2925; Case C-18/88, *RTT v. GB-INNO-BM* [1991] ECR I-5941; Case C-179/90, *Port of Genoa* [1991] ECR I-5889; Case C-320/91, *Paul Corbeau* [1993] ECR I-2533.

[5] Fifteenth Report on Competition Policy, Point 259 (concerning Germany, Belgium, France and Ireland); Nineteenth Report on Competition Policy, Point 228 (concerning Italy); Twentieth and Twenty-first Reports on Competition Policy, Points 356 and 326 (concerning Denmark). Commission Decision 90/616/EEC on Dutch PTT, OJ 1990 L10/47, subsequently appealed to Court of Justice in *Netherlands-Koninklijk PTT Nederland NV and PTT Post BV* v. *Commission* (Cases 48/90 and 66/90) [1992] ECR I-565.

cross-border communications are essential for the commercial and social life of the Community, it is a matter of concern to the Community that service performance for cross-border services is so unreliable'.

(4) *Divergences – cohesion.* A Community objective is 'to ensure its own cohesion'. Wide divergences existing in the postal sector do not contribute to the achievement of this objective. Regions with unreliable postal services are disadvantaged in respect of their communications and goods delivery requirements.

(5) *Market distortions.* There is concern that the scope of exclusive rights granted to national postal administrations is often larger than required to assure the provision of universal service.

The Green Paper sets out three main policy objectives:

(1) To ensure provision of universal postal service throughout the Community at prices affordable to all through the establishment (inasmuch as it was needed in Member States individually), of a set of reserved services which would confer some special and exclusive rights, in order to maintain the resources necessary for the undertaking of the public service mission in good conditions; at the same time, consistent with this objective, to have the largest possible part of the sector operating in free competition.

(2) To have common obligations for the universal service operators of the Community in respect of the special and exclusive rights granted to them by the reserved services in order to enable them to provide universal services, in particular with regard to the quality of service provided.

(3) To make any necessary efforts towards Community cohesion through appropriate harmonization measures.

Within these policy objectives, the Green Paper identifies four key regulatory issues. These are:

(i) to establish a set of universal services[6];

(ii) to develop a definition of reserved services[7] linked to the definition of universal service;

(iii) to ensure compatibility of other Member States' commitments with Community legislation and policies; and

(iv) to separate regulatory and operational functions.

---

[6] Universal service refers to postal services which are available to all consumers without discrimination, service should be identical for similar users and should be guaranteed without interruption except in the case of *force majeure.*

[7] Reserved services are monopoly services reserved to certain operators. The Green Paper states that 'such reserved services are normally the subject of some special and exclusive rights (these rights in effect curtailing the normal operation of the free market). In the postal sector, the reserved service operator in each Member State is the postal administration'.

The Green Paper also sets out obligations for providers of universal service. These are:

(a) access conditions to universal services should be the same for everyone;
(b) tariffs of universal services should be related to average costs;
(c) inter-administration compensation (i.e. the system of charging between postal administrations – terminal dues) should reflect delivery costs; and
(d) service standards for universal service should be set and performance monitored.

The Green Paper also sets out criteria in connection with harmonization and cohesion to ensure that the postal services sufficiently meet the postal needs of the whole Community.

The Green Paper thus sets out a framework for discussion on future Community policy based on liberalization. It perceives universal service to be of paramount importance to the liberalization process, whilst also noting that fulfilment of the universal service objective could justify the creation of a set of universal services and the grant of certain special or exclusive rights to the universal service provider. However, such reserved services should be the minimum necessary to guarantee full universal service provision. The Green Paper emphasizes the need to provide definitions of universal service and reserved service. The Green Paper itself does not provide definitions; it merely notes their absence at Community level and refers to problems of definition and to obligations inherent in these types of service.

The Green Paper envisages liberalization of express courier services, parcel services, magazine and newspaper services and self-posting. 'New' services which are significantly different from standard postal services should be placed in the non-reserved sector. It also envisages the liberalization of cross-border traffic to be followed by the liberalization of direct mail.

The Green Paper on postal services has had a determining influence on the preparation of further texts by the Commission and continues to be a key reference document.

Guidelines for the development of Community postal services: response to the Green Paper

The Guidelines were published in response to the public consultation process set in motion by the Green Paper. According to the Guidelines, a large majority of the 200 written contributions endorsed the Commission's approach to the postal sector. There was overall agreement that a common definition of universal service is a fundamental requisite for a Community policy on postal services. The majority of contributors also endorsed the need for a set of reserved services. Other areas of consensus referred to the need to improve the quality of services; the creation of a more rational system of terminal dues; and the need to separate regulatory and operational functions.

The Commission proposed that Community action, particularly legislative action, should be taken to develop the Guidelines in pursuit of a genuine single market in postal services. Legislation should be directed towards the following areas:

- Directives on the definition of the universal service, dealing with regulatory issues such as the division between operational and regulatory functions, a description of the obligations of universal service providers in respect of tariff principles, non-discriminatory access to networks etc., and defining services which could be reserved at Community level;
- a Directive on service quality in respect of quality standards, monitoring systems and principles to be applied to complaints concerning cross-border traffic; and,
- a directive on commercial harmonization and technical standardization.

It was clear from the Guidelines that whilst there was a high level of consensus from contributors to the consultation process under the Green Paper, many issues were still unresolved. For example, on the common definition of universal service, there was division of opinion as to whether parcel post should be included in the universal service. On the reserved services to be maintained, there were divergent views on whether direct mail and cross-border mail should be retained in this sector. On service quality standards, opinions differed as to whether service quality standards should be imposed only on the reserved sector or on the entire universal service. Similarly, different views were put forward in connection with cross-subsidization. Whilst most contributors indicated the need for rationalization of terminal dues, there was reluctance to advise aligning terminal dues with actual costs too quickly, and many contributors expressed concern about international aspects of terminal dues, particularly the role of the Universal Postal Union (UPU) in the area. The issue of remail was also mentioned by some contributors as an issue requiring resolution.

Council Resolution on the development of Community Postal services

In response to the Commission Guidelines, the Council adopted a resolution[8] endorsing the general approach contained in the Green Paper and Guidelines. The resolution invited the Commission to propose the measures necessary to ensure the implementation of a Community policy on postal services.

---

[8] Council Resolution of 7 February 1994 on the development of community postal services, OJ 1994 C48/3.

Case-law

Case-law of the European Court of Justice, together with individual Commission Decisions, has resulted in significant advances in establishing limits to the exclusive rights of State monopolies in the postal sector and has been instrumental in developing Community postal policy.

To date, cases specific to postal services on the interpretation of Article 90 have been in connection with limitations placed by national postal monopolies on private express courier services. They involve two Commission Decisions relating to Dutch and Spanish couriers (the express delivery services cases) and a key ruling of the Court in the *Corbeau*[9] case.

Express delivery services cases

The issue of express delivery services was examined at length in two Commission Decisions concerning such services in the Netherlands[10] and Spain[11].

In both cases, the Commission examined the compatibility with Articles 90 and 86 of national laws which limited competition with the national postal administrations on the express delivery market by imposing discriminatory conditions on private express delivery services, thereby placing them at a competitive disadvantage on that market. The Commission analysed the service itself. It concluded that the basic postal service and the express postal service constitute two separate markets. 'The essential difference lies in the value added by the providers to the transport of the goods concerned.'[12]

In the Netherlands case, the Commission established a list of services associated with express couriers. These services comprise:

- home collection of packages;
- delivery to the addressee in person;
- possibility of changing destination or addressee in transit;
- acknowledgment of the receipt given to sender;
- keeping track of packages;
- individual attention given to customers; and
- *à la carte* service offered[13].

The Spanish case added an additional category:

- guarantee of mail delivery by a specified date[14].

---

[9] See footnote 4 above.

[10] Commission Decision of 20 December 1989 concerning the provision in the Netherlands of express delivery services (90/16/EEC), OJ 1990 L10/47.

[11] Commission Decision of 1 August 1990 concerning the provision in Spain of international express courier services (90/456/EEC), OJ 1990 L233/19.

[12] Point 3, paragraph 1, Decision 90/16/EEC; Point 2, paragraph 1, Decision 90/345/EEC.

[13] Point 3, Decision 90/16/EEC.

[14] Point 2, Decision 90/345/EEC.

These categories served as a reference in the *Corbeau* case[15] and have subsequently been reproduced in legislation on postal services (discussed later in this chapter).

The Commission also considered the markets for the basic postal service and for express delivery services. It concluded that the services were not interchangeable. Express delivery services met the requirements of business customers who needed a guarantee that their mail would be delivered within a given time, whereas the basic service met the needs of the general public where the price of the service was as important as the speed. The additional services provided by express delivery services are reflected in the higher price[16].

The Commission examined the Dutch and Spanish postal administrations and found them to be undertakings to which their respective Member States had granted exclusive rights within the meaning of Article 90(1); they were subject to the competition rules under Article 86, unless they were able to show that, under Article 90(2), the application of the competition rules obstructed the performance, in law or in fact, of the particular task assigned to them[17]. Both postal administrations operated on the market for international express delivery of letters, which formed a value added market, adjacent to, but separate from, the basic postal service[18]. By virtue of the grant of exclusive rights to the Dutch and Spanish postal administrations (to collect, transport and distribute letters, to instal post boxes and to issue postage stamps within their national territories), these administrations held a dominant position on the market as regards the basic postal service[19]. According to the Commission, these exclusive rights resulted in a lack of competition on this market and it enabled the postal administrations to operate independently of their competitors who were excluded from the market and also of consumers who could not obtain the basic service elsewhere.

The Commission referred to the type of behaviour constituting an abuse of a dominant position:

> An abuse is committed where an undertaking holding a dominant position on a particular market reserves to itself or to another undertaking belonging to the same group, without any objective necessity, an *ancillary* activity which might be carried out by another undertaking as part of its activities on a neighbouring but separate market, with the possibility of eliminating all competition from such an undertaking[20]. (*emphasis added*)

[15] See footnote 4 above.
[16] Point 3, Decision 90/345/EEC, Point 4, Decision 90/16/EEC.
[17] Point 7, Decision 90/345/EEC, Point 8, Decision 90/16/EEC.
[18] Point 8, Decision 90/456/EEC, Point 9, Decision 90/16/EEC.
[19] Point 9, Decision 90/456/EEC, Point 10, Decision 90/16/EEC.
[20] Point 10, Decision 90/345/EEC, Point 11, Decision 90/16/EEC. This was established in Case 311/84. *CBEM/CLT and IPB* [1985] ECR 3278 at Point 27.

The Commission established an abuse in both cases. The Dutch law prohibited competitors of the postal administration (PTT Post BV) from providing an express service for less than a specified charge (Fl 11.9), thereby abusively reserving a large section of that market for PTT Post BV. The Commission decided that the Dutch law which reserved for PTT Post BV the express collection, transport and distribution of postal items of a specified weight and price was incompatible with Article 90(1), read in conjunction with Article 86[21]. The Dutch government successfully appealed the decision of the Commission before the Court but on procedural grounds[22].

In the Spanish case, the Commission held that the prohibition under Spanish law on private courier services from providing an international express courier service, and reserving this service exclusively to the Spanish postal administrations, similarly constituted an abuse of a dominant position under Article 86. The Commission decided that the Spanish laws which reserved for the postal service the international express collection, transport and distribution of letters were incompatible with Article 90(1) in conjunction with Article 86[23]. An additional point made in the Spanish case[24], was that the Spanish law in question, together with the behaviour of the Spanish post office, had the effect of limiting supply and technical development within the meaning of Article 86, thereby also constituting an infringement of Article 90 in conjunction with Article 86.

The abuse of limiting technical development is one which could arise in the future given the range of new postal services, including new technological services, on offer.

The Corbeau case

The case arose from a referral for a preliminary ruling to the Court by a Belgian court (the Tribunal Correctionnel de Liège) on four questions concerning the interpretation of Articles 86 and 90, with a view to verifying the compatibility of the Belgian law on the postal monopoly with these Treaty Articles. The questions were raised in the context of criminal proceedings against Mr Corbeau who had allegedly breached the Belgian legislation on the postal monopoly by delivering mail within the City of Liège. Mr Corbeau provided the service with his family. He would contractually guarantee his customers that he would deliver their mail by noon the next day following collection.

The questions related to the following issues:

(1) the extent to which the Belgian postal monopoly as defined by the Belgian law of 1965 and which prohibits actions such as those of Mr Corbeau is in conformity with Articles 90, 85 and 86;

[21] Article 1, Decision 90/16/EEC.
[22] *Kingdom of the Netherlands* v. *Commission*, Cases C-48/90 and C-66/90 [1992] ECR 565.
[23] Article 1, Decision 90/456/EEC.
[24] Point 11, Decision 90/456/EEC.

(2) the extent to which Member States must modify such a monopoly to comply with Community obligations, particularly Article 90;

(3) whether the statutory monopoly in the postal sector benefiting from the grant of exclusive rights is by virtue of Article 90(2) subject to the Treaty competition rules, particularly Articles 7 and 85 to 90; and,

(4) whether such an undertaking holds a dominant position in a substantial part of the common market within the meaning of Article 86, as a result of a statutory monopoly, or as a result of the particular circumstances.

The Court, in its own words, 'reformulated' these points into a single question. The substance of the question is whether national laws of a Member State that confer on an undertaking such as the Belgian Régie des Postes the exclusive right to collect, carry and distribute mail, and to prohibit an economic operator established in that state, by means of criminal sanctions, from offering certain specific services on that market, must be interpreted as contravening Article 90 (judgment, paragraph 7).

The Court held first that the Belgian postal administration constitutes an undertaking to which a Member State has granted exclusive rights in the sense of Article 90(1). Confirming earlier EC case-law, it held that a beneficiary of a legal monopoly in a substantial part of the common market could be considered as having a dominant position. This, however, is not in itself incompatible with Article 86 (Article 86 is concerned with the abuse of a dominant position and not its creation).

The Court then went on to say that Article 90(1) must be read in conjunction with Article 90(2). Article 90(1) provides that undertakings enjoying exclusive or special rights must not act in contravention of the Treaty and specifically the competition rules. Article 90(2) specifies that undertakings which have been entrusted with a task in the general economic interest are subject to the competition rules in so far as the application of those rules does not obstruct the performance, in law or in fact, of the task assigned to them. Member States are thus entitled to grant exclusive rights and corresponding restrictions in competition only in so far as these are necessary to ensure the performance of the particular task assigned to the undertaking.

The Court held without hesitation that the Belgian postal administration is an undertaking which has been entrusted with a task of general economic interest within the meaning of Article 90(2). This task consisted of collecting and distributing mail on behalf of all users at uniform tariffs and under similar quality conditions. It had therefore to be established to what extent competition should be restricted to permit it to carry out its task of general economic interest and in particular to benefit from economically acceptable conditions. For such an undertaking to carry out its task under acceptable economic conditions, there must be the possibility of compensation between profitable sectors and less profitable sectors. Thus it follows that there must be a restriction of competition from individual undertakings in certain profitable sectors. In the absence of this safeguard, individual undertakings could focus on the economically profitable operations and offer lower tariffs than those adopted by the holders of the exclusive rights, since unlike the latter, they would not have to offset losses in the unprofitable sectors against profits in the more profitable sectors.

The Court then proceeded to lay down a twofold test to determine the limit of exclusive rights[25]. It held that exclusive rights cannot extend to certain services defined as:

(1) specific services, distinct from the service of general interest, which meet the particular needs of economic operators, and consist of additional features that the traditional postal service does not offer such as home collection, greater speed or reliability in distribution, or the possibility of changing the destination during the course of delivery; and

(2) the services in question, by their nature and the conditions in which they are provided, including the geographical area in which they are provided, do not jeopardize the economic balance of the service of general economic interest provided by the holder of the exclusive right.

It is only if the two conditions in the test are satisfied that it will be illegal, i.e. contrary to Article 90(1), for a Member State to grant an undertaking exclusive rights over such services.

It is an indication of the significance of this case that there were interventions by three Member States, namely Spain, the UK and Ireland. The Commission presented a written submission which explicitly stated that the major issues of fact in the case should be dealt with in the national court and not before the Court of Justice. It submitted that the data in the documents before the European Court did not make it possible to conclude that there was an infringement of Article 86. Furthermore, the Treaty provisions on free movement of persons and freedom to provide services did not apply to situations which are purely internal within a Member State which was the case with Mr Corbeau.

The Court did not analyse the nature of the specific service offered by Mr Corbeau and appeared to accept that it was some form of rapid courier service. Indeed, Mr Corbeau himself acknowledged that although the service offered by him did provide more added value than the Belgian postal services' normal service, it did not constitute a real rapid courier service. The Court did not explore this area.

This is a key judgment on the interpretation of Article 90. If proceedings are started, the Member State concerned will have to demonstrate that it is necessary for the economic stability of the undertaking benefiting from exclusive rights to maintain these rights over the services in question. The

[25] Judgment, para. 19.

burden of proof is likely to rest on the Member State to show the necessity of these exclusive rights and to carry out a proportionality test based on economic criteria. This constitutes a significant development.

From a practical point of view the judgment is not entirely satisfactory. It imposes a considerable responsibility on national courts to decide whether there is an economic justification for preserving a monopoly on the basis of the two part test. There is a danger that the courts will be unable to apply this test in practice, and it leaves the way open for divergent national court interpretations which could lead to different decisions. Once the national courts have pronounced on their national law, it will be difficult for the Community institutions to intervene to create uniformity throughout the Community.

The cases which have followed *Corbeau* have still not answered the questions which arise on the compatibility of special and exclusive rights with Treaty rules. It will be necessary for several more questions to be asked of the Court to enable it to provide clearer guidance on the application of the principles developed in *Corbeau*.

Legislative measures on postal services

The *Corbeau* judgment further clarifies the interpretation of Article 90(1) and 90(2). However, the crucial task of defining the components of reserved and universal services and of consigning the various postal services into one of these categories, had still to be completed. The Commission has taken the step of issuing two proposals for postal services in the Community, namely a European Parliament and Council Directive on common rules for the development of Community postal services and a draft Commission Notice on the application of the competition rules to the postal sector.

Proposed Directive on common rules for the development of the internal market of Community postal services and the improvement of quality of service

A Common Position[26] was reached on this Directive on 29 April 1997. The Directive is designated as a harmonization measure to harmonize the rules relating to postal services within the European Union. It has as its legal basis Article 100a and not Article 90(3). Therefore its adoption involves the European Parliament and Council.

The Directive sets common rules on universal postal service provision; criteria for defining reserved services; access to the postal network; prices;

[26] Common Position (EC) 25/97 adopted by the Council on 29 April 1997 regarding a Directive of the European Parliament and of the Council Directive on common rules for the development of the internal market of Community postal services and the improvement of quality of service, OJ 1997 C-188/9.

terminal dues; the fixing of quality of service standards; the harmonization of technical standards; and the establishment of independent national regulatory bodies.

Universal service

Chapter 2 sets out provisions on universal service. Firstly, under Article 3, Member States must ensure that users have the right to a universal service defined as a postal service of specified quality provided throughout their territory at prices affordable to all users. Member States must ensure that there are sufficient collection points to meet user requirements. Universal service providers, i.e. a 'public or private entity designated by a Member State to ensure the provision of a universal service or parts thereof' guarantee collections from post boxes and delivery to the address of every person, at least five days a week except in exceptional circumstances.

The universal service will consist of a *minimum* of the following services:

- the clearance, sorting, transport and distribution of postal packages up to 2 kg;
- the clearance, sorting, transport and distribution of postal packages up to 10 kg; and
- services for registered and insured items.

The universal service covers both national and transfrontier services.

Member States must designate *one or more* postal operators to provide the universal service and inform the Commission accordingly. Furthermore, the obligations and rights assigned to the universal service provider(s) by the Member State must be published. Since most national postal administrations in the Member States are protected by special or exclusive rights, this article is of particular significance. By providing that Member States may appoint *one or more* postal operators to provide the universal service, there is the possibility that the monopoly enjoyed by the national postal administration in this area could be replaced by the sharing of universal service provision with at least one other postal operator. This raises the question of how the special or exclusive rights held by national postal administrations would operate alongside other operator. Universal service provision obligations would have to be formulated in very clear and precise terms.

Member States must ensure that the universal service offers an identical service to all users under comparable conditions.

Member States must ensure that universal service providers give users sufficiently precise information on the characteristics of the universal service offered, particularly concerning general conditions of access to services, price and quality levels.

The Directive aims to impose on universal service providers legally enforceable obligations (except in cases of *force majeure*). In the Commission's view:

> without any such obligation at national and Community levels, the existing disparities between centrally located and remote peripheral areas would be expected to widen, because private operators (for example express couriers, parcel service providers) and substitute services for letters (for example data networks, fax, e-mail) have a tendency to concentrate on centrally located, densely populated areas.[27].

Article 19 establishes on a Community-wide basis the concept of liability for failure to fulfil universal service obligations, involving user compensation, particularly in connection with loss, theft, damage or non-compliance with service quality standards. For Member States that have not been previously subject to legally enforceable duties in the provision of postal services, the Directive marks a significant departure from former responsibilities. Complainant users, for example, presumably would be entitled to establish the liability of universal service providers for gross delays in service provision.

It is not clear from the Directive how the obligations of the universal service provider and the rights of users would function, i.e. whether it would be possible for users to enforce the provision of universal service directly or whether all enforcement and compensation matters would fall under the responsibility of the regulatory authority to be established in each Member State. The latter would appear to be the intention of the Directive, but Article 19 is ambiguous on this point.

Reserved service

Chapter 3 deals with reserved services – one of the most controversial issues since it aims at limiting the scope of national postal monopolies. The definition of reserved services in the Directive is the result of a compromise. The Directive sets maximum limits to Member States' monopolies but provide that those limits may be revised in the future. It sets out two dates for revision, namely 2000 and 2003.

The definition of reserved service is linked to that of universal service. Article 7(1) of the Directive provides that the scope of the reserved service should be such as to enable Member States to provide a universal service. At present, the scope of the reserved service regarded as necessary to ensure a universal service is defined in terms of weight and prices as:

> The clearance, sorting, transport and delivery of items of domestic correspondence, whether by accelerated delivery or not, the *price* of which is less than five times the public tariff for an item of correspondence in the first weight step of the fastest standard category where such a category exists, provided that they weigh *less than 350 g.* (emphasis added)

The Directive provides that two categories of mail services – cross-border mail and direct mail – which the majority of Member States could not agree

[27] Information Note concerning postal policy. European Commission document, 1995.

to liberalize may for the time being continue to be in the reserved sector. The necessity for maintaining these two services in the reserved sector will, however, be subject to examination by the Commission before the end of 1998.

It is particularly interesting to note that the definition of reserved service in the Directive is not a fixed definition likely to last for a considerable time. The Directive expressly provides for the forthcoming revision of the definition which should result in a further limitation of postal monopolies.

Earlier versions of the Draft Directive included a definition of 'new services', which was evidently influenced by the *Corbeau* judgment. The definition referred to services different from traditional services, distinct and dissociable from services reserved to universal service providers. It did not, however, fully incorporate the *Corbeau* definition which adds that the specific services should 'not compromise the equilibrium of the services of general economic interest performed by the holder of the exclusive right' either by their nature or by the conditions under which they are offered, such as the geographical area in which they are provided. Nor was it certain that the definition of 'new services' as set out in the Directive would have been sufficient to take into account new services such as electronic postal services, or problems such as reliance on copyright provisions to prevent the emergence of new services, as happened in the *Magill* case[28]. This definition has been entirely removed from the final text of the Draft Directive. However, the provision in Article 7 that the reserved service is to be reviewed to take into account the technical, economic and social changes as well as the need to ensure the financial equilibrium of universal service providers anticipates the further liberalization of the postal sector which should cover certain new services.

Authorization procedure

The Directive provides for a system of authorization procedures which is not mandatory. Member States may, if they so wish, provide that postal operators, outside the reserved area, will be subject to a system of authorization before they start providing services. Those services which by definition will be in the competitive sector will fall into one of two categories: they will either fall within or outside the universal service. In both cases, Member States will be entitled to apply an authorization procedure to ensure that permitted postal operators (i) meet 'essential requirements' defined in the Directive as including, in particular, observance of the confidentiality of correspondence and maintenance of a secure network free of dangerous goods and (ii) contribute, in appropriate cases, to the provision of the universal service. Member States may further ask all postal operators in the universal sector to

[28] Joined cases C-241/91 P and C-242/91 P, *Radio Telefis Eireann (RTE) and Independent Television Publications Ltd (ITP)* v. *Commission and Magill TV Guide Ltd*, 6 April 1995.

contribute to a compensation fund which would help ensure the provision of a universal service.

The Directive stresses that the authorization procedures should be transparent, non-discriminatory and based on objective criteria.

Tariffs and transparency of accounts

Chapter 5 sets out tariff principles applying to universal services and provisions on transparency of accounts. The Directive imposes an obligation on the Member States applying to universal services to ensure that tariffs for the universal services are set with regard to three principles:

(1) prices must be affordable to all users;
(2) prices must be fixed in relation to costs; the Directive, however, provides that Member States may apply a uniform tariff throughout their territory;
(3) tariffs must be transparent and non-discriminatory.

The Directive also contains provisions on terminal dues which are the remuneration applied between universal service providers for the delivery of international mail. Member States must ensure that terminal dues are determined in relation to the costs of dealing with incoming international mail and in relation to the quality of the services provided by postal operators. Transitional arrangements are permitted to avoid disrupting postal markets.

The Directive imposes strict obligations on universal service providers to keep separate accounts distinguishing between the reserved and the non-reserved sectors. The Directive provides for a preferred system of cost allocation but states that other systems may be adopted subject to strict conditions including approval by the national regulator. Moreover, universal service providers must have their annual accounts audited by independent auditors.

Quality of services

Chapter 6 deals with quality standards relating particularly to deadlines for the forwarding of mail and the regularity and reliability of universal service. Standards for national services are to be set by the Member States and those for intra-Community cross-border services are to be set by the European Parliament and the Council in accordance with standards set out in the Annex to the Directive. National services standards must be compatible with those laid down for intra-Community cross border services.

The standards for national mail should have the effect of improving services in those Member States which provide slower services, which should result in an improvement to international services by all Member States. As regards intra-Community cross-border mail, provisions in the Directive do not allocate responsibilities between the different national administrations, which could result in compliant postal administrations breaching the standards as a result of delays by postal administrations in other Member States.

The Directive establishes a complaints procedure, particularly in cases involving loss, theft or damage, or non-compliance with service quality standards. A compensation/reimbursement scheme is also envisaged. Various Articles in the Directive (Articles 6, 16 and 19) establish a requirement for universal service providers to publish service and performance information. Publication of such information might provide a comparative insight into the activities of postal administrations.

The Directive further provides for the harmonization of technical standards, whose development should be in accordance with 'the interests of users'. The European Committee on Standardization (CEN) is mandated to establish technical standards for the postal sector, taking into account harmonization measures adopted at international level, particularly those developed by the Universal Postal Union (UPU).

Advisory Committee and National Regulatory Authority

The directive provides for the establishment of an *advisory committee* comprising representatives from the Member States under the chairmanship of the Commission. The Committee will assist the Commission in the preparation of new measures.

Advisory committees have been established pursuant to several Directives (on telecommunications, railway access, energy strategy and public procurement). The European Single Act of 1987 recommended their use in a declaration annexed to this Act. The declaration requests that the advisory committee procedure should be given 'a predominant place in the interests of speed and efficiency on the decision-making process, for the exercise of powers of implementation conferred on the Commission within the field of Article 100A of the EC Treaty'. Although the nature of the Member States' representatives on these advisory committees has not been specified in any of the Directives which uses them, in practice representation would tend to be from the relevant national government ministry.

The Directive provides that each Member State must appoint a *national regulatory authority* for the postal sector that is 'legally and operationally independent of the postal operators'. The primary task of these authorities is to ensure that the obligations established by the Directive are respected. However, they may also be entrusted with ensuring that national and Community competition rules in the postal sector are respected. The regulator would thus have the authority to supervise and monitor the postal monopoly in Member States. The powers of the regulator appear to be strengthened by Article 19, which suggests that the regulator may also be primarily responsible for ensuring that a universal service provider fulfils its obligations and for ensuring compensation for users for failure by the universal service provider to fulfil these obligations. Article 23 makes provision for the creation of a *review body* to 'monitor the evolution of the sector'. It is to comprise up to five 'independent experts' appointed by the Commission. Their task is to collect information on developments in the postal sector, particularly in connection with economic, social and technological aspects, as well as on quality of service.

Timetable

The Commission envisages that the Directive will enter into force shortly after its publication in the *Official Journal of the European Communities* (on the 20th day following such publication). Implementation by the Member States is intended to be within twelve months of this date.

Commission draft Notice on the application of the competition rules to the postal sector and in particular on the assessment of certain State measures relating to postal services[29]

The draft Notice is intended to be complementary to the Directive. The Commission published its initial draft in the *Official Journal* in December 1995, inviting comments. This attracted many comments including criticisms that the Notice went beyond the scope of the planned Directive and in fact obliged Member States to limit their postal monopolies more drastically than the Directive. Although a Notice is not legally binding it is very influential[30] and postal administrations and governments felt that the Commission was trying to introduce competition more rapidly than Member States were prepared to accept. In so far as the Notice tried to impose new obligations on Member States including a further limitation of their postal monopoly, the appointment of an independent regulator and separation of accounts – at a time when it was not certain that the Directive would be adopted – the Commission's draft appeared to go beyond the scope of a Notice which should be limited to explaining the law and should not purport to create new obligations.

Since the December 1995 proposal, the Commission has prepared new drafts. At the time of writing, the latest draft, dated June 1997, has not yet been published in the *Official Journal* and might be subject to further amendments. It, however, includes many changes over the original draft and appears to be close to the version likely to be adopted by the Commission.

The revised text takes into account the Common Position reached by the Council on the proposed Directive and therefore many of the main criticisms attached to the initial draft must now fall. The revised Notice does not seek to impose new obligations but sets out how the Commission would apply the competition rules in the light of the Directive. As the Commission Notice on

---

29 Draft Notice from the Commission on the application of the competition rules to the postal sector and in particular on the assessment of certain State measures relating to postal services. OJ 1995 C322/3.

30 See *Grimaldi* v. *Fond des Maladies Professionelles* [1989] ECR 4407.

competition rules in the telecoms sector which complemented various regulatory measures, this Notice purports to accompany the underlying postal Directive and to explain how the Commission will apply the competition rules in a more liberalized environment created by the Directive.

The Commission explains that the Directive does not deal with all competition matters: first, there are State measures that are not dealt with in the Directive and that could be in conflict with the competition rules. These State measures presumably include the various authorizations and decisions taken by postal ministries on a day-to-day basis. Second, the Directive does not deal with the behaviour of postal administrations which might also be anti-competitive. The purpose of the Notice is to indicate when State measures or postal administrations' action might constitute an infringement of the competition rules.

The Notice states that the Commission will take full account of the definition of permitted reserved services in the Directive. However, it further explains that whilst compliance with the definition raises a presumption of compatibility with Treaty rules and in particular with the Article 90(2) derogation, the Commission will examine any possible infringements on a case-by-case basis. It will examine, in particular, the situation prevailing in any Member State and take into account the fact that postal services may be further opened to competition pursuant to national law. In addition, compliance with the Directive might not protect a Member State from infringing the competition rules if, for example, its postal administration is totally inefficient and third parties are prohibited from offering a competing service.

The Commission Notice starts with a long preface which explains the background to the Commission's approach. It explains that the postal sector is currently facing significant changes as a result of technological progress and the development of competing sources of communication. In order to meet the challenge of external competition the postal administrations must themselves become more competitive by allowing more competition within the postal sector. At the same time, the Commission says that it is fully aware of the social role that postal administrations have to play and the need to ensure that competition will not endanger their task of general economic interest. The Commission's concern is therefore to allow as much competition as possible in the interest of consumers whilst ensuring that postal administrations fulfil their tasks of general interest, i.e. provide a service of good quality to all users at affordable prices.

The Notice deals with important issues relating first to the definition of the relevant market and second with different types of abuses.

The relevant market

The Commission considers that Member States continue to form separate geographic markets for much of their postal activities, namely for the domestic delivery of both domestic and international mail.

With regard to products and services the Commission identifies a large number of separate markets which could play an important part in its analysis of competition cases.

The Commission considers that the general letter service which includes the delivery of items of correspondence to the addresses shown on the items must be distinguished from other services such as direct mail, express mail, document exchange and self provision of mail. Each of those services would normally constitute a separate market. The Commission recognizes that direct mail is treated in the Directive as falling within the reserved area, as any other items of correspondence which meet certain price and weight thresholds, but it takes the view that it is a distinct market. This approach could become particularly relevant when this activity is open to competition.

The Commission further considers that the general letter service comprises different activities which may all constitute separate markets; these are clearance (i.e. collection of mail), sorting, transport and distribution. This is because business customers actively seek means of obtaining these various services from alternative sources. According to the Commission, there is a separate demand for each of those services by business customers as opposed to private consumers who demand a full service from clearance to distribution. The Commission nevertheless recognizes that where the general letter service falls within the reserved area, it constitutes one market which is not divided into these various activities unless Member States choose to open some of the activities to competition.

These market definitions will be important in any competition case to examine alleged infringements by the postal administrations of the competition rules.

Abuses of dominance

The Commission indicates in the preface that it wishes to fight three principal forms of abuse: (i) the extension by a post office of its dominant position in one sector into another sector, for example, the extension of a dominant position in the general letter service market into the express mail market; (ii) unfair discrimination by a postal administration, for example, the grant of preferential terms to its own subsidiary operating in the competitive sector; (iii) price cartels in the field of cross-border services.

The Commission then examines other forms of abuse mostly derived from the case-law of the Court. These include the following:

- *Inefficient service*: The Commission indicates that the provision of a service that does not satisfy the needs of customers or is seriously inefficient and fails to take advantage of technical developments may constitute an abuse, particularly if third party operators are prohibited from providing substitute services. This position of the Commission may be regarded as a bold application of the case-law of the Court, in particular, *Höfner* and *Port of Genoa*, to the postal sector.

- *Tying*: the Commission stresses the importance for postal administrations to keep separate accounts for their activities in the reserved and non-reserved sectors and to make sure that in respect of services which are part-reserved and part-liberalized, the reserved element should be open to all users of the service, irrespective of whether or not the non-reserved component is purchased.
- *Unfair cross subsidization*: The Commission recalls the principle that postal administration may not normally allocate part of the cost of running a competitive service to their reserved activity. An exception exists for cultural mail and for services to the socially, medically and economically disadvantaged and other similar justifiable cases. It is not entirely clear, however, what the full scope of this exception will be.
- *Predatory pricing*: The Commission lays down guidelines on what it regards as acceptable pricing by opposition to predatory pricing in the competitive sector. As in the telecoms sector the Commission makes a significant departure from existing case-law by abandoning the *Akzo* test for a new test. The Commission provides that:

  > the price of competitive services offered by the operator should be at least above the [total] [incremental] costs of provision plus an appropriate proportion of the overhead cost of the operator. Normally, the appropriate proportion would be based on the turnover of the services involved.

  The words in square brackets suggest that the Commission may not have entirely determined its position on this. In any event, new cases might be necessary to confirm whether this test may validly apply to the postal sector.
- *Refusal to grant access or grant of access on a discriminatory basis*: The Commission lays much emphasis on access. Postal administrations should 'provide non-discriminatory access to customers or intermediaries at appropriate public points of access in accordance with the needs of these users'. This statement could have wide practical repercussions for postal operators. The Notice, however, provides that postal administrations do not have to grant access to intermediaries in the reserved sector if they have not previously done so. Of particular importance to the Commission is the offer of access on non-discriminatory terms to customers and intermediaries.
- *Illegal state aid*: The Commission states that it will carefully review the grant of any aid in the postal sector. It recalls its decision regarding alleged State aid granted to the French Post Office and the judgment of the Court of First Instance which confirmed its decision: State aids might be compatible with Treaty rules and, in particular Article 90(2), if they do not go beyond the cost of carrying out universal service obligations and do not cross-subsidize activities in the competitive area. The

Commission, however, stresses the importance for postal operators to adopt proper cost accounting systems, so that the legality of the State aid can be verified.

Conclusion

The greatest advances in Community postal services law has come, in a first stage, from the Court. In cases such as *Corbeau*[31], as well as other key rulings on Articles 86 and 90 such as *Höfner*[32] and *The Port of Genoa*[33], the Court has lent enormous weight to the Commission's efforts to curb the exclusive rights of State monopolies in the postal sector.

While there is nothing to stop Member States continuing to grant special or exclusive rights to postal undertakings, the effect of recent jurisprudence has been to introduce obligations of accountability and justification for special and exclusive rights limiting the activities of competitors.

The development of EC case-law has been instrumental to the progress made in the proposed Directive and the draft Notice.

As this is highlighted both in the Directive and the Notice the definition of permitted monopolies will evolve as will the definition of the universal service to take into account technological, social and economic changes. Furthermore, the Commission indicates in the Notice its intention to apply the competition rules on a case-by-case basis taking into account the evolution in any particular Member State. Although this approach seems logical, it undermines to a certain extent the stated aim of the Notice which is to bring further clarity and legal certainty. It is likely to be difficult, in practice, to know precisely, at any one time and in respect of any specific case, what restrictions of competition will be permitted.

---

[31] See footnote 4 above.

[32] *Ibid.*

[33] *Ibid.*

Chapter 13

COMPETITION AND EXCLUSIVE RIGHTS IN THE EU TRANSPORT SECTOR

INTRODUCTION

The application of Community law to the transport sector is slightly more complicated than for other sectors in that transport was given limited exceptions from the Treaty in the early days of the European Community. However, it is now well established that the competition rules of the Treaty do apply to transport. Transport was excluded from the implementing regulation for Articles 85 and 86 (Regulation 17)[1] by Regulation 141/62[2], but not from Articles 85 and 86 themselves which fully apply. Furthermore, different transport sectors are subject to block exemptions and to special sectoral procedural regulations, pursuant to Article 87 for rail, road and inland waterway[3], sea transport[4] and air transport[5].

Criticising the regulatory organization of the transport sector, Sir Leon Brittan in his role as Competition Commissioner stated in early 1992 that 'for a long time the transport sector was organised on the basis of regulation rather than competition'[6]. He commented in particular that governments have regulated market access, capacity and prices. 'It is not surprising, therefore, that many Community markets were marked by monopoly or oligopoly supply, high prices and market sharing[7]'.

This chapter deals particularly with the air, rail and sea transport sectors in which the existence of special or exclusive rights are being curtailed as the

---

[1] Council Regulation 17/62, OJ 1962 13/204, 1959–62.
[2] Council Regulation 141/62, OJ 1962 124/2751, 1959–62.
[3] Council Regulation 1017/68, OJ 1968 L175/68.
[4] Council Regulation 4056/86, OJ 1986 L378/4.
[5] Council Regulations 3975/87, OJ 1987 L374/1; 82/91, OJ 1991 L10/7 (Groundhandling); 83/91, OJ 1991 L10/9 (Computer Reservations Systems); and 84/91, OJ 1991 L10/14 (Slot Allocation).
[6] Speech to the Centre of European Policy Studies (CEPS) on 'Competition Policy and Transport', 1992.
[7] *Ibid.*

sectors continue to be opened to full competition. A key thread running through these sectors is the concept of access to essential facilities which, for example, in the rail transport sector is being pursued through legislative measures, while in sea transport the Court has taken a significant lead in interpreting Community law in several major judgments concerning ports.

Air transport

The regulatory framework

Introduction

Community policy on deregulation in sectors such as telecommunications, energy, transport, postal services and broadcasting aims to introduce new market operators to compete with public monopolies. Liberalization in the air transport sector conforms with this policy. What is now happening in these sectors is that public enterprises having special or exclusive rights are having these rights more narrowly defined so that they provide services of general economic interest without restricting access to networks to other operators at a reasonable cost.

A significant development in the jurisprudence of the Community led the way for a subsequent three-stage liberalization package. A Court ruling in an important case, *Nouvelles Frontières*[8], firmly removed any doubt as to whether the general Treaty rules do apply to transport. Thus, competition in the air transport sector is governed by the Treaty competition rules. In the air transport field, Articles 85 and 86 are implemented by Regulation 3975/87 (as amended)[9]. However, this Regulation applies only to air transport itself, and not to other related services such as groundhandling[10], which fall within the scope of the general implementing regulation, Regulation 17[11]. Various block exemption Regulations apply to ancillary services such as computer reservation systems[12].

Liberalization of air transport within the European Union arose out of a three-stage package of legislative measures introducing competition in international transport between Member States. This occurred between 1987 and the end of 1992.

First package

The first package was adopted in 1987 and consisted of two Regulations adopted under Article 87 which related to the implementation of Articles 85

[8] Joined Cases 209 to 213/84, *Ministère public* v. *Asjes et al (Nouvelles Frontières)* [1986] ECR II-1425.
[9] Council Regulation 3975/87, OJ 1987 L374/1, amended by Council Regulation 2410/92, OJ 1992 L240/18.
[10] Groundhandling is also subject to the Directive on groundhandling (Council Directive 96/67/EC of 15 October 1997 on access to the groundholding market at Community airports. OJ 1996, L 272/45).
[11] Council Regulation 17/62, OJ 1962 13/204, 1959–62.
[12] See footnote 5 above.

and 86. Council Regulation 3975/87[13] laid down procedures for the application and enforcement of these articles and Council Regulation 3976/87[14] enabled the Commission to grant block exemptions for various categories of agreements and restrictive practices[15]. However, both Regulations limited the Commission's powers to the field of international air transport between Community airports. Consequently, Community competition law could only be applied to domestic air transport within a single Member State and international air transport between Community airports and third countries was applied by use of cumbersome procedures laid down in Articles 88 and 89 on transport.

There were a further two legislative measures. Council Directive 87/601/EEC[16] on fares for scheduled air services between Member States set in place a system of approval of fares in an effort to breach the resistance of the Member States to lower air fares. Its effects were limited, but airlines were freer to set prices and discounts were permissible subject to conditions.

Council Decision 87/602/EEC[17] on the sharing of passenger capacity between air carriers on scheduled air services between Member States and on access for air carriers to scheduled air service routes between Member States provided for a degree of capacity sharing, and made available limited third and fourth freedom rights[18].

Second package

The second package was adopted in 1990 and consisted of three legislative measures, the effect of which was to revoke Directive 87/601/EEC[19] on air fares and Decision 87/602/EEC[20] on capacity sharing adopted at the first stage, and to amend the framework block exemption, Regulation 3976/87[21].

---

[13] Council Regulation 3975/87 of 14 December 1987 laying down the procedure for the application of the rules on competition to undertakings in the air transport sector, OJ 1987 L374/1.

[14] Council Regulation No.3976/87 of 14 December 1987 on the application of Article 85(3) of the Treaty to certain categories of agreements and concerted practices in the air transport sector, OJ 1987 L374/9.

[15] As a result of the Framework Block Exemption Regulation, the Commission adopted specific block exemptions in connection with slot allocation, tariff consultation, revenue sharing, capacity planning, groundhandling, computer reservation systems and consultation on cargo rates.

[16] Council Directive 87/601/EEC of 14 December 1987 on fares for scheduled air services between Member States, OJ 1987 L374/12.

[17] Council Decision 87/602/EEC of 14 December 1987 on the sharing of passenger capacity between air carriers on scheduled air services between Member States and on access for air carriers to scheduled air service routes between Member States, OJ 1987 L374/19.

[18] *Ibid.* A third freedom right means 'an air carrier having the right to put down, in the territory of another State, passengers, freight and mail taken up in the State in which it is registered' (Article 2(d)). A fourth freedom right means 'an air carrier having the right to take on, in another State, passengers, freight and mail, for off-loading in its State of registration' (Article 2(d)).

[19] See footnote 16 above.

[20] See footnote 17 above.

[21] See footnote 14 above.

Council Regulation 2342/90[22] on fares for scheduled air services extended approval provisions contained in Directive 87/601, further liberalized conditions for discounts, introduced limited fifth freedom rights[23] and established a consultation and arbitration scheme.

Council Regulation 2343/90[24] on access for air carriers to scheduled intra-Community air service routes and on the sharing of passenger capacity between air carriers on scheduled air services between Member States, extended third and fourth freedom rights between airports or airport systems in one Member State and those in another Member State where such airports or airport systems are open for traffic between Member States or for international services (Article 4), subject to limitations. Limited fifth freedom rights were possible, subject to conditions. Capacity sharing percentages were also increased.

Council Regulation 2344/90[25] amended the block exemption Regulation, Regulation 3976/87. The amendment introduced a provision for consultations on cargo rates.

Third package

The third and final stage of the air liberalization package has been completed since 1 January 1993, with the exception of restrictions on cabotage which was fully liberated on 1 April 1997. The main features of Commission policy in this area suggest a Community competitive regime operating to enable oligopolies to be challenged by competition from other operators. Public enterprises holding special or exclusive rights should have their monopoly rights narrowly defined and such rights should be compatible with the provision of a service of general economic interest, for example, by providing access to networks at a reasonable cost. Whilst airlines are generally free to set their fares, Member States do have the ability to intervene but only to stop excessively high or low prices. Moreover, price competition should be maximised and price fixing prevented, although tariff consultations in the air transport sector may be permitted subject to an Article 85(3) exemption. Price competition in air transport is expected to result in lower fares since cartel price fixing would no longer bind the carriers.

---

[22] Council Regulation 2342/90 of 24 July 1990 on fares for scheduled air services, OJ 1990 L217/1.

[23] Fifth freedom rights are defined as 'the right of an air carrier to undertake the air transport of passengers, freight and mail between two States other than the State in which it is licensed' (Article 2(f)).

[24] Council Regulation 2343/90 of 24 July 1990 on access for air carriers to scheduled intra-Community air service routes and on the sharing of passenger capacity between air carriers on scheduled air services between Member States, OJ 1990 L217/8.

[25] Council Regulation 2344/90 of 24 July 1990 amending Regulation 3976/87 on the application of Article 85(3) of the Treaty to certain categories of agreements and concerted practices in the air transport sector, OJ 1990 L217/15.

This stage comprised five legislative measures. Council Regulation 2407/92[26] on the licensing of air carriers; Council Regulation 2408/92[27] on access for Community air carriers to intra-Community air routes; Council Resolution 2409/92[28] on fares and rates for air services. The competition regulations implementing Articles 85 and 86 were also amended, i.e. Council Regulation 2410/92[29] on procedures for the application of the competition rules to air transport undertakings and Council Regulation 2411/92[30] on the general air transport block exemption.

Licensing of air carriers

Council Regulation 2407/92 on the licensing of air carriers[31] establishes the requirements for the granting and maintenance of operating licences by Member States for Community air carriers. Companies applying for licences must have their principal place of business, and, if any, their registered office located in that Member State; and its main occupation must be air transport, either solely, or combined with any other commercial operation of aircraft or repair and maintenance of aircraft (Article 4(1)).

A Member State must accept onto its national register, without discriminatory fee or delay, aircraft owned by nationals of other Member States, and transfers from aircraft registers of other Member States, provided that they conform to the requirements of the Regulation (Article 8(4)). Procedures for the granting of operating licences must be made public by the Member States concerned and the Commission should also be informed.

Fares and rates for air services

Council Regulation 2409/92 on fares and rates for air services[32] concerns the criteria and procedures to be applied for the establishment of fares and rates on intra-Community air services. The Regulation has a general provision that Community air carriers are free to set fares (Article 5(1)). The Regulation, however, permits Member States concerned, to decide 'at any moment' to withdraw a basic fare which, when examined in the context of the whole fare structure for the particular route and the competitive market situation, is excessively high. Member States involved may also prevent, in a non-

[26] Council Regulation 2407/92 of 23 July 1992 on licensing of air carriers, OJ 1992 L240/1.

[27] Council Regulation 2408/92 of 23 July 1992 on access for Community air carriers to intra-Community air routes, OJ 1992 L240/8.

[28] Council Regulation 2409/92 of 23 July 1992 on fares and rates for air services, OJ 1992 L240/15.

[29] Council Regulation No.2410/92 of 23 July 1992 amending Regulation 3975/87 laying down the procedure for the application of the rules of competition to undertakings in the air transport sector, OJ 1992 L240/18.

[30] Council Regulation 2411/92 of 23 July 1992 amending Regulation 3976/87 on the application of Article 85(3) of the Treaty to certain categories of agreements and concerted practices in the air transport sector, OJ 1992 L240/19.

[31] See footnote 26 above.

[32] See footnote 28 above.

discriminatory manner, further fare reductions in a market where such reductions vary significantly from normal seasonal price movements and which result in widespread losses among all air carriers concerned with the services in question (Article 6).

Article 4 requires air carriers operating within the Community to inform the general public, on request, of all air fares and cargo rates. The Regulation empowers the Commission to investigate the level of air fares on the basis of a complaint from a party having a legitimate interest (Article 7(2)).

The Regulation, subject to safeguards in the interests of consumers, establishes the framework for liberalization of air fares by permitting Community air carriers to set their fares freely. This, in practice, means that Member States' involvement in this area is greatly reduced. Member States may become involved to ensure that there is neither excessive nor predatory pricing by air carriers. Moreover, the Commission has the final say on whether a safeguard taken by a Member State is justifiable, since it has the power to examine the safeguard measures taken by the Member State in question.

Access of Community air carriers to intra-Community routes

Council Regulation 2408/92 on access for Community air carriers to intra-Community air routes[33] governs the access to traffic rights on intra-Community air routes. Soon after it entered into force, the Commission issued a Decision in the *Viva Air*[34] case (referred to in detail later in this chapter), which turned on its interpretation, particularly Articles 3(1) and 8(1). The Regulation sets out provisions under which Member States may impose a public service obligation in respect of scheduled air services to airports serving, for example, less developed or peripheral regions, 'thin' routes to regional airports if such route is considered vital for the economic development of the region in which the airport is located, or where other methods of transport are unable to ensure an adequate and uninterrupted service (Article 4).

'Public service obligation' is defined as meaning 'any obligation imposed upon an air carrier to take, in respect of any route which it is licensed to operate by a Member State, all necessary measures to ensure the provision of a service satisfying fixed standards of continuity, regularity, capacity and pricing, which standards the air carrier would not assume if it were solely considering its commercial interest' (Article 2(o)).

The Regulation takes into account exclusive concessions granted under law or contract on domestic routes. Such concessions are permitted to continue for a maximum of three years provided that other forms of transport are unable to ensure an adequate and uninterrupted service (Article 5).

---

[33] See footnote 27 above.

[34] Commission Decision 93/347 of 28 May 1993 on a procedure relating to the application of Regulation 2408/92 – *Viva Air*, OJ 1993 L140/51.

Article 8(1) establishes that the Regulation should not affect a Member State's right to regulate without discrimination on grounds of nationality or identity of the air carrier, the distribution of traffic between the airports within an airport system. Article 8(2) provides that 'the exercise of traffic rights shall be subject to published Community, national, regional or local operational rules relating to safety, the protection of the environment and the allocation of slots'.

The Regulation also contains provisions under which a Member State may, in certain circumstances, refuse access to small aircraft operating on new routes between regional airports, where the number of passengers is low (Article 6). Refusals or limitations to access may also be imposed by Member States, subject to conditions and circumstances involving serious congestion and/or environmental problems (Article 9).

During the third liberalization phase, amendments were also made to the Competition Regulation implementing Articles 85 and 86 (Council Regulations 2410/92 and 2411/92).

Application of the competition rules to the air transport sector

Council Regulation 2410/92, amending Regulation 3975/87 laying down the procedure for the application of the rules on competition to undertakings in the air transport sector[35], extends the Commission's powers to take measures for the application of Articles 85 and 86 to undertakings in the air transport sector in situations where trade between Member States may be affected. The original regulation, Regulation 3975/87, was limited in its application to competition in respect of international transport between Community airports. However, it did not empower the Commission to intervene in competition matters *within* a Member State. Therefore the Commission's powers did not extend to domestic transport nor to transport between a Member State and a third country. As a result of the Regulation, such transport is now also subject to the competition rules under Articles 85 and 86.

This measure, therefore, was particularly important in the context of the third air liberalization package for the creation of a single market in air transport.

Air transport block exemption

Council Regulation 2411/92, amending Council Regulation No 3976/87 on the application of Article 85(3) of the Treaty to certain categories of agreements and concerted practices in the air transport sector[36], gave the Commission power to adopt block exemptions in the air transport sector. However, this Regulation expired on 31 December 1992 and the new Regulation permits the continuation of exemptions, particularly concerning

[35] See footnote 29 above.
[36] See footnote 30 above.

agreements on joint planning and co-ordination of airline schedules, tariff consultation, joint operations on new, less busy scheduled air services, slot allocation at airports and airport scheduling, and common development and operation of computer reservation systems (Article 2(2)).

This Regulation extends Regulation 3976/87 to domestic air transport and to air transport between a Member State and a third country, as well as international air transport. It entered into force on 27 August 1992.

Slot allocation

The allocation of slots at airports is closely related to the issue of access. In January 1993 Council Regulation 95/93 on common rules for the allocation of slots at Community airports[37] was adopted to regulate the allocation of slots at Community airports. The main issue of contention with regard to slot allocation concerns so-called 'grandfather' rights, i.e. where an airline by virtue of its use of a slot in a particular season, has the right to continue to use it in the next equivalent period. Thus a system of virtual ownership of rights to slots has emerged. The Commission was concerned that this practice militated against new entrants to the airline market.

Regulation 95/93 attempts to reconcile provision for 'grandfather' rights with provisions to allow new entrants into the market. A 'new entrant' is defined (in Article 2(b)) as being:

> (i) An air carrier requesting slots at an airport on any day and holding or having been allocated fewer than four slots at the airport on that day, or
> (ii) An air carrier requesting slots for a non-stop service between two Community airports where at most two other air carriers operate a direct service between these airports or airport systems on that day and holding or having been allocated fewer than four slots at that airport on that day for that non-stop service.

Air carriers holding more than 3 per cent of the total available slots on that day at a particular airport, or more than 2 per cent of the total slots available on that day in an airport system of which that airport forms part, are not to be considered new entrants at that airport (Article 2(b)).

The Regulation sets out conditions for airport co-ordination under which two categories, that of 'co-ordinated airport' and 'fully co-ordinated airport' are established[38]. Airport co-ordination must conform to the principles of transparency, neutrality and non-discrimination (Article 3(2)). Where capacity problems exist, the Regulation sets out a procedure for an airport to be

---

[37] Council Regulation 95/93 of 18 January 1993 on common rules for the allocation of slots at Community airports, OJ 1993 L14/1.

[38] 'Co-ordinated airport' means an airport where a co-ordinator has been appointed to facilitate the operations of air carriers operating or intending to operate at that airport (Article 2(f)). A 'fully co-ordinated airport' means a co-ordinated airport where, in order to land or take off, during the periods for which it is fully co-ordinated, it is necessary for an air carrier to have a slot allocated by a co-ordinator (Article 2(g)).

designated as a co-ordinated airport. Member States must ensure that a capacity analysis is carried out in situations where '(i) air carriers representing more than half of the operations at an airport and/or the airport authority consider the capacity is insufficient for actual or planned operations at certain periods, or (ii) when new entrants encounter serious problems in securing slots, or (iii) when a Member State considers it necessary' (Article 3(3)). The analysis and the method underlying it must be made available to interested parties. If, following consultation with air carriers using the airport regularly, the representative organizations, airport authorities, air traffic control authorities and passenger organizations, it emerges that there is no short-term possibility of resolving capacity problems, the Member State is required to ensure that the airport is designated as fully co-ordinated for the periods during which congestion occurs. This designation should be lifted when sufficient capacity to meet actual or planned operations becomes available at the airport (Articles 3(4) and 3(5)).

The process of slot allocation is set out in Article 8. The system gives precedence to 'grandfather' rights, subject to provisions set out in Article 10 concerning a slot pool (which, *inter alia*, provides for unutilized slots to be withdrawn and placed into the slot pool). Article 8(1)(a) provides that 'subject to the provisions of Article 10, a slot that has been operated by an air carrier as cleared by the co-ordinator shall entitle that air carrier to claim the same slot in the next equivalent scheduling period'. Where all slot requirements cannot be met to the satisfaction of the air carriers concerned, 'preference shall be given to commercial air services and in particular to scheduled services and programmed non-scheduled services' (Article 8(1)(b)). Moreover, the co-ordinator must take account of priority rules established by the air carrier industry, as well as guidelines recommended by the co-ordination committee taking account of local conditions, provided that such guidelines respect Community law. Where a requested slot cannot be granted, the co-ordinator must inform the air carrier requesting it of the reasons and indicate the nearest alternative slot (Article 8(2)). There are also provisions laid down in the Regulation for the transfer of slots.

As this Regulation stands, it does little to change the former IATA system[39]. The IATA system operated on the basis of 'grandfather' rights with some provisions for new entrants. The operation of the Regulation is, however, being monitored. The Commission is mandated to issue a report on its operation three years after its entry into force. Article 13 sets out a non-exhaustive list of elements to be considered in the report. These are:

- structure of the airline industry;
- progress made by the industry in reducing the non-use of slots;
- size of the slot pool each season at selected airports;

[39] Before the entry into force of Regulation 95/93, Community airports used a system established by the International Air Transport Association (IATA).

- volume of unsuccessful applications for slots each season at selected airports;
- number of new entrants applying for slots each season at selected airports;
- use of dispute procedures established under Article 8 of the Regulation.

Whilst the Regulation is not radical in its changes from the IATA system, it does, however, reinforce the principle of transparency and non-discrimination in the process of allocating slots. It also assists new entrants through its provisions requiring airlines to relinquish slots that they do not use.

On the basis of a Commission proposal the Council must decide whether to continue or to revise this Regulation by 1 July 1997 (Article 14). At the time of writing the Commission is examining the issue of whether allocated slots can be sold. At present, they can be exchanged. A major concern is that the sale of slots could lead to anticompetitive effects, particularly to the reinforcement of dominant positions.

Application of Article 90 to air transport

Introduction

In the air transport sector, the Commission has not used Article 90(3) in a Commission Directive of general application; it has, however, issued or threatened to issue Decisions based on Article 90(3) addressed to individual Member States, for example, it successfully used its powers under Article 90(3) as a lever to put to an end breaches of Article 90 concerning the conduct of two Member States – Germany and Belgium – which required under their national legislation that civil servants travelling on government business should use their national airlines, i.e. Lufthansa and Sabena respectively[40]. In both cases, the governments concerned responded by abolishing the national laws, thereby rendering a formal Article 90(3) Decision unnecessary. In other cases, the Commission has actually used Article 90(3) in individual cases to put to an end breaches of Article 90(1). In the Spanish transport Decision[41], a Spanish law that granted lower air and sea transport fares solely to Spanish nationals resident in the Canary and Balearic Islands was held by the Commission to be in breach of Article 90(1). In a more recent Commission Decision taken under Article 90(3), the Commission has declared as incompatible with Articles 90(1) and 86 the operation of a discount system for landing fees at Brussels airport[42]. This case links other recent cases on access to an essential facility at ports with an analogous issue

---

[40] Commission Press Release IP/90/730 and Commission Press Release IP/91/494.

[41] *Re Transport to the Canary and Balearic Islands* [1988] 1 CMLR 331.

[42] Commission Decision of 28 June 1995 relating to a proceeding pursuant to Article 90(3) of the Treaty, OJ 1995 L216/8.

of access in the air transport field, and is discussed in detail later in this chapter.

It is likely that the granting of special or exclusive rights in air transport will continue. The Commission is pressing the Council to permit it to negotiate airline agreements on behalf of the fifteen Member States in respect of third countries. The Council is reluctant to hand over such powers to the Commission. Subsidiarity, too, is being used by the Member States as a means of eroding the Commission's power, thus the Commission is keen to maintain its powers under Article 90(3) to strengthen its hand in controlling anti-competitive practices in a sector where airlines have been traditionally owned or controlled by the Member States and have been granted special or exclusive rights.

The question of what is meant by undertakings granted special or exclusive rights in the airline sector was dealt with in the *Sterling Airways-SAS Denmark* Decision[43] in which the Commission said that an airline designated particular routes by a Member State held special or exclusive rights within the meaning of Article 90(1). *Ahmed Saeed*[44] was a landmark ruling of the Court on the application of the EC competition rules to airline tariffs, in connection with Articles 85, 86 and 90. The *Ahmed Saeed* case established for the first time that the competition rules applied in principle to air transport between Member States and third countries. It also contains important elements on the application of Article 90 to airline tariffs.

Tariffs: Ahmed Saeed

In a referral for a preliminary ruling from the German federal court (the Bundesgerichthof) the Court was asked the following question:

> Is the approval of tariffs for scheduled air flights by the competent authority of a Member State incompatible with the second paragraph of Articles 5[45] and Article 90(1) of the EEC Treaty and therefore void, even if the Commission has not objected to such tariff approval (Article 90(3))?

The Court's conclusion on the interpretation of Articles 5 and 90 EC Treaty was threefold:

- national authorities are prohibited in particular by Article 90(1) from *encouraging* the conclusion of agreements on tariffs contrary to Article 85(1) and Article 86 of the EC Treaty. Article 90(1) applies even if the Commission has not acted by way of Decisions or Directives pursuant to Article 90(3);

---

[43] *Sterling Airways-SAS Denmark*, Tenth report on competition policy [1981] Point 135.

[44] Case 66/86, *Ahmed Saeed Flugreisen and Silver Line Reisebüro GmbH* v. *Zentrale Zur Bekämpfung Unrautern Wettbewerbs e.V.* [1989] ECR 803.

[45] The competition rules established by Articles 85 and 86 concern the behaviour of undertakings and not measures of the authorities in the Member States (Article 5).

- national authorities are prohibited from *approving* tariffs resulting from such agreements; and
- a limitation on the application of the competition rules may be permitted but only in so far as it is indispensable for the performance of a task of general interest which air carriers are required to perform, and provided the nature of that task and its economic impact on the tariff structures are clearly established.

Access

Viva Air[46]

The Commission's tendency to uphold the general principle of free access to Community routes is clear in this case. The Commission investigated a complaint from Viva Air challenging the legality of a decision of the French authorities to refuse it a licence to operate on the Paris–Madrid route through Charles de Gaulle Airport. The justification of the French authorities for the refusal was that one airline could not provide medium-haul international services from both Paris Orly and Paris Charles de Gaulle Airports. (They considered Viva Air to be part of the Iberia Group which already operated on the Madrid–Paris (Orly) route).

The Commission considered the French authorities' linking of Viva Air and Iberia, concluding that Viva Air had its own operating licence, own staff, own commercial image and own fleet. Thus, Viva Air should be regarded as separate from Iberia, since Article 2(b) of the Regulation simply defines a Community air carrier as being one with a valid operating licence issued by a Member State without any reference to the question of control.

The Commission concluded that the French authorities had misapplied the provisions of Article 8(a) of Regulation 2408/92 and were therefore incorrect to refuse Viva Air authorization to the requested traffic rights.

It is clear from this case that the Commission will uphold the principle of free access to routes within the Community and that it will strictly interpret any exceptions to this principle. Moreover, refusals to grant access must be duly motivated and based on transparent, objective and non-discriminatory criteria.

TAT Decisions

In these cases, the Commission issued Decisions which required France to open up competition on certain air routes, following an investigation into allegations by TAT European Airlines (a subsidiary of British Airways) that the French authorities had refused it an operating licence for the Paris–

[46] See footnote 34 above.

Marseille, Paris–Toulouse[47] and Paris-London routes from Paris (Orly) Airport[48].

On the Paris–London route, the Commission decided that the French authorities could no longer continue to refuse Community air operators air transport rights on this route departing from and arriving at Orly Airport.

The Commission Decision draws a distinction between discrimination on grounds of the carrier's *nationality* and discrimination on grounds of its *identity*. Discrimination on grounds of nationality is embodied in Article 7 of the Treaty. Regulation 2408/92 went beyond this to add an additional principle of non-discrimination on the basis of the air carrier's identity. According to the Commission:

> the principle of non-discrimination on the grounds of the air carrier's identity prevents a Member State, without objective grounds, from withholding from one or more carriers, whether already operating on a route or applying to do so, the traffic rights granted to a limited number of other carriers who are or can be precisely identified to operate the same service on comparable terms.

On this assessment, the Commission considered that the French authorities had breached the principle of non-discrimination on the basis of the carrier's identity. The Commission noted that the French authorities maintain an exclusive concession of Air Inter on the same routes to and from Orly Airport under Article 5 of Regulation 2409/92. Refusal to permit any Community airline other than Air Inter to serve Orly Airport as part of its operation of scheduled air services between the Paris airport system and Toulouse and Marseille, was 'clearly a case of discrimination to the advantage of Air Inter alone in the allocation of traffic between the airports within the Paris airport system'. The Commission also held that France could not rely on the derogation provided by Article 5 of the Regulation. This derogation allows maintenance of exclusive concessions in favour of one airline only where there is no other way of ensuring adequate transport services between two points within the same Member State. The Commission held that the facts showed that this was not the case. France could therefore not maintain its restriction of competition on the three routes at issue.

Landing fees: Brussels National Airport[49]

The Commission in a recent Decision has held a system of discounts granted on landing fees at Brussels National Airport to be incompatible with Article 90(1) read in conjunction with Article 86.

---

[47] Commission Decision 94/290/EC of 27 April 1994 on a procedure relating to the application of Council Regulation 2408/92 – TAT–Paris (Orly)–London, OJ 1994 L127/22.

[48] Commission Decision 94/291/EC of 27 April 1994 on a procedure relating to the application of Council Regulation 2408/92 – TAT–Paris(Orly)–Marseille and Paris–(Orly)–Toulouse, OJ 1994 L127/32.

[49] Commission Decision of 28 June 1995 relating to a proceeding pursuant to Article 90(3) of the Treaty, OJ 1995 L216/8.

The case arose out of a complaint to the Commission by British Midland alleging that a system of discounts granted at Brussels National Airport which increase in accordance with an airline's traffic volume, favours carriers with a high volume of traffic at that airport, thereby placing small carriers at the airport at a competitive disadvantage. The discount system is enshrined in Belgian law pursuant to a Royal Decree of 1989 which lays down the fees payable for the use of Brussels National Airport.

The Commission found that the Brussels Airways Authority, a public enterprise benefiting from the grant of exclusive rights under Belgian law, holds a dominant position (in its capacity as airport authority) on the market in aircraft landing and take-off services for which the fee in question is charged. In examining whether that dominant position had been abused, the Commission noted that the system of discounts had the effect of applying dissimilar conditions to airlines for equivalent transactions linked to landing and take-off services, which placed some airlines at a competitive disadvantage. Moreover, the threshold to be attained to qualify for a discount (BF 5 million in monthly fees) was so high that only a carrier based at the airport could benefit, to the disadvantage of other Community carriers. In addition, the stepped discount system, under which the cost reductions offered increased by more than a proportional basis to the number of landings and take-offs, further accentuated the differences between heavy traffic carriers and all other carriers. The system also disadvantaged regional carriers. The airline to benefit most from the discount system was the Belgian national carrier, Sabena. On the Brussels–London route, on which Sabena and British Midland compete using the same type of carriers, Sabena benefited from an 18 per cent discount on take-off and landing fees for an equivalent service, to the detriment of its competitor, British Midland.

The Commission held that the application of dissimilar conditions by the airway authorities to commercial partners for equivalent transactions, which placed some airlines at a competitive disadvantage, amounted to an abuse of a dominant position under Article 86. Moreover, since the system operated by the airway authorities was established pursuant to a State measure, such a measure constituted an infringement of Article 90, taken together with Article 86(c).

Limit to the application of Articles 85, 86 and 90: Eurocontrol[50]

In this significant case, the Court ruled that Articles 86 and 90 of the Treaty are to be interpreted as meaning that an international organization such as Eurocontrol does *not* constitute 'an undertaking' within the meaning of those Articles. It thus follows that Articles 86 and 90 do *not* apply to Eurocontrol, and, in a wider sense, neither should they apply to other international organizations which would fulfil the criteria of the *Eurocontrol* case.

---

[50] Case C-364/92 *SAT Fluggesellschaft GmbH and European Organisation for the Safety of Air Navigation (Eurocontrol)* [1994] ECR I-0043.

The case was referred from the Belgian Court de Cassation for a preliminary ruling on the question of whether Eurocontrol constituted an undertaking within the meaning of Articles 86 and 90 of the Treaty. The subject matter of the case involved a dispute concerning recovery of route charges with the company SAT Fluggesellschaft GmbH ('SAT').

SAT argued that Eurocontrol is an undertaking within the meaning of Articles 86 and 90 on the basis that organising and collecting route charges by Eurocontrol constituted economic activities that could be carried out by bodies governed by private law. Furthermore, SAT claimed that air navigation control, too, is an economic activity, on the grounds that in some Member States private undertakings exercise this type of control. Four Member States (France, Germany, Greece and the UK) intervened in the case on Eurocontrol's behalf, arguing that Eurocontrol should be regarded as a public authority acting in the exercise of its power, i.e. in a supervisory capacity intended to ensure public safety, and that the collection of route charges merely constituted the consideration for the air and navigation services provided by the contracting States. Thus they submitted that Eurocontrol is not an undertaking within the meaning of the EC competition rules.

The Court concluded that taken together, Eurocontrol's activities:

> by their nature, their aims and the rules to which they are subject, are connected with the exercise of powers related to the control and supervision of airspace which are typically those of a public authority. They are not of an economic nature justifying the application of the Treaty rules on competition.

Accordingly, the Court ruled that 'an international organization such as Eurocontrol does not constitute an undertaking subject to the provisions of Articles 86 and 90 of the Treaty'.

This case leaves it open for an international (or possibly even national) organization to seek to rely on the Eurocontrol criteria on the exercise of certain types of power as being those which are typically associated with a public authority, and which are not of an economic nature, therefore falling outside the scope of application of Articles 86 and 90.

Ancillary services

The Commission is also interested in ensuring open access to ancillary services such as groundhandling and computer reservation systems.

Groundhandling

The Council issued a Directive on groundhandling on October 1996. The Directive is based on Article 84(2) of the Treaty. The Commission had originally drafted a Directive based on Article 90(3) but this controversial project was eventually abandoned in favour of a Council Directive involving full scrutiny by the European Parliament and the Council of Ministers. The Directive provides for the gradual opening up of groundhandling to competition. Groundhandling covers many services including: passenger handling (all services to passengers such as checking tickets, registering baggage); baggage handling (such as loading and unloading from the aircraft, transporting to the sorting area); freight and mail handling; aircraft services (such as cleaning); aircraft maintenance; catering services.

Until recently these services were subject to exclusive or special rights held by a monopoly in most Member States. The purpose of the Directive is to introduce competition by allowing many more operators to have access to groundhandling. The Directive does this by allowing first *third party handling* (i.e. the right for a company to provide groundhandling services to third parties) and second, *self-handling* (i.e. the right for an airport user to carry out groundhandling services for itself).

The Directive provides for the gradual introduction of competition so that, in a first stage, only the largest airports dealing with the largest traffic will have to allow certain forms of third party handling and self-handling. By 2001 all airports in the Community whose annual traffic is not less than 2 million passenger movements or 50,000 tonnes of freight will be subject to the Directive.

The Directive provides for certain limited exemptions in the event of lack of space or capacity.

In order to ensure transparency and fair competition the Directive contains unbundling provisions which require that accounts for groundhandling services and other services be clearly separated. The Directive further prohibits the cross-subsidization by an airport authority of groundhandling activity with the revenues arising from the management of the airport.

Computer reservation systems (CRS): legislation

Community legislation on computer reservation systems (CRS) has been substantially motivated by the need to ensure that airlines have free market access to this important distribution network. In fact, the Commission's first formal competition decision pursuant to the air transport competition rules related to the refusal of Sabena to provide access to its CRS to a competitor, also a new market entrant, that was offering cut-price fairs in competition to Sabena on a particular air route. Sabena was fined for infringing Article 86 of the Treaty[51]. A block exemption regulation[52] subsequently adopted requires

[51] Commission Decision of 4 November 1988 relating to a proceeding under Article 86 of the EEC Treaty – *London European – Sabena*, OJ 1988 L317/47.

[52] Commission Regulation 83/91 of 5 December 1990 on the application of Article 85(3) of the Treaty to certain categories of agreements between undertakings relating to computer reservation systems for air transport services, OJ 1991 L10/9, based on Framework Council Regulation 3976/87 of 14 December 1987 on the application of Article 85(3) of the Treaty to certain categories of agreements and concerted practices in the air transport sector (as amended), OJ 1987 L374/9.

CRS to provide access to airlines on non-discriminatory and reasonable terms.

A computer reservation system, as defined under Community legislation[53] refers to a computerised system which contains information concerning, *inter alia*, air carriers' schedules, availability, fares and related services. It may also have facilities through which reservations may be made or tickets may be issued to the extent that some or all of these services are made available to subscribers.

CRS have developed to a high level of technical sophistication in recent years. The establishment of CRS involves high investment costs. Consequently their legal structure tends to be organised on the basis of joint ventures between groups of airlines. The complexity of flight information available on CRS, not to mention supplementary services such as car rental, hotel information and ticket booking, make CRS a formidable commercial force in the global provision of air transport services. Inherent in their oligopolistic structure, however, exists the potential for anti-competitive abuses. Such abuse can include refusal to grant access to a competitor; prejudicial ranking of information in favour of airlines having ownership in particular CRS; unfair contract terms, such as onerous termination conditions; attaching supplementary conditions to CRS participation that are unconnected with the CRS agreement; better quality information provision on owner-airlines flight details to the detriment of non-owning participant carriers. Thus the Community was prompted to legislate in this field.

The legislation on CRS is intended to prevent restrictive or other anticompetitive practices in connection with CRS. The Commission's concern that competition between different CRS should not be distorted is reflected both in a code of conduct for CRSs and in the conditions attached to a block exemption for CRSs in air transport services. The Commission takes the view that a refusal by parent carriers to provide the same information on schedules, fares and availability to systems other than their own and to accept bookings made by those systems can seriously distort competition between CRS.

Regulatory control of CRS at Community level has so far focused on the airline sector, since the majority of existing CRS concern air transport services. CRS have, however, expanded into other sectors, for example, the rail sector and car rental services. There is also case-law (Commission decisions and judgments of the Court) on the application of the competition rules to CRS.

The principal legislation governing CRS in the European Union consists of a Commission Regulation on the application of Article 85(3) of the Treaty to certain categories of agreements between undertakings relating to computer reservation systems for air transport services and concerted practices in the

---

[53] Council Regulation 2299/89 of 24 July 1989 on a code of conduct for computerised reservation systems, OJ 1989 L220/1, as amended by Council Regulation 3089/93, OJ 1993 L278/1.

air transport sector (as amended)[54] – the block exemption; and a Council Regulation on a code of conduct for computer reservation systems (as amended).

The Code of Conduct Regulation[55] establishes a mandatory code of conduct which sets out the responsibilities and obligations of computer reservation system owners, participants and subscribers, to each other and to consumers.

The Block Exemption Regulation exempts from Article 85(1) of the Treaty, pursuant to Article 85(3), agreements between undertakings whose purpose is at least one of the following: to purchase or develop a CRS in common; to create a 'system vendor'; or, to regulate the provision of distribution facilities by the system vendor or by distributors. Conditions are attached to the exemption: an obligation not to engage directly or indirectly in the development, marketing or operation of another CRS; an obligation on the system vendor to appoint parent carriers or participating carriers as distributors in respect of all or certain subscribers in the defined area of the common market; an obligation on the system vendor to grant a distributor exclusive rights to solicit all of certain subscribers in the defined area of the common market; or an obligation on the system vendor not to allow distributors to sell distribution facilities provided by other system vendors.

Provisions on access contained in the Block Exemption Regulation require a system vendor to permit any air carrier the opportunity to participate, on an equal and non-discriminatory basis, in its distribution facilities within the available capacity of the system, subject to any technical constraints beyond the control of the system vendor (Article 3(1)). Access provisions also prohibit the system vendor from attaching unreasonable conditions to contracts with participating carriers or requiring acceptance of supplementary conditions which under normal commercial usage have no connection with participating in its CRS. Moreover, a system vendor is not allowed to make participation in its CRS conditional upon a participating carrier not participating in another system. This Regulation also prohibits discrimination between CRS members, sets out conditions for provision of accurate and non-misleading information; provides requirements for non-discriminatory displays of flight information as between airlines; and establishes conditions for reciprocity in connection with third country airlines.

The issue of access to a CRS in one transport-related sector by an operator in another transport sector, for instance, access to an air transport CRS by a railway operator, is likely to become increasingly relevant in the future. While

---

[54] Commission Regulation 3976/87, OJ 1987 L374/9; Commission Regulation 2672/88 of 26 July 1988 on the application of Article 85(3) of the Treaty to certain categories of agreements between undertakings relating to computer reservation systems for air transport services, OJ 1988 L239/13; as amended by Commission Regulation (EEC) of 5 December 1990, OJ 1991 L10/9; as amended by Commission Regulation 1618/93 of 25 June 1993, OJ 1993 L155/23; as amended by Commission Regulation 3652/93 of 22 December 1993, OJ 1993 L333/37.

[55] Council Regulation 3089/93, OJ 1993 L278/1.

participants can, in principle, rely on EU competition law to obtain access and non-discriminatory treatment to any CRS, it has yet to be tested in Community law whether a carrier is under all circumstances entitled to obtain equal access to a CRS which provides distribution facilities for another transport mode.

Community legislation on CRS, in practice, applies largely to the two major CRS systems operating in the European Union: Amadeus, a joint venture whose members include the carriers Air France, Iberia, Lufthansa and SAS; and Galileo (now Galileo International) whose members include Aer Lingus, Alitalia, British Airways, Covia, KLM, Olympic, Sabena, Swissair and TAP.

Computer Reservation System (CRS): case-law

The Commission has specifically considered anti-competitive practices in respect of CRS. European Court of Justice jurisprudence, whilst not directly ruling on CRS, has established precedents that could probably be applied to anti-competitive practices with regard to CRS. For example, the Court has held that a company commits an abuse where it is in a dominant position on a particular market and, in the absence of objective necessity, reserves to itself an ancillary activity on a neighbouring but separate market and eliminates competition in that market[56]. An abuse is also committed even if competition is not eliminated but merely restricted or distorted[57]. Thus the refusal by a CRS to accept bookings not made on its own CRS could constitute an abuse of a dominant position in breach of Article 86 unless the CRS in question was able to show that its conduct was objectively necessary.

London European – Sabena[58]

This first Commission Decision on CRS arose out of a complaint by London European Airways to the Commission against the Belgian national flag carrier, Sabena, alleging that Sabena had refused London European access to its Saphir computer reservation system, and that this refusal was motivated by the lower priced fares being offered by London European.

On examination, the Commission found Sabena to be in breach of Article 86 of the Treaty on the grounds that the Saphir CRS held a substantial share of the market for the provision of CRS to travel agents in Belgium and that Sabena held a dominant position on that market. Sabena had, moreover, abused its dominant position by refusing London European access to the Saphir CRS.

ACRISS[59]

In addition to the block exemption and individual exemptions concerning CRS in the airline sector, and the *London European – Sabena* Commission

[56] See *Telemarketing* [1985] ECR 3261.
[57] Case C-18/88, *RTT* v. *GB-INNO-BM SA* [1991] ECR 5941.
[58] See footnote 51 above.
[59] Notice pursuant to Article 19(3) of Council Regulation 17 – ACRISS, OJ 1993 C149/9.

Decision, the Commission has also examined a computer reservation system in the car rental sector. The Association of Car Rental Industry Systems Standards (ACRISS) notified the Commission of a computer reservation system agreement which included a code of conduct, car classification system, special equipment code and standard rental voucher relating to car rental in the EC.

The Commission found that the ACRISS car rental computer reservation system agreement was consistent with the main features contained in EC legislation on CRS in the airline sector, i.e. fairness between competitors; non-discrimination as between participants (applicable for both ACRISS members and non-members); transparency; and a comprehensive, accurate and non-misleading presentation of information on screen displays. The Commission thus took a favourable view of the agreement and issued a 'comfort letter'.

Amadeus/Sabre[60]

The Commission has authorised a co-operation agreement between two major CRS, Amadeus (a joint venture whose main members include Lufthansa, Air France and Iberia) and Sabre (owned by American Airlines). Whilst the Commission considered that the agreement would have the effect of reducing competition between the two parties, it nevertheless took the view that the agreement would enhance distribution in Europe and globally. The exemption was granted subject to guarantees by the parties and their main carriers, particularly to ensure that competition was not restricted on the markets covered by Amadeus and Sabre.

Sabre/Air France and Iberia[61]

Sabre, the CRS owned by American Airlines, issued a complaint to the Commission that the subsidiaries of Air France and Iberia had refused to participate to a sufficient level in its distribution facilities. Air France and Iberia are major shareholders in the Amadeus CRS. Sabre was concerned that it could not effectively compete in the national markets in which it operated if it could not distribute the services of the domestic airlines. Air France and Iberia took the view that their own CRS which covered over 80 per cent of their respective national markets, gave them all the distribution facilities they needed. They alleged that Sabre, the internal reservation system of its owner, American Airlines, discriminated against them by providing a better quality of service for American airlines than for other carriers. The parties eventually agreed to a settlement under which the domestic subsidiaries of Air France and Iberia agreed to participate in Sabre.

The Commission, in drafting the new code of conduct for CRS and the revised block exemption for CRS, took on board the issues arising out of

---

60 Commission Press Release IP(91) 784, 2 August 1991.

61 European Commission Competition Report [1993] at point 238.

these complaints and consequently strengthened the requirements for parent carriers to participate in competing CRS and the non-discriminatory operation of 'hosted' CRS.

The Commission is focusing attention upon a particular link in the CRS distribution chain, i.e. travel agents. The Commission is concerned that travel agents, as the direct link between CRS and the consumer, have considerable potential for anti-competitive practices. For example, an exclusivity clause in a travel agent's contract with a particular system might be designed to prevent a travel agent from participating in other CRS. Consequently, the information provided by that travel agent to the consumer would be limited to the extent that it relied on the information contained in that particular system and would not reflect the full range of travel options available. Moreover, travel agents tend to be given concessionary rates on CRS equipment in return for generating business for that particular system. This kind of tie-in can mitigate against fair competition by effectively binding the travel agent to that particular computer reservation system. The Commission is examining the financing of CRS in the air transport sector.

Interlining

Regulation 84/91[62] on the application of Article 85(3) to the co-ordination of capacity, tariff consultations and slot capacity, considers that consultations on passenger and cargo tariffs may contribute to the generalised acceptance of interlineable fares and rates to the benefit of air carriers and air transport users. However, such consultations should not exceed their lawful purpose which is to facilitate interlining.

Interlining agreements are commercial arrangements whereby airlines authorise IATA signatories to sell their services. This in practice means that passengers have the opportunity of travelling to and from a destination using different airlines, while holding a single ticket. Airlines also recognise each other's authority to change tickets to ensure that passengers can change reservations, routes or air lines after the ticket has been issued. The Commission approves interlining since it enhances flexibility and generally benefits airlines, travel agents and passengers. Nonetheless, interlining has a potential for anti-competitive abuses. This might be the case, for example, where interlining is refused to a new entrant on a route operated by a dominant competitor carrier. The Commission has expressed its views on interlining in a press release[63].

> Where an airport has a large share of a particular market between two airports, a refusal to interline may hinder the maintenance or development of competition on that route or on connecting routes, and thus constitute an abuse of that dominant position. It is important, in particular for small airlines and new entrants to air

[62] OJ 1991 L10/14.
[63] Commission Press Release IP(93)84, 15 May 1990.

> transport, to be able to offer their passengers access to the network of the dominant airline in order to be able to compete effectively with that carrier. The Commission nevertheless finds it reasonable that the dominant carrier could restrict interline sales of its services to those made at its own tariff level, and ask the traveller to pay any supplement which may be necessary. The Commission intends to ensure, however, that a dominant airline allows sales at any fare at which the other airline needs to interline if it wants to compete effectively.

The Commission also intervened under Articles 85 and 86 to require national airlines to grant interlining facilities to smaller airlines running parallel services with the larger airlines[64].

Sea transport

Abuse of monopoly rights

Introduction

The Commission and the Court have, respectively, in a series of Decisions and judgments, made it clear that a company in a monopoly position, should not abuse its position by obstructing access to the market or by unfairly discriminating against third parties.

The Commission case-law on maritime transport in the context of Article 90 is particularly interesting because it introduces and defines for the first time the concept of essential facility. It has done so in two decisions of the same date, one was taken under Article 86 alone, namely *Sea Containers* v. *Stena Sealink*; the other was taken under Articles 86 and 90, namely *Rødby Port.* Essential facility was defined as 'a facility or infrastructure, without access to which competitors cannot provide services to their customers'[65].

Sea Containers v. Stena Sealink[66]

The Commission issued a Decision at the end of 1993 regarding a request for interim measures in the case of *Sea Containers* v. *Stena Sealink Ports and Stena Sealink Line* for an alleged infringement of Article 86.

Sea Containers alleged that Stena Sealink Ports had abused its dominant position as owner and operator of the port of Holyhead in breach of Article 86 by failing to permit Sea Containers access on a *reasonable basis* to an essential facility or infrastructure and that it had relied on its exclusive rights to protect its commercial interests as a ferry operator without objective justification. According to Sea Containers, Stena Sealink had not separated its

---

[64] *Lufthansa/Air Europe*, (European Commission Competition Report [1990] at point 106. *British Midland* v. *Aer Lingus*, OJ 1992 L96/34.
[65] *Stena Sealink*, para. 66.
[66] Commission Decision of 21 December 1993 relating to a proceeding pursuant to Article 86 of the EC Treaty – *Sea Containers* v. *Stena Sealink* – interim measures, OJ 1994 L15/8.

role as port operator at Holyhead from that of ferry operator. The Commission agreed that there was a *prima facie* case of abuse by Stena Sealink.

It found that Sealink had not offered access to Holyhead on non-discriminatory terms to Sealink up to October 1993. At this point, however, Sealink made an offer to Sea Containers permitting it to begin a service in 1994 at the sailing times it considered as essential for the operation of a commercially viable service. Sea Containers subsequently accepted this offer, which the Commission regarded as reasonable in the context of a port with a limited capacity. This acceptance enabled the Commission to reject the application by Sea Containers for interim measures against Stena Sealink. The interim decision cannot, however, be ignored because it lays down important principles which were confirmed in the other Decision of the same day, namely *Rødby Port.*

Rødby Port[67]

Like *Sealink*, the Rødby Port Decision concerned access to an essential facility or infrastructure which had relied on exclusive rights to protect its own commercial interests as an operator in the absence of objective justification.

Two companies, Euro-Port AS and Scan-Port GmbH, subsidiaries of Stena Rederi AB, brought a complaint to the Commission following a refusal by the Danish transport minister to allow them either to build a new terminal in the immediate vicinity of Rødby Port, or to have access to the existing terminal at the port to operate a ferry service between Rødby Port in Denmark and Puttgarden in Germany.

In this case, the owner and manager of the port was not a private enterprise, but a public undertaking (DSP) which held the status of a department of the Danish transport ministry. In order to use the port terminals, operators had to be authorised by the transport ministers acting on a proposal from DSP. Moreover, DSP also held the exclusive rights to organise rail traffic in Denmark, as well as operating ferry services between Denmark and neighbouring countries (but did not hold any exclusive rights to such services). The Rødby/Puttgarden route was jointly operated by DSP and Deutsche Bundesbahn (now Deutsche Bahn), a German public undertaking. No other companies provided ferry services on this route.

Unlike *Sealink*, the Commission examined the case under Articles 86 and 90. It applied the narrow definition of 'relevant market', based on the Court's judgment in the *Port of Genoa*[68], i.e. that the organization on behalf of a third party of port operations in a *single* port can constitute a relevant market within the meaning of Article 86. In this case, the relevant market was defined as the market for the organization of port services in Denmark operating on the Rødby/Puttgarden route for passengers and vehicles.

---

[67] Commission Decision of 21 December 1993 concerning a refusal to grant access to the facilities of the Port of Rødby (Denmark), OJ 1994 L55/52.

[68] Case C-179/90, *Merci di Convensionale Porto dei Genova SpA* [1991] ECR I-5889.

The Commission also considered that there was no valid alternative transport with the same advantages as those offered by Rødby Port for sea transport between Eastern Denmark and Germany and the rest of Western Europe, in so far as alternative forms of transport, such as air transport, were more expensive and limited in their freight carrying capacity. Figures showed that around 71 per cent of passengers and 88 per cent of lorries travelling by sea between Denmark and Germany on the Rødby/Puttgarden route were dealt with by Rødby Port.

Consequently, the Commission considered the market for maritime services between the two ports to constitute a neighbouring but separate market which might be affected by the behaviour of an undertaking on the market for organization of port services.

The Commission concluded that DSP was a public undertaking, which, by virtue of the exclusive rights granted by the State in its capacity as port authority, held a dominant position in Denmark on the market for the organization of port operations in connection with the transport of passengers and vehicles by ferry on the Rødby/Puttgarden route.

The Commission decided that the refusal by Rødby Port to allow Euro-Port to operate from Rødby had the effect of eliminating a potential competitor from the Rødby/Puttgarden route, thereby strengthening the dominant position of DSP and Deutsche Bundesbahn on that route. The Commission concluded that an undertaking owning or managing, and using, an essential facility without which its competitors would be unable to offer their services to customers, and which refuses to grant them access to such a facility, is abusing its dominant position. Consequently, the Commission decision stated that 'an undertaking that owns or manages an essential port facility from which it provides a maritime transport service may not, without objective justification, refuse to grant shipowners wishing to operate on the same route access to that facility without infringing Article 86'.

On the issue of the Euro-Port request to build a new terminal, or to operate from existing port facilities, the Commission held that Article 90(1) prohibits Member States from placing public undertakings and undertakings to which they grant exclusive rights, in a position which would inevitably lead to an infringement of Article 86. Where a Member State has refused such access and has further strengthened the effect of the refusal by also refusing to authorise the construction of a new port, this constitutes a State measure contrary to Article 90, read in conjunction with Article 86.

The Danish authorities argued that technical constraints existed which prevented Scan-Port from using the Rødby/Puttgarden route. The Commission rejected this argument, together with the Danish authorities' assertion that a new service could result in saturation of existing port facilities. The Commission said that even on a saturated market there could be quality improvements in goods or services, or price reductions resulting from competition, which would benefit the customer, and that this could lead to increased demand which could be met by expanding the port.

The Commission ruled out the application of the exemption provided for in Article 90(2) because the application of the competition rules did not impede the performance of a task of general economic interest.

Port of Genoa[69]

The *Sealink* and *Rødby Port* cases clearly illustrate that the Commission is taking an increasingly strict line on the exclusive rights of monopolies. In *Port of Genoa*, the Court has also indicated that it will not permit undertakings in a dominant position to obstruct access to the market or permit less favourable access for a competitor. In this case, the Court ruled that Italian legislation granting exclusive rights to Italian ports in respect of loading and unloading operations at ports was in breach of Article 90(1) in that it required the staff to be Italian workers[70] In Italy, all loading, unloading and transfer of goods in a port were reserved to the port authorities whose employees were required to be Italian nationals. These rules were enshrined in the Italian Navigation Code, and their breach would result in criminal sanctions. A shipping operator, Siderurgica Gabrielli, sought to use its own loading facility and staff at the Port of Genoa. Permission to unload was refused on the grounds that it would be contrary to the Italian Navigation Code. It was forced to use the port's facilities which were less technologically developed and which led to delays which would not have occurred had the shipping operator been permitted to use its own facilities. The shipping company took a case against the port company to court. The Italian court referred a question to the European Court of Justice, asking whether the exclusive rights of the Port of Genoa were compatible with Article 90(1) and the competition rules of the Treaty.

The Court established that the Port of Genoa (a single port) constituted a substantial part of the common market and that the port company, Merci di Converzionali Porto di Genova held a dominant position in that market. The Court held that a Member State would infringe Articles 90 and 86 of the Treaty if a company to which it granted exclusive rights were led by the mere exercise of these rights to exploit its dominant position in an abusive manner. This would, in particular, be the case if the company was led to demand payment for services not requested, or to invoice disproportionate amounts, or to fail to use modern technology. The point concerning failure to use modern technology is particularly relevant since it marks an advance in the jurisprudence of the Court in limiting the scope of monopolies. Thus the Court established that there had been a breach of EC competition law in a case where a monopoly fails to meet market demand, whether for technical or other reasons. In this case, the Court placed a strict interpretation on the

---

[69] *Ibid.*

[70] Article 7 of the Treaty prohibits discrimination on grounds of nationality.

ways in which a body holding a monopoly can exercise this monopoly without being liable for an abuse of a dominant position.

Corsica Ferries[71]

In a further case concerning discriminatory conduct by, or involving, undertakings which hold exclusive rights, the Court has given judgment in an action brought by Corsica Ferries against the Genoa Pilots Association. Corsica Ferries had raised objections to an Italian government approved monopoly on pilot services in the Port of Genoa, as set out in the Italian legislation governing piloting services, the Navigation Code.

Under this code only vessels flying the Italian flag were entitled to obtain a cabotage licence. The tariff applicable to vessels with a cabotage licence was 30 per cent lower than the tariff for vessels without a licence. A reduction of 50 per cent was granted to vessels carrying out regular scheduled services between Italian ports on a fixed route and making at least one stopover per week at the Port of Genoa. Further reductions were granted to vessels of over 2,000 tonnes carrying out cabotage and using piloting services a specified number of times each month.

The Court ruled that under Council Regulation 4055/86/EEC[72] on the free movement of maritime services between Member States and Member States and third countries, it is contrary to Community law to charge, for identical piloting services, a different tariff depending on whether an undertaking is providing transport services between two Member States, or between ports located in the national territory.

It also ruled that it is contrary to Articles 86 and 90(1) of the Treaty for a national authority to authorise an undertaking to which exclusive rights have been granted over a substantial part of the common market to apply such discriminatory tariffs. These discriminatory tariffs resulted from an abuse of a dominant position and therefore fell within the prohibition set out in Article 90(1), when read in conjunction with the second paragraph of Article 5 of the Treaty which requires Member States to abstain from any measure that could jeopardise the attainment of the objectives of the Treaty.

The above cases and Decisions are of considerable significance in so far as they further extend the limitations on monopolies held by companies operating port services and reinforce that users of ports can take successful legal actions challenging abuses by monopoly operators – be they private or public – at Community ports and, by analogy, this could presumably extend to facilities in the rail, air transport, energy, telecommunications and broadcasting sectors. The Commission relied extensively on jurisprudence established

---

[71] Case C-18/93, *Corsica Ferries Italia SRL* v. *Corpo dei Piloti del Porto di Genova* [1994] ECR I-1783.

[72] Council Regulation 4055/86 of 22 December 1986 applying the principles of freedom to provide services to maritime transport between Member States and between Member States and third countries, OJ 1986 L378/1.

in *Port of Genoa*[73] and *Corsica Ferries*[74] in its recent decision in the air transport sector in connection with landing fees at Brussels National Airport[75].

Rail transport

The regulatory framework

Introduction

Community attention to the railways was effectively insignificant until the mid-1980s when developments were begun to align rail transport costs with those of road transport. However, it was not until the end of 1990 that a serious attempt was made to adopt legislation to establish a Community-wide rail policy. The key legislation, a framework Directive on the development of the Community's railways, was adopted in 1991[76].

In common with other sectors such as telecommunications and energy, and maritime transport, a major issue in the rail transport field is that of access. In 1991 the Council adopted a Directive on the development of the Community railways (the Access Directive)[77]. This Directive was designed to clarify access to railway networks by railway companies. However, the Directive itself raised certain questions which themselves need clarification and have led to new Council Directives on the allocation of railway infrastructure capacity and the charging of infrastructure fees[78], and on the licensing of railway undertakings[79].

The Access Directive

The Access Directive arose out of a draft document prepared by Commissioner Van Miert, when Commissioner for Transport, in 1989. When the text was presented to the Council it was initially resisted, but, following amendments, was adopted, but not before the Commission (through DG IV) had put pressure on the Council by threatening to impose competition among the railways by means of an Article 90(3) Directive. As this was around the time of the adoption of an Article 90(3) Directive in the telecommunications field, this threat was taken seriously and prompted the Council to adopt the Access Directive, subject to important amendments.

---

[73] See footnote 68 above.
[74] See footnote 71 above.
[75] See footnote 49 above.
[76] Council Directive 91/440/EEC of 29 July 1991 on the development of the Community's railways, OJ 1991 L237/25.
[77] *Ibid.*
[78] Council Directive 95/19/EC of 19 June 1995 on the allocation of railway infrastructure capacity and the charging of infrastructure fees, OJ 1995 L143/75.
[79] Council Directive 95/18/EC of 19 June 1995 on the licensing of railway undertakings, OJ 1995 L143/70.

The Commission was concerned that the railways might want to reserve railway traffic opportunities to themselves; therefore the Access Directive aimed to facilitate the entry of new European market entrant railway companies, based on the principle of access to an essential facility.

For the first time, the concept of separation of infrastructure activities from the supply of rail transport services was introduced into the sector. The Access Directive provided a mechanism, albeit limited, given that it applies only to international groupings and to international intermodal freight operators, to reduce State rail monopoly powers over the Community's railways. Traditionally, the rail transport sector had developed as a single operation in which infrastructure and services were controlled by a single operator. The Access Directive diffused, but did not abolish, this monopoly influence.

The Directive set out objectives to ensure:

- the management independence of railway undertakings;
- unbundling of management of railway operation and infrastructure from transport services, including mandatory separation of accounts and institutional separation remaining optional;
- access to Member States' networks for international groupings of railway undertakings and for railway undertakings involved in international combined freight transport (which could be independent operators).

The Directive defines 'railway undertakings' as either public or private undertakings, having as their main business the provision of freight transport services and/or passengers, provided that the undertaking *ensures traction.* Thus the Directive applies to both public and private railway companies. Whilst, from the commercial standpoint, the mere existence of open access does not imply automatic opportunities for profitability for private operators, it does, nonetheless, open the option for new operators to enter the market under improved competitive conditions.

Directive 91/440 defined access as being the right to pick up and unload passengers in a Member State. Transit was defined as the right to go through a Member State without picking up and unloading passengers.

The Directive obliges Member States to see to it that railway companies adjust their activities to the market and manage them in a way that ensures 'efficient and appropriate services at the lowest possible cost for the quality of the service required'.

The Directive also specifies that the public service obligations entrusted to railway companies by the state must, similar to commercial activities, be managed in accordance with the principles applicable to commercial companies.

Railway companies may form international groupings with one or more railway companies from a different Member State. An 'international grouping' is defined as 'any association of at least two railway undertakings . . .'. It would appear that this would not be applicable to a link-up between the railway undertaking and an infrastructure manager. It has to involve two railway undertakings which ensure traction. Railway undertakings are defined by Member States. An 'association' was left wide in the Directive and would cover any type of joint venture. The definition of 'railway undertaking' gives Member States a certain flexibility of definition. It allows them to permit the creation of new railway undertakings, hence making it possible for Member States to break up railway monopolies and to create alternative railway companies. The interpretation of 'railway undertaking' was left broadly to the Member States to determine, leaving it open for Member States to adopt divergent criteria and authorization procedures; the Commission could, however, review the Member States' implementing legislation in order to ensure its compliance with the Directive. An 'international grouping' can include any type of company as long as it includes at least two railway undertakings. They may establish their internal organization (subject to compliance with the Directive's provisions on the separation of infrastructure management from transport operations). They are also free to control the supply and marketing of services and to fix the price of such services (subject to adherence to the Council Regulation governing public service obligations in the sector)[80].

Such international groupings have access and transit rights in the Member State of establishment and transit rights only in other Member States for the provision of international services between the Member States of the companies comprising the grouping and the Member State of establishment. The Directive also provides open access for individual operators of international intermodal freight services.

The Access Directive does not specify that all members of an international grouping must be railway companies. The only requirement is that such groupings must comprise at least two railway companies from more than one Member State. The future trend might well be for other interests to become involved in railway activities by becoming part of an international grouping.

One issue left uncertain in Directive 91/440 was whether the definition of 'railway undertaking' meant that a new undertaking requesting access should actually own its means of traction. It would appear that ownership is not essential, i.e. a railway undertaking must have a means of access to traction but need not necessarily own it. Therefore it would appear that traction could be leased.

Infrastructure managers determine user fees for the use of infrastructure by railway companies and international groupings. User fees must be calculated in a manner which does not discriminate between railway companies.

---

[80] Council Regulation 1191/69 of 26 June 1969 on action by Member States concerning the obligations inherent in the concept of a public service in transport by rail, road and inland waterway, OJ 1969 L156/1. Amended by Council Regulation 1893/91, OJ 1991 L169/1.

The Directive does not apply to railway companies which provide urban, suburban or regional services only, i.e. mainly domestic services. The Directive applies to joint ventures and has been in force since 1 January 1993[81].

Licensing of railway undertakings

The Directive on the licensing of railway undertakings[82] establishes the criteria to be applied by Member States concerning the issue, renewal or amendment of licences for railway undertakings established in the Community for the provision of services referred to in Article 10 of Directive 91/440[83].

It was thought that the Directive on the licensing of railway undertakings would help to clarify certain aspects of Directive 91/440/EEC, for example whether traction could be leased or rented. The Directive in its final form did not provide such clarification. In fact it avoided doing so. Article 2 defines 'railway undertaking' as meaning 'any private or public undertaking the main business of which is to provide rail transport services for goods and/or passengers with a requirement that the undertaking must ensure traction'. An earlier version of the text had added the further provision that the undertaking must ensure traction 'with traction equipment owned, leased or rented'. This would have clarified Article 3 of Directive 91/440 by providing that traction may be leased or rented. However, the adopted Directive does not contain this phrase and uncertainty remains on this point.

It defines 'licence' as 'an authorization issued by a Member State to an undertaking by which its capacity as a railway undertaking is recognised. That capacity may be limited to the provision of specific types of services'. This would suggest that an undertaking which is not a railway undertaking but has the capacity of becoming one by specialising in certain railway services could be granted an operating licence. Railway undertakings established in the Community have the right to apply for an operating licence. Member States can refuse to grant licences or to maintain their validity in cases of non-compliance with the Directive's requirements. The Member States must designate a licensing authority to grant railway operating licences (Article 3). A railway undertaking which meets the Directive's requirements would be entitled to an operating licence and to apply for access to railway infrastructure. However, Article 4(4) limits the rights of the licence by providing that the operating licence does not of right confer any access rights: 'Such a licence shall not in itself entitle the holder to access to the infrastructure'. In

---

81 The Directive is currently under amendment, with the aim of extending the access rights to railway infrastructure established under the Directive, and to establish rights of access for railway undertakings to operate international and cabotage services for goods transport, combined transport of goods and international passenger transport services. Amended proposal for a Council Directive amending Directive 91/440/EEC on the development of the Community's railways. COM(97) 34, 1997.

82 See footnote 79 above.

83 See footnote 76 above. Article 10 refers to access and transit rights granted to international groupings for activities involving the international combined transport of goods.

addition the licence granted to the operator must be appropriate for the services to be provided (Article 4(45)).

Before commencing the provision of rail transport services, a railway undertaking must be able to demonstrate to the licensing authorities of the Member State concerned that it will, at any time, be able to meet the requirements relating to good repute, financial fitness, professional competence and that it is adequately insured (Articles 5 to 9).

The Directive sets out provisions on the validity of the licence (Articles 10 to 13). Member States are obliged to make the procedures for granting licences public and to inform the Commission of such procedures (Article 15(1)) and decisions of licensing authorities are subject to judicial review (Article 15(3)).

The Commission is mandated to put forward a report after two years of application of the Directive with a view to possibly widening the scope of the Directive. Member States must comply with the Directive's provisions by the end of June 1997.

Allocation of railway infrastructure and charging of fees

The Directive on the allocation of railway infrastructure capacity and the charging of infrastructure fees[84] sets out the principles and procedures applicable to the allocation of railway infrastructure capacity and the charging of infrastructure fees for railway undertakings established in the Community and the international groupings that they form and where they carry out the services referred to in Article 10 of Directive 91/440/EEC[85].

Under its provisions, Member States must designate the authority responsible for the allocation of infrastructure capacity in compliance with the requirements of the Directive (Article 3). This authority does not have to be independent of the owner of the infrastructure, but it must particularly ensure that railway infrastructure is allocated on a fair and non-discriminatory basis. The authority must also ensure that the allocation procedure is efficiently organised and that it applies market principles, subject to derogations in connection with public service obligations and with the granting of special or exclusive rights (Articles 4 and 5).

Article 4 provides that the responsible authority may take the necessary measures to ensure that in the allocation of railway infrastructure capacity, priority is given to rail services which are provided in the interests of the public (as defined in Regulation 1191/69 as amended by Council Regulation 1893/91 concerning the obligations inherent in the concept of a public service in transport by rail, road and inland waterways). There is also provision for a derogation which may be granted in respect of services operating on infrastructure designed for certain specific services, notably high-speed or

---

[84] See footnote 78 above.
[85] See footnote 76 above.

freight lines; however, the competition rules contained in Articles 85 and 86 as well as Article 90 on public undertakings, must be respected.

As regards services provided in the public interest, the Directive requires that Member States should compensate the infrastructure manager for any financial loss incurred as a result of imposing a certain infrastructure capacity allocation (Article 4(2)). Pursuant to Article 5, the responsible authority may grant special or exclusive rights with regard to infrastructure allocation but this must be done on a non-discriminatory basis to operators of certain types of services in each area; moreover, it must be indispensable to ensuring adequate public services or efficient use of infrastructure capacity or to allowing the financing of new infrastructures, and comply with EC competition rules, particularly Articles 85, 86 and 90.

On the charging of infrastructure fees, infrastructure-related accounts of an infrastructure manager should balance under normal business conditions over a reasonable time period. Therefore, income from infrastructure fees, together with contributions from the Member State, should balance with the infrastructure expenditure (Article 6).

A non-discrimination clause contained in Article 7 provides that there should be no discrimination between different railway undertakings and between national and international services in the charging of services of an equivalent nature in the same market. The Member State should set out rules to determine the infrastructure fees after consultations with the infrastructure manager. The following principles should be applied to such rules:

- the fees should be paid directly to the infrastructure manager;
- the fees should be fixed in accordance with the nature and time of the service, market situation and quality of the infrastructure;
- fees should be charged on a non-discriminatory basis; and
- there should be adequate prior notification by the infrastructure manager to the users of any changes in infrastructure charged or in quality or capacity of the infrastructure concerned.

This Directive clarifies that infrastructure capacity is allocated on a fair and non-discriminatory basis, and it introduces market considerations both in the allocation of railway infrastructure capacity and in the charging of fees for use of the infrastructure. This Directive requires to be implemented by the Member States by the end of June 1997.

Application of the competition rules to rail transport

Introduction

Community policy for the railway sector tends to attempt to reconcile the principles contained in the competition rules with the development of rail transport as a technically advanced, consumer-beneficial network at national and trans-

European level, as well as an environmentally sustainable transport mode. This has led to a number of cases involving agreements which would normally be prohibited under Article 85(1), being granted exemptions under Article 85(3). These exemptions, however, have been made subject to stringent conditions governing, in particular, non-discriminatory access to an essential facility.

A number of important Commission and Court decisions have been taken in the railway sector. Although these cases are not based on Article 90, they are worth mentioning here because:

(1) they involve the concept of access to an essential facility and the conditions under which the Commission is prepared to grant an exemption (*Eurotunnel*; *ENS*; *ACI*; combined tariff structure);
(2) they involve controversial definitions of the market by the Commission (*Eurotunnel*; *ENS*; *HOV SUZ*; *UIC*);
(3) they set out the scope of Regulation 1017/68 as applied by the Commission and as defined by the Court (*UIC*).

These cases reveal a tension between the Commission and the Court:

- the Commission has tried unsuccessfully to limit the application of Regulation 1017/68 in order to control more agreements; the Court condemned the Commission's limited application of Regulation 1017/68 in *Eurotunnel*;
- the Commission has also tried to limit the effect of the bottlenecks and other restriction of competition resulting from the creation of joint ventures involving transport through the Channel tunnel. It has tried to do so by imposing conditions on the joint venture partners. The Court, however, overturned the Commission decision in the *Eurotunnel* case; and the railways have challenged the conditions which the Commission imposed upon them in the *ENS* case.

As these cases demonstrate the railways have been able, pursuant to Regulation 1017/68, to enter into agreements which in the absence of the Regulation would have been regarded as anti-competitive. This tailor-made regulation has often provided them with an exemption from the full application of the competition rules as efficient as that contained in Article 90(2). Question marks, however, remain on the full scope of the derogation permitted by Regulation 1017/68, which might only be resolved following new cases before the Court.

Conditional exceptions from the competition rules

Eurotunnel usage agreements

The Channel Tunnel Group Limited and France Manche SA notified a contract between themselves and BR and SNCF[86] to the Commission with a

[86] Notice pursuant to Article 12(2) of Council Regulation 1917/68 concerning Notification IV/32490 – Eurotunnel, OJ 1998 C292/2.

view to obtaining a declaration of non-applicability of Article 2 of Regulation 1017/68[87] (which regulates the application of the competition rules to rail, road and inland waterway transport).

The agreement in question concerned the grant by Channel Tunnel Group Limited and France Manche SA ('the Concessionaries') to the railways of the right to use 50 per cent of the capacity of the fixed link together with obligations on BR and SNCF ('the Railways') to run passenger and freight trains. The terms of the contract included the obligation on the Concessionaries to make available a fixed link across the Channel by a specified date and fulfilling certain conditions and on the Railways to maintain a set quality and frequency of service. The agreement also dealt with the methods of payment for the usage of the fixed link. Payment by the Railways was to be based, in part, on the number of passengers and tonnage carried by the trains through the fixed link.

The Concessionaries stated that the aim of the agreement was to achieve optimum use of the fixed link through the equal distribution of capacity between the Concessionaries and the Railways. It was also claimed that the allocation of 50 per cent of the capacity of the fixed link to the Railways was justified as the Railways were the only undertakings able to operate a direct train service and to make the best use of the fixed link. However, the remaining 50 per cent of capacity could be made the subject of contracts with third parties, for example for shuttle services, and the railways would be free to conclude agreements with third parties, including other potential operators of through trains.

The parties therefore considered that the contract should be exempted from the prohibition in Article 2 of Regulation 1017/68 as it improved the quality of transport services, promoted greater stability in the meeting of transport needs, increased the productivity of undertakings, boosted technical and economic progress in the interests of users and promoted competition in the transport market.

The Concessionaries emphasised that the rail service and shuttle service were distinct transport services concerning respectively the markets for passenger and freight transport by train and the markets for transport of accompanied motor vehicles by specially designed rail shuttles.

In February 1989 the Commission decided not to oppose the usage agreement[88], and, pursuant to Regulation 1017/68[89] the agreement was exempted for three years from November 1988. Meanwhile, the Commission examined the granting of the exemption for a longer period.

However, the exemption decision was not renewed at the time, since in the intervening period Council Directive 91/440/EEC[90] (the Access Directive) on the development of the Community's railways was adopted in July 1991.

---

[87] See footnote 3 above.
[88] European Commission Competition Report [1989] at point 57.
[89] See footnote 3 above.
[90] See footnote 76 above.

The Directive introduced rights of access to railway infrastructure from 1 January 1993, subject to conditions, and distinguished between infrastructure management and rail transport activities.

A full Commission Decision renewing the three-year exemption granted in November 1988 was issued in December 1994[91] but subject to stringent conditions including an obligation to relinquish a quarter of their guaranteed right to use the Tunnel. In the Decision the Commission recognised the Channel Tunnel 'as constituting an essential facility for railways wishing to provide transport services between the UK and the continent'.

The relevant market for provision of infrastructure capacity through the Tunnel was defined as being geographically limited to the Tunnel and its access areas. Referring to the precedent of the Court in *Port of Genoa*[92] and its own decision in *Sea Containers* v. *Stena Sealink*[93], the Commission held this market to constitute a substantial part of the common market. One of the many controversial aspects of the case was that the Commission asserted a separate market in the provision of infrastructure (i.e. the Tunnel) as distinct from the market for international rail services, so that the former fell under Regulation 17/62 and the latter under Regulation 1017/68.

The Commission analysed restrictions of competition as a result of the usage contract in the light of the new Directive 91/440.

It held that the transport markets would be divided between Eurotunnel, which would operate the shuttles, and BR and SNCF which would operate international trains carrying passengers and freight. The Commission concluded that by dividing the markets between BR, SNCF and Eurotunnel, the usage contract had as its object or effect the appreciable prevention or restriction of competition contrary to Article 85(1).

A further restriction or prevention of competition was deemed present in the provision of hourly paths for rail transport in the Channel Tunnel. The Concessionary, Eurotunnel, owned all of the hourly paths. Under the contract, half of the Tunnel capacity was reserved for shuttle services and the other half for international passenger and freight transport. In addition, BR and SNCF were at all times entitled to 50 per cent of the Tunnel capacity to operate international trains – or, in fact, 100 per cent of the hourly paths available for that category of transport – unless they relinquished part of their entitlement. According to the Commission these arrangements thereby prevented *other railway companies* from obtaining from the infrastructure managers the hourly paths required to operate international passenger or freight trains in competition with BR and SNCF. This thus distorted competition on the market for the provision of hourly paths for rail transport in the Tunnel and on the transport markets.

---

[91] Commission Decision of 13 December 1994 relating to a proceeding under Article 85 of the Treaty and Article 53 of the EEA agreement – *Eurotunnel*, OJ 1994 L354/66.

[92] See footnote 68 above.

[93] See footnote 66 above.

The Commission, therefore, decided that it could only grant a long-term exemption (for a period of thirty years) subject to very specific obligations on the parties:

- BR and SNCF were prohibited from withholding their agreement to the sale by the infrastructure managers to other railway companies of the hourly paths needed to operate international passenger and freight services;
- BR and SNCF could only retain, for the period up to 31 December 2006, 75 per cent of the hourly capacity of the Tunnel in each direction which is reserved for international passenger and freight trains. Over the same period other railway undertakings and groups of undertakings should have a guaranteed availability of the remaining 25 per cent;
- the conditions imposed should not prevent BR and SNCF, during that period, from using more than 75 per cent of the hourly capacity if the other railway companies do not use their 25 per cent, and, correspondingly, railway companies other than BR and SNCF should not be prevented from using during the period more than 25 per cent of the hourly capacity, if BR and SNCF do not use the 75 per cent of capacity reserved to them;
- the proportion of paths reserved to BR and SNCF should be subject to re-examination by the Commission before 31 December 2006.

The Commission Decision also imposed the obligation on BR and SNCF to inform the Commission on an annual basis from 1 January 1995 of the number of hourly paths, by hour, and in each direction, calculated on a monthly basis. Similarly, the Channel Tunnel Group Limited and France Manche SA must inform the Commission annually from 1 January 1995 of the average number of applications for hourly paths submitted by railway undertakings or groups of railway undertakings other than BR and SNCF which could not be met because no paths were available.

Following the Commission Decision, BR and SNCF brought an action before the Court of First Instance against the Commission[94]. The parties sought the annulment of the Commission Decision 94/894/EEC, condemning the conditions imposed by the Commission. BR and SNCF considered the conditions imposed to be illegal and prejudicial to their interests on the grounds that they have been obliged to renounce, to the benefit of third party railway companies, a quarter of the usage rights of the Tunnel. The parties applied for an interim order to suspend the Decision but this was refused by the Court on the grounds that there was no obvious urgency since no other railway company was seeking to use the Tunnel.

[94] Registered in the Court as cases T-79/95 R and T-80/95 R, *Société Nationale de Chemins de Fer Française (SNCF) and British Railways Board (BR)* v. *Commission of the European Communities.*

Judgment on the substance of the case was rendered in October 1996[95] annulling the Commission decision on the grounds of incorrect appreciation of facts by the Commission. The Court of First Instance held that, contrary to what the Commission maintained, the usage contract did not prevent Eurotunnel from making capacity available to third parties to operate international rail services from Eurotunnel's remaining 50 per cent of capacity. Furthermore, the Court did not consider that the usage contract contained restrictive provisions.

ENS European Passenger Services

In January 1993 a notification was made to the Commission from European Night Services Limited (ENS), British Rail (BR), Deutsche Bundesbahn (DB), NV Nederlandse Spoorwegen (NS), Société Nationale de Chemins de Fer Française (SNCF) and Société Nationale de Chemins de Fer Belge (SNCB) for a declaration that Article 2 of Regulation 1017/68 did not apply to several agreements concerning the carriage of passengers by rail through the Channel Tunnel, or, alternatively, the exemption of the agreements under Article 5 of that Regulation[96]. These five national railways were parties to an agreement to operate a new night passenger service. All the railways except SNCB had taken part in the creation of the joint venture ENS. ENS was set up to run a new service via the Channel Tunnel over the following routes: London–Amsterdam; London–Frankfurt/Dortmund; Glasgow/Swansea–Paris; and Glasgow/Plymouth–Brussels.

ENS was designed to meet the needs of three categories of travellers; business travellers receiving high levels of quality and service and arrival times early enough to get to business; leisure travellers expecting luxury services; and leisure travellers who were more price-sensitive. ENS was responsible for the issuing of rail tickets on each route in its own name.

Each of the railways would provide a number of services including haulage services over its network (provision of locomotive, train crew and paths); cleaning and on-board services; and passenger handling services on the commercial stops at stations on the railway network. In addition, BR and SNCF would provide haulage services through the Tunnel consisting of making available appropriate locomotives, train crews and paths.

The joint venture would require specialised rolling stock suitable for running on different rail systems. For this, the participating railways, BR, DB, NS and SNCF, through ENS, entered into a sale and lease-back agreement under which the lessors agreed to lease the rolling stock to ENS for an initial period of twenty 20 years.

In its legal assessment, the Commission examined the relevant market and concluded that there were two service markets, i.e. one was the market for the transport of business travellers, and the other was the market for the

[95] Joined cases T79/95R and T80/95R, *Société Nationale de Chemins de Fer Française (SNCF) and British Railways Board (BR)* v. *Commission of the European Communities* [1996] ECR II-1491.

[96] Commission Notice pursuant to Article 12(2) of Regulation 1017/68, OJ 1993 C149/10.

transport of leisure travellers. In its definition of the geographic market, the Commission confined its scope to the routes actually served by ENS. The Commission examined the joint venture ENS and concluded that it was a co-operative joint venture and fell within the scope of Article 85 of the Treaty. Similarly, the operating agreements between ENS and each of the railways constituted agreements within the meaning of Article 85.

The Commission analysed restrictions on competition between the parent companies and the possible effects on third parties. It reiterated that under Article 10 of Directive 91/440/EEC[97], international groupings of railway undertakings providing passenger transport have a right of access to the railway infrastructures of their Member States of establishment, as well as transit rights on the infrastructure of other Member States in order to permit them to provide international services. It pointed out that such opportunities are open to existing railways, but also to new companies including subsidiaries of existing ones. The Commission concluded that the parent companies of ENS held a dominant position in the supply of rail services in their Member States, particularly regarding special locomotives for the Channel Tunnel, and that ENS would have direct access to those services under the operating agreements concluded with the other parties. In such cases the special relationships between the parent companies and ENS might place other operators at a competitive disadvantage. In view of the commercial strength of the parent companies, the formation of ENS might impede market access by transport operators trying to compete with it.

The Commission was also concerned that joint ventures strengthen the links between the parent companies and thus weaken the competition between them.

The Commission held that the ENS agreement was not an agreement whose sole object and effect was to apply technical improvements or achieve technical co-operation. It was also a commercial arrangement and therefore fell short of the criteria for exemption under Article 3 of Regulation 1017/68[98].

The Commission took the view that any joint venture between oligopolies, particularly in relation to the use of an essential facility such as the Channel Tunnel, would normally fall under Article 85(1) and raise questions under Article 86. The Commission is reluctant to see a series of monopolies replaced by a network of joint ventures between the railway companies exercising dominance. There was a real risk that the railways would not compete with each other and would close the market when the Access Directive[99] and subsequent directives on licensing[100] and on allocation and infrastructure charges[101] intended to open it.

---

[97] See footnote 76 above.
[98] See footnote 3 above.
[99] See footnote 76 above.
[100] See footnote 79 above.
[101] See footnote 78 above.

The Commission did conclude, however, that the formation of ENS was likely to contribute to economic progress for several reasons: a high quality transport service would be introduced to the Channel Tunnel; ENS offered a service substitutable for air transport by business travellers which had not formerly existed, therefore, the ENS services could increase competition between transport modes; and the agreement should contribute to the success of the Channel Tunnel by encouraging traffic between the UK and the continent. Users would also receive direct benefit from ENS through new high-quality transport services, as well as benefiting from competition between these new services and air transport. The Commission also accepted that the financial risks were substantial and would be difficult to bear by a single company. The pooling of the know-how acquired by each of the railways on its own territory was also important for the success of the project.

Accordingly, the Commission was prepared to grant an exemption, but strictly limited in time and subject to stringent conditions, to help ensure that restrictions remained within indispensable limits and that other rail transport companies competing with ENS could operate.

The Commission granted an exemption to Article 85(1) for a period of eight years ending on 31 December 2002. The following conditions were attached:

- the railways involved in the ENS agreements must supply to any international grouping of railway undertakings or any transport operator wishing to operate night passenger services through the Channel Tunnel the same necessary rail services as those they have agreed to supply to ENS;
- these services consist of the provision of the locomotive, train crew and path on each national network and in the Channel Tunnel;
- the railways must supply these services on their networks on the same technical and financial terms as they allow to ENS.

While the joint venture partners would be required to provide a competitor with all the services necessary to provide night services on some or all of the routes served by ENS, the competitor would have to provide its own coaches and other rolling stock.

The parties to the Decision have taken an action to the Court [102] against the Commission seeking an annulment of the Decision or requiring the Commission to modify its Decision. ENS alleges that the Commission has committed errors of law and of procedure in reaching its decision. It further alleges that the conditions imposed are excessively burdensome and interfere with the commercial freedom of the railways in a manner disproportionate to the competition restrictions identified by the Commission, which run

---

[102] Action brought on 22 November 1994 by European Night Services Ltd against the Commission of the European Communities (case T-374/94), OJ 1994 C392/8. Action brought on 22 November 1994 by European Passenger Services Ltd against the Commission of the European Communities (case T-375/994), OJ 1994 C392/9. The cases are pending at the time of writing.

contrary to the principles of proportionality and legal certainty. In the alternative, if such conditions were justified and legal, an eight-year exemption is too short when taking into account the financial, technical and other risks facing the railway companies. ENS alleges that in imposing such conditions the Commission has exceeded its competence and misused its powers by extending to the railway undertakings the obligations that Directive 91/440/EEC[103] places upon Member States.

Eurotunnel agreements on rail transport of goods: ACI [104]

The Commission has granted an exemption from the applicable competition rules under Council Regulation 1017/68[105], subject to strict conditions, in connection with two agreements on the rail transport of goods through the Channel Tunnel.

BR, SNCF and Intercontainer CV (IC), a co-operative company jointly owned by twenty-four railway companies engaged in intermodal freight in Europe, sought a declaration of inapplicability of Article 2 of Council Regulation 1017/68 to two agreements entered into between them. The first agreement concerned Allied Continental Intermodal Services Ltd (ACI), a company established by BR, SNCF and IC, to market intermodal rail services through the Channel Tunnel and the provision of related operations between the UK and the continent.

The second agreement, known as the 'Wagon Provision' Agreement, was between British Rail (BR) and, Fret Manche Europe (FME) (a company formed jointly by SNCF and Intercontainer) whereby BR and FME were to provide specialised railway wagons for the movement of freight for hire by ACI. ACI, in return, agreed to sub-hire any long-term surplus wagons to BR or FME in preference to other operators provided that those wagons were to be used on routes through the Channel Tunnel.

The Commission took the view that the agreement could qualify for an exemption for five years subject to the following conditions:

- shippers and combined operators were to have the opportunity to buy rail traction services from BR and SNCF on the same non-discriminatory basis as ACI; and
- the surplus wagons were to be sub-let on a non-discriminatory basis, but it was acceptable to give priority to any Channel Tunnel operator.

Commission Decision on Combined Tariff Structure[106]

The Commission conditionally exempted an agreement drawn up by the members of the 'Inter Unit Subsidiary Committee' (UIC) which included the

[103] See footnote 76 above.
[104] Commission Decision under Council Regulation 1917/68, OJ 1994 L224/28.
[105] See footnote 3 above.
[106] Commission Decision of 24 February 1993 relating to a proceeding under Article 85 of the EEC Treaty – Tariff Structures in the Combined Transport of Goods; OJ 1993 L73/38.

railway companies of the twelve Member States[107]. The agreement established a common tariff structure for the sale of rail haulage within the Community and in its original form was challenged by the Commission as contrary to Article 85. In January 1992 an amended version of the agreement was accepted by the Commission as valid for five years from 1 March 1992. It was underlined by the Commission that the principle of income neutrality must be enforced, i.e. the agreement must not entail any increase or decrease in the income of the railways.

Combined transport services are rarely sold directly to consignors. The sale is usually made by specialised operators which are sometimes subsidiaries of railway companies. The contract usually involves the dual supply of a locomotive and access to the rail network.

Council Directive 91/440/EEC[108] permits equitable access for all railway undertakings to the infrastructure of the Member States for the purpose of operating international combined transport goods services. This Directive only applies to those undertakings which can also provide rail haulage. It thereby excludes the majority of combined transport operators, who are still obliged to purchase rail haulage from railway companies.

The relevant market was defined by the Commission as rail transport services provided by railway companies to combined transport services.

In considering the restriction of competition, the Commission held that the agreement between the railway companies had a dual purpose; the structuring of sale prices and methods for determining such prices; and the establishment of a flat charge rate on the incidental costs set at a rate of 1 per cent of the total haulage price.

The Commission said that the agreement would have an indirect effect on price levels by preventing negotiation between operators and companies and also by preventing each railway company from adopting its own tariff structure to attract business. The agreement thereby had a restrictive effect on *actual* competition and on *future* competition with the operation of Directive 91/440. This restriction on negotiation also applied to the fixed incidental costs.

The Commission, however, considered that the agreement fixing a common rate for incidental costs might facilitate transactions between haulage suppliers and operators by eliminating unnecessary procedures and invoices. Even after the implementation of Directive 91/440, co-operation between the railways would still be necessary. The Commission therefore found that the restrictions contained in the agreement were necessary.

Furthermore, there would be no elimination of competition. Competition on the various routes to which the agreement applied would be maintained and would be further increased by the effect of Directive 91/440.

---

[107] The case occurred prior to the accession of Austria, Finland and Sweden to the European Union.

[108] See footnote 76 above.

The Commission granted the exemption, subject to the following conditions:

- the rate of remuneration was to be optional and to correspond to the services actually provided;
- before any modifications were made to the tariff structure, the representatives of the operators were to be consulted and any modification subject to a year's delay;
- the railway companies were to ensure that there was no discrimination where different tariff structures existed at a national and international level;
- the agreement was to be applied equally to all enterprises;
- there was to be no generalised agreement or concerted action on the level of tariffs;
- prior to withdrawal by any company, reasonable notice was to be given to ensure stability in investment in rolling stock;
- the principle of income neutrality was to be applied.

Anticompetitive restrictions on sale of rail tickets by travel agents: UIC decision[109]

On 25 November 1992 the Commission adopted a decision against the International Union of Railways ('UIC') for infringement of the provisions of Article 85 of the Treaty. The proceeding related to the conditions laid down by the UIC for the appointment of travel agents authorised to issue tickets for the carriage of passengers by rail and the conditions under which agents could sell these tickets. This decision deals with important competition issues, in particular, the definition of an agent, the meaning of recommendation by an association, the application of Regulation 1017/68[110] to an agreement by railway operators and the conditions that the latter can impose on their travel agents.

International rail travel is carried out on the basis of co-operation between the railway operators of the Member States, with the ticket price of an international journey being the sum of the various ticket prices for travel through the territory of the Member States involved. The railway operators receive the money due after completion of the journey. In addition to selling tickets directly to the public, railway companies also sell tickets through networks of agencies. The UIC issues a series of leaflets which regulate the relationship between the railway companies. Leaflet number 130 states that the appointment of a travel agency in a Member State can only be made by the national railway company, subject to special exceptions.

---

[109] Commission Decision of 25 November 1992 relating to a proceeding under Article 85 of the EEC Treaty – distribution of railway tickets by travel agents, OJ 1992 366/47.

[110] See footnote 3 above.

Acting in concert, the railways have broadly followed the recommendations of the UIC and have set a single rate of commission to be paid to the agencies for ticket distribution as well as uniform conditions for payment of the commission. The provision obliges the railways to sell the tickets at the official fares indicated in the tariffs and, in certain Member States, forbids the passing on by the travel agencies of part of their commission to their customers. The agency is also prohibited from promoting other competing forms of transport, whether by advertising or by advice.

The UIC argued that the competition rules were not applicable to the agreement on the grounds that: the appointed agents did not assume the risks of performance of the transport contract; the railway companies co-operate rather than compete; and agents can only negotiate and conclude contracts on behalf of the railways and cannot pass on a percentage of their commission.

The Belgian government had disputed the application of Article 85(1), stating that the relationship between a tour operator and a travel agent was one of principal and agent. However, the Commission considered that the travel agents in question were independent agents in accordance with the Court's ruling in *VVR* v. *Sociale Dienst*[111], as they also sell a wide range of additional services. The Commission added that railway companies sell through a large number of distributors, some agents, some other railway companies. The Commission also found that Council Regulation 17/62[112], as opposed to Regulation No 1017/68 which offers more procedural safeguards and a wider scope for exemption, applied to the agreement as it could not be argued that agreement between the agencies and the railway companies directly related to the provision of transport services.

The Commission found the agreement to be anti-competitive in several ways and UIC was ordered to bring the infringement to an end within a twelve-month period and a fine of ECU 1 million was imposed.

The UIC brought an action to the Court of First Instance[113] seeking an annulment of the Commission Decision (Decision 92/568/EEC). In a judgment given in June 1995, the Court of First Instance declared the Commission Decision void on procedural grounds in that it had misapplied the law by applying Regulation 17 instead of Regulation 1017/68[114]. This amounted to a breach of an essential procedural requirement which had the effect of depriving UIC of the procedural safeguards to which it was entitled under the application of Regulation 1017/68. This judgment was upheld on appeal to the Court of Justice[115].

The *UIC* case is important to the understanding of the scope of application of Regulation 1017/68. The Court rejected the Commission's narrow

[111] Case 311/85 [1987] ECR 3801.

[112] See footnote 1 above.

[113] Case T-14/93, *Union Internationale de Chemins de Fer* v. *Commission of the European Communities.*

[114] See footnote 3 above.

[115] *Commission* v. *UIC*, Case C-264/95 P, 11 March 1997, not yet reported.

interpretation which reserved the application of the Regulation to cases *directly* related to the provision of transport services. Had the Commission challenged the UIC system pursuant to this Regulation, it is not certain however that the railways would have won by convincing the Court that their agreement met the conditions for exemption provided in the Regulation. Question marks remain on the scope of this exemption. The frontier between permitted and prohibited restrictive agreements in the railway sector is unclear except in the most blatant cases involving, for example, blatant price discrimination as in the *HOV SVZ/MCN* decision[116].

Conclusion

Treaty obligations and secondary legislation in the transport sector have been reinforced and developed by specific case-law of the Court and the Commission.

A predominant issue common to different transport modes concerns access to facilities and infrastructure owned or controlled by monopolies holding special or exclusive rights. The principle of access to essential facilities continues to evolve. It is contained both in legislation, e.g. the railway access Directive, air transport access regulations, and in case-law. While ownership of an essential facility is a valid legal and commercial right, this right must be exercised with caution. Ownership/control over an essential facility may not be used to exclude access by a competitor in a manner contrary to the competition rules. Competitors availing themselves of an essential facility are, however, entitled to charge on a commercial basis for its use. The Court and the Commission have held a range of facilities to be essential facilities to which access must be granted by the dominant owner/controller to competitors on a non-discriminatory basis, subject to objective justification for refusal of access. This chapter has shown that there are several examples of Community intervention to establish access to facilities in ports, railways and air transport activities.

Progress towards increased competition has been significant first and foremost in air transport. Restrictions of competition as a result of exclusive or special rights granted to public or private transport operators nevertheless remain significant and the Commission retains Article 90 as an instrument to restrain anti-competitive practices by these operators; it is conceivable that the continued monitoring of the liberalization of all transport sectors will involve the strategic use of Article 90.

---

116 Commission decision on combined transport *HOV/SVZ/MCN*, OJ 1994 L104/34. This was a decision prohibiting a clear infringement of Articles 85 and 86 by DB.

INDEX

Abuse,
  Article 86, 59–60, 65–93
  by state alone, 71–3
  dominant position *see* Dominant position
  exclusive rights, 28
  long-term contracts, 68
  prices *see* Unfair prices
  state-related undertakings, 65–97
  transit facilities, 216
  with state intervention, 68–71
  without state intervention, 66–8
Access,
  agreements, 269–70
  air routes, 344–5, 350–1
  airport slot availability, 168, 184, 346–8
  definition, 171–2
  denial, 84–6, 174
  facilities *see* Essential facilities
  infrastructure, 154, 170, 171, 172, 174, 184
  origin of concept, 167
  pharmaceuticals, 167
  railways, 365–8
  telecommunications, 230, 268–70
  telephony, 251
  television broadcasting, 302
  third parties *see* Third parties
  unbundling, 175
  see also Discrimination
Accounts,
  cross-subsidies, 93
  telecommunications, 242
  transparency *see* Transparency
  unbundling *see* Unbundling
Administrative decisions, appeals, 15
Advertising, television *see* Television advertising
Agreements,
  access, 269–70
  Eurocheque, 94
  measures, 15
Air routes,
  access, 344–5, 350–1
  air traffic control, 61
  capacity sharing, 341, 342
  Commission Decisions, 54, 350–2
  domestic routes, 341, 345
  dominant position, 67
  exclusive concessions, 344, 351
  fares, 12, 67, 184, 341, 342, 343–4, 349–50
  France, 350–1
  general economic interest, 23, 26, 67
  nationality discrimination, 345, 351
  non-viable, 24, 26, 28, 67
  public service obligation, 344
  route charges, 352–3
  small aircraft, 345
  special and exclusive rights, 10, 349
  unfair prices, 12, 67, 184
Air traffic control, charges, 61
Airports,
  capacity, 346–7
  catering, 168
  co-ordination, 346–7

Airports (*cont.*)
cross-subsidies, 354
dominant position, 352
essential facilities, 168–9
fuelling, 169
grandfather rights, 346, 347
groundhandling, 168–9, 353–4
International Air Transport Association (IATA), 347–8
Italy, 169
landing fees, 12, 53, 69, 348, 351–2
slots,
availability, 168, 184, 346–8
sale, 338
Ancillary services,
aviation, 169–70, 353–60
cross-subsidies, 92, 354
dominant position, 60, 76–81, 312–13
measures, 13
motor insurance, 13, 138–9
railways, 170
television advertising, 167
Appeals,
administrative decisions, 15
authorizations, 205
Commission Decisions, 54
type approval, 113, 114, 119
Article 7, television broadcasting, 304
Article 30,
dominant position, 118–19
free movement of goods, 109–13
normative measures, 121, 124
terminal equipment *see* Terminal equipment
waste oil, 110, 111
Article 36, exemptions *see* Exemptions
Article 37,
adjustment of state monopolies, 122
monopolies, 119–28
purpose, 121–2
removal of discrimination, 122–4
scope, 122–4
state agencies, 121, 124
whether redundant, 124–5
Article 52, freedom of establishment, 159–61
Article 56, justifications, 133
Article 59,
cross-border element, 131–2
economic reasons, 136
general applicability, 130–3
monopolies, 136–52
nationality discrimination, 138
restrictions, 138–40
scope, 136–7
services rendered for remuneration, 130–1
Article 85,
competition rules, 93–7
general economic interest, 93–7
indispensability test, 97, 183–4
necessity test, 95
telecommunications, 263–5
Article 86,
abuse, 59–60, 65–93
characteristic abuses, 73–4
competition rules, 60–3
criteria, 60
discrimination, 59
economic activities, 60–3
exclusive rights, 59–60
new interpretation, 60
potential effect on trade, 64–5
scope, 59–60
telecommunications, 263–5
television,
advertising, 305, 312–13
broadcasting, 306
see also Dominant position
Article 90,
access to essential facilities, 165–75
aviation, 348–9
Commission Decisions, 53–7
defences, 39–40
derogation, 17–40, 127
direct effect, 38–9, 156, 305
electricity supply, 39
energy, 210–16
European Commission powers, 41–57
key concepts, 165–87
Member States obligations, 7–15, 43, 46, 50–1
new interpretation, 1–4, 18–21, 107–9
open network provision (ONP), 257–8
original interpretation, 1–2, 18, 21, 37, 106–7
postal services, 39
telecommunications, 39–40, 262
telephony, 39
television broadcasting, 303–15
Article 92, state aids, 155–9
Article 222, public ownership, 153–4
Association of Car Rental Industry Systems Standards (ACRISS), 357–8
Auctions, mobile telephony, 239–40
Austria, state monopolies, 123
Authorizations,
appeals, 205

cable television, 289
electricity generation, 204–5
fees, 232
postal services, 330–1
satellites, 232
telecommunications, 245, 252–4
Aviation,
aeronautical authorities, 69
air traffic control, 61
airports *see* Airports
ancillary services, 169–70, 353–60
Article 90, 348–9
Belgium, 348
block exemptions, 341, 342, 345–6
cabotage, 342
competition rules, 340, 345
computerized reservation systems (CRS), 169, 184, 354–9
discrimination, 351
exclusive rights, 339–60
Germany, 348, 349–50
interlining, 85–6, 170, 359–60
landing fees *see* Landing fees
liberalization, 340–8
licensing, 343
routes *see* Air routes

Bangemann Group Report, 220
Banking,
Eurocheque agreement, 94
general economic interest, 24
Spain, 155–8
state aids, 24, 155–8
transfer of funds, 24
Belgium,
airport landing fees, 12, 53, 69, 348, 351–2
aviation, 348
cable television, 287, 289
certificates of origin, 100–1
compulsory pricing, 111–12
entertainment agencies, 132, 135
Flemish language and culture, 299–300
GSM, 262
mobile telephony, 239
pluralism, 288
postal services, 29, 324–7
telecommunications, 186, 223–4, 265
television broadcasting, 286–9, 299–300
tobacco products, 111–12
water supply, 25–6
Block exemptions,
aviation, 341, 342, 345–6
computerized reservation systems (CRS), 354–5, 356
Bovine insemination, monopolies, 33–4, 69–70, 110, 111, 114–15, 117
Broadcasting,
advertising *see* Television advertising
cable *see* Cable television
exclusive rights, 74–5, 140, 277–316
Greece, 74–5
harmonization, 299, 300
nationality discrimination, 13–14, 74–5, 185
Netherlands, 142–3, 145–6, 182–3, 307–9
policy development, 297–303
product element, 106
satellites *see* Satellites
television *see* Television broadcasting
Burden of proof,
indispensability test, 183–4
reserved services, 180
special and exclusive rights, 36–8, 327

Cable television,
authorizations, 289
Belgium, 287, 289
Commission Directives, 51, 233–6
cross-subsidies, 236
Denmark, 294
explanatory memorandum, 234
Germany, 290–1
new services, 234
restrictions, 234–5, 245
telephony, 234, 236
see also Television broadcasting
Cabotage,
aviation, 342
Italy, 148
sea transport, 148
Cassis de Dijon, free movement of goods, 101–2
Channel Tunnel,
European Night Services, 91–2, 170, 375–8
transport of goods, 378
usage agreement, 371–5
Co-operation procedure, television broadcasting, 281
Coal, long-term contracts, 96–7, 214
Commission of the European Communities *see* European Commission
Commission Communications,
digital video broadcasting, 300–2
information society, 220

Commission Communications (*cont.*)
  non-binding effect, 50
  services of general interest, 60–1, 181–2
Commission Decisions,
  air and sea transport fares, 53, 348
  air routes, 54, 350–2
  appeals, 54
  Article 90, 53–7, 348
  complainants, 55–7
  computerized reservation systems (CRS), 357–9
  courier services, 42, 72, 322–4
  defences, 54
  insurance, 53
  landing fees, 351–2
  Member States, 43, 55
  Merger Regulation, 162
  railways, 371–82
  repressive powers, 44
  satellites, 264–5, 309–10
  state-related undertakings, 42, 55
  telecommunications, 262, 263–5
Commission Directives,
  Article 90 obligations, 43
  cable television, 51, 233–6
  consultation, 49
  dual competence with Council, 46–8
  electricity supply, 93, 191–2
  energy, 191
  enforcement, 43
  gas supply, 93, 191–2
  inter-operability, 187
  lack of procedural safeguards, 49
  legal basis, 44–8
  mobile telephony, 51, 237–8
  normative measures, 50
  postal services, 49, 160–1, 176–7
  preventative powers, 44
  principal features, 51–3
  procedure, 48–50
  railways, 49, 93
  satellites, 51, 52, 230, 231–3
  telecommunications, 51, 52–3, 171, 227–33, 240–6
  telephony, 229, 230, 240–1
  telex, 230
  terminal equipment, 45–8, 51, 52, 108, 114, 227–9
  transparency, 44, 51, 158, 184, 196, 331
Commission Green Papers,
  copyright, 316
  information society, 316
  mobile telephony, 225
  pluralism, 298–300
  postal services, 82, 171–2, 318–21
  programme industry, 297–8
  satellites, 224–5
  telecommunications, 221–3, 224–7, 236, 243
  television broadcasting, 278, 297–300
Commission Notices,
  concentrations, 163–4
  energy, 216
  postal services, 50, 53, 93, 172, 333–7
  telecommunications, 268–70
Commission White Papers, growth, competition and employment, 220
Commission Working Documents,
  electricity supply, 201–2
  energy, 192–4
  see also European Commission
Common Position,
  open network provision (ONP), 247–8
  television broadcasting, 281
Companies,
  aids *see* State aids
  group restructuring, 163–4
  mergers *see* Merger Regulation
  monopolies *see* Monopolies
  state-related *see* State-related undertakings
Competition rules,
  Article 85, 93–7
  Article 86, 60–3
  aviation, 340, 345
  dominance *see* Dominant position
  enforcement *see* European Commission
  general economic interest, 25
  guidelines, 92–3, 266–7, 333–7
  mergers *see* Merger Regulation
  monopolies, 3–4
  non-discrimination, 185
  postal services, 325, 333–7
  railways, 370–82
  state-related undertakings, 7–8, 13, 37, 59–97
  telecommunications, 261–70
Complaints,
  European Commission, 55–7
  postal services, 332
Computers,
  computerized reservation systems (CRS), 169, 184, 354–9
  interface, 173
Concentrations,
  Commission Notices, 163–4
  media industry, 298–300
  Merger Regulation, 163–4

Confidential information, statistics, 196
Consolidators, postal services, 172
Consumer protection, public interest, 134
Contracts,
  exclusive purchasing *see* Exclusive purchasing obligations
  long-term, 68, 96–7, 214
  maintenance, 229
Copyright,
  Commission Green Papers, 316
  listing magazines, 173–4, 313–15
Council of Europe, television broadcasting, 279–80
Council of Ministers,
  digital video broadcasting, 302
  Directives, 46–8
Courier services,
  Commission Decisions, 42, 72, 322–4
  dominant position, 72
  freedom to provide services, 42
  liberalization, 320
  Netherlands, 53, 54, 72, 322–4
  prices, 31
  Spain, 42, 53, 72, 322–4
  value added services, 322–3
Cross-subsidies,
  airports, 354
  ancillary services, 92, 354
  cable television, 236
  dominant position, 92–3
  electricity supply, 206
  loss making activities, 30
  postal services, 30, 32, 92, 158, 179, 321, 336
  separation of accounts, 93
  telecommunications, 244–5, 256
Cultural policy,
  pluralism *see* Pluralism
  public interest, 134, 308
  television broadcasting, 288, 299

Defences,
  Article 90, 39–40
  Commission Decisions, 54
  smooth transition towards competition, 149–50
Denmark,
  cable television, 294
  ferry services, 53, 72, 166, 361–3
Deregulation,
  energy, 194
  liberalization, 186–7
Derogation,
  Article 90, 17–40
  meaning, 17–18
  mobile telephony, 239
Desama Report, 198
Direct effect,
  Article 90, 38–9, 156, 305
  freedom to provide services, 132–3
Directives,
  animal health, 115
  Council of Ministers, 46–8
  definition, 44, 48
  electricity,
    supply, 196, 203–9
    transit, 196
  European Commission *see* Commission Directives
  gas transit, 195–6
  groundhandling, 353–4
  interconnection, 254–6
  legal basis, 44–8
  meeting aims, 115–16
  open network provision (ONP), 245–52, 254–6
  postal services, 327–33
  railways, 170, 186, 365–70
  telecommunications *see* Telecommunications
  telephony, 249–50, 259
  television,
    advanced standards, 302–3
    advertising, 280
    broadcasting, 277–89
Discrimination,
  air carriers, 351
  Article 86, 59
  behaviour of monopoly, 138–40
  essential facilities, 85
  existence of monopoly, 140–4
  freedom to provide services, 306
  mere possibility, 80–1
  nationality *see* Nationality discrimination
  necessity test, 185
  non-discrimination *see* Non-discrimination
  postal services, 336
  prices, 28
  railways, 370
  removal, 122–4
  see also Access
Dominant position,
  absence of valid justification, 78
  abuse,
    with state intervention, 68–71
    without state intervention, 66–8
  air routes, 67

Dominant position (*cont.*)
airports, 352
ancillary services, 60, 76–81, 312–13
Article 30, 118–19
collective dominance, 212
competitors' products, 78–9
computerized reservation systems (CRS), 357
courier services, 72
cross-subsidies, 92–3
denying access to essential facility, 84–6
downstream markets, 79, 173
essential facilities, 167–8, 173, 174–5, 360
extension, 76–81
ferry services, 362
mere possibility of discrimination, 80–1, 307
monopolies, 63–4, 76
pharmaceuticals, 167
ports, 360, 363
postal services, 42, 323–4, 325, 335–7
preventing new services, 87–92
refusal to provide indispensable services, 77
restriction of competition, 79–80
special and exclusive rights, 14, 60, 74–6
substantial part of common market, 63–4
telecommunications, 66, 76, 119, 234–5, 268, 269
telephony, 235
television,
advertising, 66, 312–13
broadcasting, 306
listings, 173–4, 313–15
unable to meet demands, 60, 81–3
unfair prices, 27, 34, 67–8
see also Article, 86

Economic activities,
Article 86, 60–3
commercial monopolies, 120–8
employment bureaux, 62–3, 130–1
pensions, 62
services rendered for remuneration, 130–1
Economic equilibrium,
general economic interest, 30, 32, 35, 39
postal services, 88, 89, 177, 330
specific services, 88
Electrical appliances, nationality discrimination, 25–6
Electricity supply,
Article 90 derogation, 39
autoproducers, 209
Commission,
Directives, 93, 191–2
Working Documents, 201–2
cross-subsidies, 206
Directives, 196, 203–9
distribution, 206
equalization charges, 211, 212
essential facilities, 172
exclusive purchasing obligations, 32–3, 94, 201, 211–14
exclusive rights, 125, 127–8
France, 111, 215–16
free movement of goods, 25, 106, 111
general economic interest, 23, 33, 35, 96–7
geographical zones, 215–16
hydroelectricity, 215–16
inter-operability, 205
Italy, 111
joint ventures, 216
necessity test, 33, 35–6
negotiated access, 207, 208–9
Netherlands, 32–3, 95, 111, 211–14
prices, 207–8
production, 204–5
single buyer system, 192, 201–4, 207–9
Spain, 111
subsidiary, 204
tendering, 202
third party access, 192, 203–4
transit facilities, 196, 216
transmission, 205–6
transparency, 206
unbundling, 206
universal service, 118, 151
see also Energy
Employees, public interest, 134
Employment bureaux,
economic activities, 62–3, 130–1
potential effect on trade, 64
social objectives, 146–7
special and exclusive rights, 10, 14
unable to meet demand, 60, 81–3
Energy,
Article 90, 210–16
coal, 96–7
Commission,
Directives, 191
Notices, 216
Working Documents, 192–4
common rules,

inherent principles, 197
key elements, 197
legal base, 197–8, 200
modified proposals, 199–200
new compromise, 200–1
response to proposals, 198–9
competition and exclusive rights, 191–217
deregulation, 194
electricity *see* Electricity supply
essential facilities, 172
exclusive rights, 191–217
free movement of goods, 193–4
gas *see* Gas supply
general economic interest, 162–3
internal market, 191, 192–4, 196–202
liberalization, 191–217
Member States, 217
oil *see* Petroleum
parallel networks, 172
single buyer system, 192, 200–4, 207–9
subsidiary, 197
tendering, 200
third party access, 196–203
treaty articles and competition, 210–1
unbundling, 200
universal service, 194

Entertainment agencies, Belgium, 132, 135

Environment,
protection, 115, 116
regulation, 39
restrictions on competition, 147–8

Equalization charges, electricity supply, 211

Equivalent effect,
certificates of origin, 100–1
free movement of goods, 99
minimum alcohol content, 101–2
monopolies, 125
requirements to be met by goods, 102, 103

Essential facilities,
airports, 168–9
computer interface, 173
copyright, 173–4
definition, 84, 165
denial of access, 84–6, 174
discrimination, 85
dominant position, 167–8, 173, 174–5, 360
electricity supply, 172
energy, 172
ferry services, 84–5, 165–6, 360–3
gas supply, 172
interlining, 85–6, 170, 359–60
management, 84–5
objective justification, 85–6
postal services, 171–2
railways, 86, 170
right of access, 84–6, 165–75
telecommunications, 171
television,
advertising, 167, 173
broadcasting, 172–3
transport, 168–70

Essential requirements,
definition, 230, 330
mobile telephony, 238
postal services, 330
telecommunications, 230, 238, 246
telephony, 230, 251

Essential services,
loss making *see* Loss making activities
refusal to provide, 77
tasks *see* General economic interest
universal *see* Universal services

Eurocontrol, 61, 352–3

Euroelectric, 198

Eurogas, 198

European Broadcasting Union (EBU), 309–11

European Commission,
Article 90 powers, 41–57
Communications *see* Commission Communications
competition guidelines, 92–3, 266–7
complaints, 55–7
Decisions *see* Commission Decisions
Directives *see* Commission Directives
enforcement powers, 42–3
Green Papers *see* Commission Green Papers
infringements by Member States, 42, 43, 45
Notices *see* Commission Notices
notification of state aids, 155, 157
opinions, 42, 45
supervisory duty, 38, 41, 44–5, 47, 176
supervisory powers, 41–2, 157, 158
surveillance duty, 44
telecommunications, 92–3, 220–7, 230–1, 266–70
Working Documents *see* Commission Working Documents

European Committee on Standardization (CEN), 332

European Conference of Postal and Telecommunications Administrations (CEPT), 232
European Convention on Human Rights, television broadcasting, 288
European Economic Area, mobile telephony, 240
European Telecommunications Standards Institute (ETSI), 249
European works, television broadcasting, 279, 280
Eurovision, 309
Excessive prices *see* Unfair prices
Exclusive purchasing obligations, electricity supply, 32–3, 94, 201, 211–14
Exclusive rights *see* Special and exclusive rights
Exemptions,
  animal health, 114–15, 117
  efficient operation of product, 115
  environmental protection, 115, 116
  European Broadcasting Union (EBU), 310
  justification, 113–18
  meeting aims of Directive, 115–16
  network safety, 114, 115
  non-economic reasons, 116–17
  public security, 116–17, 162–3
  railways, 371–82
  socio-economic reasons, 117–18
  state aids *see* State aids
  subsidiary, 117
  user safety, 114
Exports,
  exemptions *see* Exemptions
  quantitative restrictions *see* Equivalent effect
  restrictions, 110–11
  state agencies, 121

Fares, air routes, 12, 67, 184, 341, 342, 343–4, 349–50
Fees,
  air routes, 352–3
  airports, 12, 53, 69, 348, 351–2
  authorizations, 232
  piloting services, 12, 69, 70, 139, 364–5
  railway infrastructure, 367, 369–70
  terminal dues, 321, 331
  see also Prices
Ferry services,
  Denmark, 53, 72, 166, 361–3
  dominant position, 362
  essential facilities, 84–5, 165–6, 360–3
Financial services,
  cold calling, 135
  postal services, 32, 158
Finland, state monopolies, 123
France,
  air routes, 350–1
  bovine insemination *see* Bovine insemination
  commercial monopolies, 122
  electricity supply, 111, 215–16
  funeral services *see* Funeral services
  gas supply, 111
  pensions, 62
  postal services, 92, 158
  telecommunications, 176, 179
  television advertising, 283–4
Free movement of goods,
  Article 30 restrictions, 109–13
  cassis de Dijon, 101–2
  certificates of origin, 100–1
  compulsory pricing, 111–12
  derogation, 99
  electricity supply, 25, 106, 111
  energy, 193–4
  exemptions *see* Exemptions
  exports *see* Exports
  gas supply, 35, 111
  general interest test, 101–2
  goods linked to services, 106
  imports *see* Imports
  mandatory requirements, 101–2, 110, 119
  minimum alcohol content, 101–2
  monopolies over goods, 105–9
  petroleum, 111, 116–17
  proportionality test, 101, 102, 111, 113, 115, 117, 118
  quantitative restrictions *see* Equivalent effect
  requirements to be met by goods, 102, 103
  selling arrangements, 103–5
  state-related undertakings, 99–128
  terminal equipment, 2–4, 45–8, 106–9, 114
  traditional case-law, 100–5
  whisky, 100–1
Freedom of establishment,
  Article 52, 159–61
  conditions for own nationals, 160
  insurance, 159–60
Freedom to provide services,
  courier services, 42
  direct effect, 132–3

discrimination, 306
justifications, 133–6, 145–52
proportionality, 182–3
restrictions, 133–6, 145–52
state-related undertakings, 129–52
telecommunications, 143–4
television broadcasting, 277, 288, 305, 306, 308
Funeral services,
general economic interest, 22, 23
potential effect on trade, 65
product element, 106
special and exclusive rights, 10, 64
substantial part of Common Market, 64
unfair prices, 12, 27, 67–8

Gambling, regulation, 24
Gas supply,
Commission Directives, 93, 191–2
Directives, 195–6
essential facilities, 172
exclusive rights, 125, 127–8
France, 111
free movement of goods, 35, 111
general economic interest, 35
natural gas transit, 195–6
necessity test, 35–6
see also Energy
General economic interest,
adjustment of monopolies, 123
air routes, 23, 26, 67
Article 85, 93–7
Article 90 derogation, 17–40
banking, 24
Commission Communications, 60–1, 181–2
competition rules, 25
criteria, 23
economic equilibrium, 30, 32, 35, 39
electricity supply, 23, 33, 35, 96–7
energy, 162–3
funeral services, 22, 23
gas supply, 35
inland waterways, 23, 181
insurance, 24
meaning, 21–4
necessity test, 3, 19, 20, 29–30, 33, 35–6, 183, 185
obligations, 29–30
performing rights, 22, 23–4
postal services, 23, 29–32, 325–6
requisite conditions, 22–3
review by court, 24
specific services, 87–8, 326
specific tasks, 22
state aids *see* State aids
sufficient revenue, 30
telecommunications, 23
telephony, 23
television broadcasting, 23, 95–6
universal service, 181
waste oil, 23
water supply, 23, 26
Geographical zones,
electricity supply, 215–16
no effect on trade, 65
special and exclusive rights, 10
Germany,
aviation, 348, 349–50
cable television, 290–1
co-insurance, 131, 133
employment bureaux *see* Employment bureaux
synchronous digital hierarchy (SDH), 254, 262
telecommunications, 254, 262
television broadcasting, 289–91
Global system for mobile telecommunications (GSM),
Belgium, 262
Ireland, 262
Italy, 13, 53, 73, 239, 262
licensing, 238, 239
Spain, 262
Goods,
free movement *see* Free movement of goods
linked to services, 106
meaning, 105–6
Greece,
insurance, 53, 54, 159–60
petroleum, 122, 126, 127
television broadcasting, 74–5, 139–40, 149, 172–3, 305–7
Green Papers *see* Commission Green Papers
Groundhandling, airports, 168–9, 353–4
GSM *see* Global system for mobile telecommunications
Guidelines,
competition rules, 92–3, 266–7, 333–7
postal services, 320–1, 333–7
telecommunications, 92–3, 266–7

Harmonization,
broadcasting, 299, 300
ONP *see* Open network provision

Harmonization (*cont.*)
postal services, 318, 320
telecommunications, 245–56

Imports,
absolute bans, 111
discrimination *see* Nationality discrimination
exclusive rights, 123, 125
exemptions *see* Exemptions
monopolies, 3, 122
parallel, 26
quantitative restrictions *see* Equivalent effect
restrictions, 110–11
state agencies, 121
Indispensability test,
Article 85, 97, 183–4
burden of proof, 183–4
Information society, 220, 243, 316
Infrastructure,
access, 154, 170, 171, 172, 174, 184
alternative *see* Cable television
railways, 154, 170, 186, 366, 367
telecommunications, 171, 186, 225–7, 236
Inland waterways, general economic interest, 23, 181
Insurance,
co-insurance, 131, 133, 136
freedom of establishment, 159–60
general economic interest, 24
Germany, 131, 133
Greece, 53, 54, 159–60
loss adjusters, 13, 138–9
motor vehicles, 13, 138–9
taxation, 135, 158
Integrated service digital network (ISDN), open network provision (ONP), 249
Intellectual property,
copyright, 173–4, 313–15, 316
patent agents, 136
public interest, 134
television listings, 173–4, 313–15
Inter-operability,
Commission Directives, 187
electricity supply, 205
networks, 151–2, 187, 222, 226–7
telecommunications, 226–7, 254–6, 258
Interconnection,
Directives, 254–6
mobile telephony, 238
open network provision (ONP), 238, 243, 254–6
prices, 256
telecommunications, 242–3, 254–6, 258
telephony, 242
Interlining, aviation, 85–6, 170, 359–60
Internal market, energy, 191, 192–4, 196–202
International Air Transport Association (IATA), 347–8
International Energy Agency, 116
International Union of Railways (UIC), 380–2
Ireland,
GSM, 262
petroleum, 111, 116–17, 214–5
Italy,
airports, 169
cabotage, 148
commercial monopolies, 122
electricity supply, 111
GSM, 13, 53, 73, 239, 262
motor insurance, 13, 137–8
piloting tariffs, 12, 69, 70, 139, 364–5
ports, 363–5
telecommunications, 26–7
television broadcasting, 1–2, 303–5
tobacco products, 79

Joint ventures,
computerized reservation systems (CRS), 358
electricity supply, 216
satellites, 293–5, 309–10
telecommunications, 264–5
television broadcasting, 289–97
Jurisdiction, television broadcasting, 282, 284–5
Justifications,
absence, 78
Article 56, 133
exemptions, 113–18
lack of spare capacity, 148–9
network security, 151–2
no duplication of requirements, 135
non-discrimination, 135
objective, 85–6, 223–4
proportionality, 135–6
public interest, 133–6, 142–3
pursuit of Community objective, 148
reasons,
environmental, 147–8
socio-cultural, 145–6
socio-political, 146–7
restrictions on services, 133–6, 145–52

smooth transition to competition, 149–50
universal service, 150–1

Landing fees,
discounts, 348, 351–2
unfair prices, 12, 53, 69
Leased lines, open network provision (ONP), 248–9
Liabilities, universal service, 329
Liberalization,
aviation, 340–8
courier services, 320
deregulation, 186–7
energy, 191–217
privatization, 186
telecommunications, 227–45, 298
telephony, 231
Licensing,
aviation, 343
mobile telephony, 238, 239, 40
railways, 368–9
satellites, 232, 285, 286
synchronous digital hierarchy (SDH), 254, 262
telecommunications, 226, 229–30, 241–2, 252–4, 258
television,
broadcasting, 279, 285, 286
listings, 313–15
Local authorities, service provision, 121
Long-term contracts,
abuse, 68
purchase of coal, 96–7, 214
terminal equipment, 68
Loss making activities,
cross-subsidies, 30
non-viable air routes, 24, 26, 28, 67

Mandatory requirements, free movement of goods, 101–2, 110, 119
Measures,
agreements, 15
ancillary services, 13
exclusive rights, 20–1
monopolies, 14
nature and form, 15
normative, 121
public authorities, 12, 22, 24
quantitative restrictions *see* Equivalent effect
relevant, 13
special and exclusive rights, 13–15
state-related undertakings, 12–13, 47, 48, 50–1
unfair prices, 12, 68
validity, 19–20
Media industry,
broadcasting *see* Broadcasting
cinematographic works, 280, 297
concentrations, 298–300
electronic newspapers, 299
pluralism *see* Pluralism
programmes, 297–8
television *see* Television broadcasting
Member States,
abuse,
by state alone, 71–3
intervention, 68–71
non-intervention, 66–8
Article 90 obligations, 7–15, 43, 46, 50–1
Commission Decisions, 43, 55
energy, 217
gambling regulation, 24
meaning, 8
measures *see* Measures
normative measures, 121
public powers, 61
state agencies, 121
television advertising, 281–2
treaty infringements, 42, 43, 45
see also Public authorities
Merger Regulation,
Commission Decisions, 162
concentrations, 163–4
group restructuring, 163–4
state-related undertakings, 161–4
telecommunications, 261, 262–3
television broadcasting, 289–97
turnover thresholds, 161, 163
Mobile telephony,
auctions, 239–40
Belgium, 239
Commission,
Directives, 51, 237–8
Green Papers, 225
derogation, 239
digital services, 238
essential requirements, 238
European Economic Area, 240
GSM, 13, 53, 73, 238–40, 262
interconnection, 238
licensing, 238, 239, 240
meaning, 237
monopolies, 237
open network provision (ONP), 238
radiotelephony, 13, 53, 73, 238–40

Mobile telephony (*cont.*)
restrictions, 238
SIMLock features, 240
standards, 238
see also Telephony
Monopolies,
Article 37, 119–28
Article 59, 136–52
bovine insemination, 33–4, 69–70, 110, 111, 114–15, 117
commercial, 120–8
competition rules, 3–4
definition, 120
Directives *see* Commission Directives
discriminatory behaviour, 138–40
dominant position, 63–4, 76
equivalent effect, 125
exclusive rights *see* Special and exclusive rights
extension, 223–4
free movement *see* Free movement of goods
imports, 3, 122
measures, 14
mere existence, 19, 28–32, 108, 140–4, 185
mobile telephony, 237
new service, 90–2
part of monopoly, 76
post *see* Postal services
public undertakings, 126–8
revenue producing, 17, 21, 22, 25, 127
sea transport, 360–5
selling arrangements, 104–5
special rights, 124
state-related *see* State-related undertakings
telecommunications, 20, 153–4
telephony, 20
terminal equipment, 20, 46–8
third parties, 18
toleration of infringement, 83
unable to meet demands, 60, 81–3, 335
value added services, 91
Motor insurance, loss adjusters, 13, 138–9

National regulatory authorities *see* Regulation
Nationality discrimination,
air routes, 345, 351
Article 59, 138
broadcasting, 13–14, 74–5, 185
electrical appliances, 25–6
parallel imports, 26
ports, 363
state agencies, 121
state-related undertakings, 8, 18
television broadcasting, 27, 304
Necessity test,
Article 85, 95
commercial monopolies, 127
discrimination, 185
electricity supply, 33, 35–6
gas supply, 35–6
general economic interest, 3, 19, 20, 29–30, 33, 35–6, 183, 185
postal services, 29–30
proportionality, 183
state aids, 157
Netherlands,
broadcasting, 142–3, 145–6, 182–3, 307–9
courier services, 53, 54, 72, 322–4
electricity supply, 32–3, 95, 111, 211–14
pluralism, 142, 145, 146, 182–3, 307–8
television advertising, 292, 307, 308
television broadcasting, 292–3, 307–9
Networks,
access to infrastructure, 154, 170, 171, 172, 174, 184
inter-operability, 151–2, 187, 222, 226–7
ONP *see* Open network provision
parallel, 172
safety exemptions, 114, 115
security, 151–2
technical integrity, 152
telecommunications *see* Telecommunications
Non-discrimination,
competition rules, 185
justifications, 135
special and exclusive rights, 11
Normative measures,
Article 30, 121, 124
Commission Directives, 50
Norms *see* Standards

Objective justification,
essential facilities, 85–6
extension of monopoly, 223–4
Objectives,
Community, 148
legitimate, 182–4
social, 146–7
Obligations,
Article 90, 7–15, 43, 46, 50–1
exclusive purchasing, 32–3
general economic interest, 29–30

petroleum, 193
postal services, 328–9
public service, 182, 194, 199, 204, 226, 344
railways, 374
universal service *see* Universal service
Oil,
petroleum *see* Petroleum
waste *see* Waste oil
Open network provision (ONP),
Article 90, 257–8
Common Position, 247–8
directives, 245–52, 254–6
framework Directive, 246–50
integrated service digital network (ISDN), 249
interconnection, 238, 243, 254–6
leased lines, 248–9
mobile telephony, 238
packet-switched data services (PSDS), 249
telecommunications, 225, 238, 243, 245–52, 254–6
telephony, 249–50, 251–2, 260

Packet-switched data services (PSDS), open network provision (ONP), 249
Parallel imports, nationality discrimination, 26
Patent agents, services, 136
Pensions, economic activities, 62
Performing rights, general economic interest, 22, 23–4
Petroleum,
free movement of goods, 111, 116–17, 214–15
Greece, 122, 126, 127
Ireland, 111, 116–17, 214–5
obligations, 193
see also Energy
Pharmaceuticals, access, 167
Piloting tariffs, ports, unfair prices, 12, 69, 70, 139, 364–5
Pluralism,
Belgium, 288
Commission Green Papers, 298–300
Netherlands, 142, 145, 146, 182–3, 307–8
Pornography, television broadcasting, 279, 281, 286
Ports,
Denmark, 53, 72, 166, 361–3
dominant position, 360, 363
ferries *see* Ferry services
Italy, 363–5
nationality discrimination, 363
piloting tariffs, 12, 69, 70, 139, 364–5
relevant markets, 361
special and exclusive rights, 14, 60
substantial part of common market, 64, 363
Postal services,
advisory committee, 332
Article 90 derogation, 39
authorizations, 330–1
Belgium, 29, 324–7
case law, 322–7
Commission,
Directives, 49, 160–1, 176–7
Green Papers, 82, 171–2, 318–21
Notices, 50, 53, 93, 172, 333–7
Common Position, 327
competition rules, 325, 333–7
complaints, 332
consolidators, 172
cross-subsidies, 30, 32, 92, 158, 179, 321, 336
Directives, 327–33
discrimination, 336
dominant position, 42, 323–4, 325, 335–7
economic equilibrium, 88, 89, 177, 330
essential facilities, 171–2
essential requirements, 330
exclusive rights, 317–37
express delivery *see* Courier services
financial services, 32, 158
France, 92, 158
general economic interest, 23, 29–32, 325–6
guidelines, 320–1, 333–7
harmonization, 318, 320
inefficiency, 335
inner circle activities, 30–2
monopolies, 20, 29–32, 50
necessity test, 29–30
new services, 90–1, 180, 320, 330
obligations, 328–9
outer circle activities, 31–2
policy, 317–21
prices, 331
quality of service, 321, 331–2
regulation, 332–3
relevant markets, 334–5
reserved services, 177, 178, 179–80, 319, 320, 321, 329–30, 336
specific services, 87–91, 150–1, 326
standards, 332

Postal services (*cont.*)
state aids, 336–7
terminal dues, 321, 331
third parties, 29
transparency, 331
unable to meet demands, 82
unfair prices, 336
Universal Postal Union (UPU), 321, 332
universal service, 176–81, 317, 319, 320, 321, 328–9
unreserved services, 177, 178, 179
Predatory pricing *see* Unfair prices
Prices,
compulsory pricing, 111–12
courier services, 31
discrimination, 28
electricity supply, 207–8
interconnection, 256
postal services, 331
railways, 378–80
tariff rebalancing, 241, 243–4
telecommunications, 222, 241, 243–4
telephony, 251
transparency, 193, 196
unfair *see* Unfair prices
Privatization,
liberalization, 186
utilities, 153, 216
Professional standards, public interest, 134
Proportionality,
free movement of goods test, 101, 102, 111, 113, 115, 117, 118
freedom to provide services, 182–3
justifications, 135–6
legitimate objectives, 182–4
necessity test, 183
public interest, 182–3
Public authorities,
Eurocontrol, 61, 352–3
meaning, 8
measures, 12, 22, 24
unfair prices, 27, 68–9
see also Member States
Public interest,
consumer protection, 134
cultural policy, 134, 308
employees, 134
financial services, 135
intellectual property, 134
justifications, 133–6, 142–3
professional standards, 134
proportionality, 182–3
tasks *see* General economic interest
taxation, 135
Public mission, television broadcasting, 310–12
Public powers, Member States, 61
Public security, exemptions, 116–17, 162–3
Public service,
meaning, 181–2
obligations, 182, 194, 199, 204, 226, 344
Public undertakings,
commercial monopolies, 126–8
meaning, 8–9

Quantitative restrictions *see* Equivalent effect
Quotas, television broadcasting, 281, 282

Radio *see* Broadcasting
Radiotelephony *see* Mobile telephony
Railways,
access, 365–8
ancillary services, 170
Channel Tunnel,
European Night Services, 91–2, 170, 375–8
transport of goods, 378
usage agreement, 371–5
Commission,
Decisions, 371–82
Directives, 49, 93
competition rules, 370–82
conditional exemptions, 371–82
definition of undertakings, 366, 367
Directives, 170, 186, 365–70
discrimination, 370
essential facilities, 86, 170
exclusive rights, 365–82
infrastructure, 154, 170, 186, 366, 367
infrastructure fees, 367, 369–70
Inter Unit Subsidiary Committee, 378
international groupings, 366–7
International Union of Railways (UIC), 380–2
licensing, 368–9
obligations, 374
prices, 378–80
relevant markets, 373
ticket sales, 380–2
travel agents, 380–2
value added services, 91–2
Regulation,
environment, 39
gambling, 24
mergers *see* Merger Regulation
postal services, 332–3
telecommunications, 230, 256, 269

utilities, 187
see also Deregulation
Relevant markets,
ports, 361
postal services, 334–5
railways, 373
Reserved services,
definition, 329–30
post, 177, 178, 179–80, 319, 320, 321, 329–30, 336
tying, 336
Restrictions,
Article 59, 138–40
cable television, 234–5, 245
competition, 79–80
exports, 110–11
free movement of goods, 109–13
freedom to provide services, 133–6, 145–52
imports, 110–11
justifications, 133–6, 145–52
mobile telephony, 238
quantitative *see* Equivalent effect
telecommunications, 230, 244–5, 246
television,
advertising, 141–2, 307–8
broadcasting, 279
waste oil, 110, 111
Revenue,
produced by monopolies, 17, 21, 22, 25, 127
sufficient, 30

Satellites,
authorizations, 232
Commission,
Decisions, 264–5, 309–10
Directives, 51, 52, 230, 231–3
Green Papers, 224–5
domestic satellite services (DSS), 285
joint ventures, 293–5, 309–10
licensing, 232, 285, 286
non-domestic satellite services (NDSS), 285
special and exclusive rights, 233
television broadcasting, 279, 282, 284–6, 293–5
very small aperture satellites (VSAT), 264
Sea transport,
cabotage, 148
exclusive rights, 360–5
ferries *see* Ferry services
monopolies, 360–5
piloting *see* Piloting tariffs
ports *see* Ports
Selling arrangements,
free movement of goods, 103–5
monopolies, 104–5
special and exclusive rights, 104–5
Services,
ancillary *see* Ancillary services
couriers *see* Courier services
cross-border element, 131–2
definition, 130
entertainment agencies, 132, 135
ferries *see* Ferry services
financial, 32
freedom *see* Freedom to provide services
funerals *see* Funeral services
general *see* General economic interest
indispensable, 77
linked to goods, 106
local authorities, 121
new, 87–92, 180, 320, 330
patent agents, 136
post *see* postal services
preventing new services, 87–92
public service, 181–2
rendered for remuneration, 130–1
reserved, 177, 178, 179–80, 319, 320, 321, 329–30, 336
specific *see* Specific services
telecommunications *see* Telecommunications
television *see* Television broadcasting
tourist guides, 132, 136
universal *see* Universal service
unreserved, 177, 178, 179
value added *see* Value added services
Significant market power,
telecommunications, 251, 255–6
Single buyer system,
electricity supply, 192, 201–4, 207–9
energy, 192, 200–4, 207–9
Social security, non-economic in nature, 61
Spain,
air and sea transport fares, 53, 348
banking, 155–8
courier services, 42, 53, 72, 322–4
electricity supply, 111
GSM, 262
public credit institutions, 155–6
state aids, 155–8
television broadcasting, 295–7
Special and exclusive rights,
air routes, 10, 349

Special and exclusive rights (*cont.*)
burden of proof, 36–8, 327
dominant position, 14, 60, 74–6
employment bureaux, 10, 14
exclusive rights,
Article 86, 59–60
aviation, 339–60
broadcasting, 74–5, 140, 277–316
electricity, 125, 127–8
energy, 191–217
gas, 125, 127–8
imports, 123, 125
marketing, 123–4
meaning, 9–10
measures, 20–1
postal services, 317–37
railways, 365–82
sea transport, 360–5
several, 28, 119, 140
telecommunications, 219–76
television, broadcasting, 74–5, 304, 306–7
television advertising, 304
third parties, 86
transport, 339–82
funeral services, 10, 64
geographical zones, 10
groundhandling, 353–4
meaning, 9–12
measures, 13–15
non-discriminatory, 11
ports, 14, 60
satellites, 233
selling arrangements, 104–5
special rights, meaning, 10–12, 229, 232
state-related undertakings, 7
telecommunications, 9–10, 14, 28, 227
telephony, 241
terminal equipment, 14, 75, 228–9
transparency, 11
waste oil, 10, 110
Specific services,
characteristics, 87–8
economic equilibrium, 88
general economic interest, 87–8, 326
new services, 89–90
postal services, 87–91, 150–1, 326
Sponsorship, television broadcasting, 280–1
Standards,
integrated service digital network (ISDN), 249
mobile telephony, 238
postal services, 332
telecommunications, 14, 119, 186, 249
television broadcasting, 279
type approval *see* Type approval
State agencies, Article 37, 121, 124
State aids,
banking, 24, 155–8
Commission supervision, 157, 158
necessity test, 157
notification, 155, 157
postal services, 336–7
social character, 155
Spain, 155–8
specific industries, 159
state-related undertakings, 8, 155–9
State-related undertakings,
abuse, 65–97
burden of proof, 36–8
characteristic abuses, 73–4
Commission Decisions, 42, 55
competition rules, 7–8, 13, 37, 59–97
complaints against, 55–7
dominant position *see* Dominant position
free movement of goods, 99–128
freedom to provide services, 129–52
measures, 12–13, 47, 48, 50–1
Merger Regulation, 161–4
monopolies *see* Monopolies
nationality discrimination, 8, 18
rights *see* Special and exclusive rights
state aids, 8, 155–9
supervision *see* European Commission
tasks *see* General economic interest
transparency, 184
unfair prices *see* Unfair prices
Statistical Office of the European Communities, 196
Subsidiary,
electricity supply, 204
energy, 197
exemptions, 117
Sweden, state monopolies, 123
Synchronous digital hierarchy (SDH),
Germany, 254, 262
licensing, 254, 262

Tariffs,
fees *see* Fees
piloting, 12, 69, 70, 139, 364–5
pricing *see* Pricing
rebalancing, 241, 243–4
unfair prices *see* Unfair prices
Taxation,
insurance, 135, 158

municipal establishment tax, 155
public interest, 135
rebates, state aids, 158
Telecommunications,
access, 230, 268–70
accounts, 242
Action Programme, 221
alternative infrastructure *see* Cable television
Article 85, 263–5
Article 86, 263–5
Article 90, 39–40, 162
audio-conferencing, 263
authorizations, 245, 252–4
Belgium, 186, 223–4, 265
cable *see* Cable television
coherent regime, 257–69
Commission,
Decisions, 262, 263–5
Directives, 51, 52–3, 171, 227–30, 241–6
Green Papers, 221–3, 224–7, 236, 243
guidelines, 92–3, 266–7
Notices, 268–70
policies, 220–7
review of services, 230–1
competition,
exclusive rights, 219–76
rules, 261–70
cross-subsidies, 244–5, 256
Directives,
Commission, 51, 52–3, 171, 227–30, 241–6
full competition, 240–3
harmonization, 245–56
interconnection, 254–6
liberalization, 227–45
services, 229–31
dominant position, 66, 76, 119, 234–5, 268, 269
efficient operation of product, 115
essential facilities, 171
essential requirements, 230, 238, 246
exclusive rights, 219–76
France, 176, 179
freedom to provide services, 143–4
general economic interest, 23
Germany, 254, 262
GSM *see* Global system for mobile telecommunications
guidelines, 92–3, 266–7
importance of sector, 220–1
information society, 220
infrastructure, 171, 186, 225–7, 236
inter-operability, 226–7, 254–6, 258
interconnection, 242–3, 254–6, 258
Italy, 26–7
joint ventures, 264–5
liberalization, 227–45, 298
licensing, 226, 229–30, 241–2, 252–4, 258
Merger Regulation, 261, 262–3
mobiles *see* Mobile telephony
monopolies, 20, 153–4
national administrations, 223
network safety, 114, 115
open network provision (ONP), 225, 238, 243, 245–52, 254–6
prices, 222, 241, 243–4
public service obligations, 226
regulation, 230, 256, 269
restrictions, 230, 244–5, 246
satellite communication *see* Satellites
significant documents, 272–6
significant market power, 251, 255–6
special and exclusive rights, 9–10, 14, 28, 227
standards, 14, 119, 186, 249
tariff rebalancing, 241, 243–4
technological innovation, 109
telex *see* Telex
terminal equipment *see* Terminal equipment
type approval, 14, 78–9, 112–13, 186, 221, 222
unbundling, 119, 186, 245, 251
unfair prices, 269
universal service, 176, 222, 225–6, 243–6, 259–61
user safety, 114
value added services, 20, 143, 144, 222, 223
voice telephony *see* Telephony
Telemarketing, television advertising, 77, 78, 167, 173, 312–13
Telephony,
access, 251
Article 90 derogation, 39
cable television, 234, 236
Commission Directives, 229, 230, 240–1
Directives, 249–50, 259
directory services, 244, 265
dominant position, 235
equipment *see* Terminal equipment
essential requirements, 230, 251
general economic interest, 23
interconnection, 242
liberalization, 231
mobiles, 51, 225, 237–40

Telephony (*cont.*)
monopolies, 20
open network provision (ONP), 249–50, 251–2, 260
prices, 251
public payphones, 252
radio *see* Radiotelephony
significant market power, 251
special and exclusive rights, 241
special market access, 251
universal service, 236, 250–2, 256, 259
Television advertising,
Article 86, 305, 312–13
commercial objects, 24
Directives, 280
dominant position, 66, 312–13
exclusive rights, 304
France, 283–4
Member States, 281–2
Netherlands, 292, 307, 308
restrictions, 141–2, 307–8
telemarketing, 77, 78, 167, 173, 312–13
Television broadcasting,
access, 302
advanced standards, 302–3
Article 7, 304
Article 86, 306
Article 90, 303–15
Belgium, 286–9, 299–300
Broadcasting Act 1990, 284–6
cable *see* Cable television
case law, 283–9
cinematographic works, 280
co-operation procedure, 281
Commission,
Communications, 300–2
Green Papers, 278, 297–300
reports, 281
Common Position, 281
concentrations, 298–300
Council of Europe, 279–80
cultural policy, 288, 299
definition, 279
digital video, 300–2
Directives, 277–89, 302–3
dominant position, 306
essential facilities, 172–3
European Broadcasting Union (EBU), 309–11
European Convention on Human Rights, 288
European works, 279, 280
Eurovision, 309
exclusive rights, 74–5, 304, 306–7
freedom to provide services, 277, 288, 305, 306, 308
general economic interest, 23, 95–6
Germany, 289–91
Greece, 74–5, 139–40, 149, 172–3, 305–7
Italy, 1–2, 303–5
joint ventures, 289–97
jurisdiction, 282, 284–5
licensing, 279, 285, 286
listing magazines, 173–4, 313–15
media services, 283
Merger Regulation, 289–97
nationality discrimination, 27, 304
Netherlands, 292–3, 307–9
passports, 279
pay-TV, 290, 294–6, 302
pluralism *see* Pluralism
pornography, 279, 281, 286
preservation of image, 78
programme industry, 297–8
public mission, 310–12
quotas, 281, 282
restrictions, 279
retransmission, 74, 287, 288, 305–7
satellites, 279, 282, 284–6, 293–5
Spain, 295–7
sponsorship, 280–1
standards, 279
telecommunications liberalization, 298, 316
teleshopping, 282, 283
see also Broadcasting
Telex,
Commission Directives, 230
forwarding agencies, 66, 115, 223
Tendering,
electricity generation, 205
electricity supply, 202
energy, 200
Terminal equipment,
Commission Directives, 45–8, 51, 52, 108, 114, 227–9
contracts,
long-term, 68
maintenance, 229
efficient operation of product, 115
free movement of goods, 2–4, 45–8, 106–9, 114
monopolies, 20, 46–8
special and exclusive rights, 14, 75, 228–9
telephone equipment, 27, 171, 224
type approval, 47, 52, 76, 113, 114, 229

Third parties,
electricity supply, 192, 203–4
energy, 196–203
monopolies, 18
new entrants, 86, 196–202
postal services, 29, 30
telephone equipment, 27
Tobacco products,
Belgium, 111–12
Italy, 79
state monopolies, 123, 126
Tourist guides, services, 132, 136
Transit facilities,
abuse, 216
electricity, 196, 216
gas, 195–6
Transparency,
Commission Directives, 44, 51, 158, 184, 196, 331
electricity supply, 206
postal services, 331
prices, 193, 196
special and exclusive rights, 11
state-related undertakings, 184
Transport,
air routes *see* Air routes
aviation *see* Aviation
essential facilities, 168–70
exclusive rights, 339–82
ferries *see* Ferry services
railways *see* Railways
see also Sea transport
Travel agents,
computerized reservation systems (CRS), 359
railways, 380–2
Type approval,
appeals, 113, 114, 119
telecommunications, 14, 78–9, 112–13, 186, 221, 222
terminal equipment, 42, 47, 76, 113, 114, 229

Unbundling,
access, 175
electricity supply, 206
energy, 200
telecommunications, 119, 186, 245, 251
utilities, 185–6
Unfair prices,
air routes, 12, 67, 184
bovine insemination, 33–4, 69–70
dominant position, 27, 34, 67–8
funeral services, 12, 27, 67–8
landing fees, 12, 53, 69
measures, 12, 68
piloting tariffs, 12, 69, 70, 139, 364–5
postal services, 336
public authorities, 27, 68–9
telecommunications, 269
Universal Postal Union (UPU), 321, 332
Universal service,
affordability, 260
Commission supervision, 176
costing, 225–6
definition, 175
electricity supply, 118, 151
energy, 194
general economic interest, 181
justifications, 150–1
key principles, 260–1
liabilities, 329
obligations, 255
postal services, 176–81, 317, 319, 320, 321, 328–9
relevant criteria, 150
telecommunications, 176, 222, 225–6, 243–6, 259–61
telephony, 236, 250–2, 256, 259
utilities, 118, 150–1, 175–81
Utilities,
electricity *see* Electricity supply
gas *see* Gas supply
post *see* Postal services
privatization, 153, 216
public ownership, 153–4
regulation, 187
remote areas, 83
telecommunications *see* Telecommunications
unbundling, 185–6
universal service, 118, 150–1, 175–81
water *see* Water supply

Value added services,
courier services, 322–3
monopolies, 91
railways, 91–2
telecommunications, 20, 143, 144, 222, 223
Voice telephony *see* Telephony

Waste oil,
Article 30 restrictions, 110, 111
general economic interest, 23
meeting aim of Directive, 115–16
special and exclusive rights, 10, 110

Water supply,
  Belgium, 25–6
  electrical appliances, 25–6
  general economic interest, 23, 26
Whisky, free movement of goods, 100–1
World Trade Organization (WTO), 220

Index compiled by John Jeffries